Bill: P - P - P 7.

Your comments are welcome!

[signature]

(T-H. Wu)

ATM Transport and Network Integrity

Tsong-Ho Wu

*Network Control and Management
    Department*
Bellcore (U.S.A.)

Noriaki Yoshikai

Multimedia Networks Laboratories
NTT (Japan)

ACADEMIC PRESS

San Diego   London   Boston   New York   Sydney   Tokyo   Toronto

ACADEMIC PRESS
525 B Street, Suite 1900, San Diego, CA 92101-4495, USA
1300 Boylston Street, Chestnut Hill, MA 02167, USA
http://www.apnet.com

ACADEMIC PRESS LIMITED
24–28 Oval Road, London NW1 7DX, UK
http://www.hbuk.co.uk/ap/

Cover design: Lapis Design

Library of Congress Cataloging-in-Publication Data

Wu, Tsong-Ho.
    ATM transport and network integrity / Tsong-Ho Wu, Noriaki
Yoshikai.
        p. cm. — (Telecommunications)
    Includes bibliographical references and index.
    ISBN 0-12-765558-1 (alk. paper)
    1. Asynchronous transfer mode. 2. SONET (Data transmission)
3. Synchronous digital hierarchy (Data transmission) I. Yoshikai,
Noriaki. II. Title. III. Series: Telecommunications (Boston,
Mass.)
TK5105.35.W8 1997   97-7520
004.6'6—dc21   CIP

Printed in the United States of America
97 98 99 00 01 IP 9 8 7 6 5 4 3 2 1

To
Shu-Jen and Yumiko;
Arthur, Mae, Sandy, Mariko, and Hiroaki

Contents

Preface   xv
Acknowledgments   xviii
Chapter 1 Broadband Services, Transport, and Network Control
  1.1 Introduction   1
  1.2 Business Drivers for ATM   3
    1.2.1 Business Drivers for the Computer and Networking Industries   4
      1.2.1.1 ATM for Desktop   4
      1.2.1.2 ATM for LANs   4
      1.2.1.3 ATM for WANs   5
    1.2.2 Business Drivers for the Telecommunications Industry   5
      1.2.2.1 Frame Relay Service   6
      1.2.2.2 ATM Cell Relay Service   7
      1.2.2.3 Switched Multimegabit Data Service (SMDS)   7
      1.2.2.4 Circuit Emulation Service   8
      1.2.2.5 High-Performance Applications   8
      1.2.2.6 Video Dial Tone (VDT) Services   9
      1.2.2.7 LAN Interconnection   10
      1.2.2.8 LAN Emulation (LANE)   10
  1.3 Existing Telecommunications Network Infrastructure   10
    1.3.1 Today's Narrowband Network Infrastructure   10
    1.3.2 Synchronous Transfer Mode Service Transport Networks   12
      1.3.2.1 Channel Format for STM Transport   12
      1.3.2.2 SONET/STM System Configuration and Operations   12
    1.3.3 Potential Limitations of SONET/STM Transport Networks   15
  1.4 Broadband Integrated Service Network Infrastructure   15
    1.4.1 B-ISDN Transport Network Reference Model   16
    1.4.2 Physical Path vs. Virtual Path   17
    1.4.3 Channel Format for ATM   18
    1.4.4 ATM Adaptation Layer (AAL)   19
    1.4.5 ATM Switching Operations   20
    1.4.6 Comparison between STM and ATM   21
    1.4.7 End-to-End Broadband Transport Path   22
  1.5 Broadband Signaling Transport Networks   23
    1.5.1 Today's SS7 Signaling Network Architecture   23
    1.5.2 Potential Limitations of SS7 Network Architecture   24
    1.5.3 Role of ATM Technology in Broadband Signaling   25
    1.5.4 Potential ATM Signaling Network Architectures   26
      1.5.4.1 Enterprise ATM Signaling Approach   26
      1.5.4.2 Public ATM Signaling Approach   27
      1.5.4.3 Signaling Network Interconnection   28
  1.6 ATM Traffic Management   28

1.6.1 ATM Traffic Management Objectives and Functions	28
1.6.2 ATM Traffic Control Functions	30
1.6.2.1 Connection Admission Control (CAC)	30
1.6.2.2 Usage/Network Parameter Control (UPC/NPC) (Traffic Policing)	31
1.6.2.3 Traffic Shaping	31
1.6.2.4 Feedback Control	32
1.6.3 ATM Network Congestion Control	32
1.6.4 Traffic Management Options and Time Scale	33
1.7 Broadband Transport Network Restoration	33
1.7.1 Network Restoration Concept and Methods	34
1.7.2 ATM Technology Impact on Network Restoration	35
1.7.3 Feature Comparisons	36
1.7.4 Multilayer Network Restoration	36
1.8 Standards Progress and ATM Network Deployment	37
1.8.1 Standards References and Progress	37
1.8.2 ATM Network Deployment	38
1.9 Summary and Challenges	40
References	41

Chapter 2   SONET/SDH Transport and Network Integrity

2.1 Introduction	45
2.2 SONET/SDH Technology Overview	46
2.2.1 SONET Frame Structure	47
2.2.2 SONET Data Rates	48
2.2.2.1 Virtual Tributaries (VTs)	49
2.2.3 SONET Layer Model	49
2.2.4 SONET Overhead Channels	50
2.2.4.1 Section Overhead	51
2.2.4.2 Line Overhead	51
2.2.4.3 STS Path Overhead	51
2.2.5. SONET Mapping and Multiplexing	52
2.3 SONET versus SDH	53
2.4 SONET Network Architecture and Systems	57
2.4.1 SONET Target Network Architecture and End-to-End Model	57
2.4.2 SONET Transport Network Architectures	61
2.4.2.1 SONET Point-to-Point System	62
2.4.2.2 SONET Rings	65
2.4.2.3 SONET Mesh Networks	66
2.4.3 SONET Operations Communications Architecture	67
2.5 SONET Network Integrity	71
2.5.1 Definitions of ITU-T Network Protection Schemes	71
2.5.2 Customer Impacts Due to Network Failures	72
2.5.3 Network Failure Detection and Propagation	73
2.5.4 Survivable SONET Network Architectures	74
2.5.4.1 APS Diverse Protection (APS/DP)	75
2.5.4.2 Dual Homing	76
2.5.4.3 Self-Healing Rings (SHRs)	76
2.5.4.4 Self-Healing Mesh Network	76

2.6 SONET Automatic Protection Switching (APS) System   76
2.7 SONET Self-Healing Rings (SHRs)   78
    2.7.1 Unidirectional Path-Switched SHR (UPSHR)   78
    2.7.2 Bidirectional Line-Switched USR (BLSHR)   79
      2.7.2.1 Four-Fiber BLSHR (BLSHR/4)   80
      2.7.2.2 Two-Fiber BLSHR (BLSHR/2)   80
    2.7.3 Application Areas for SHRs   81
    2.7.4 Ring Interworking   83
2.8 SONET DCS Reconfigurable Mesh Networks   86
    2.8.1 Path Restoration vs. Line Restoration   87
    2.8.2 Centralized DCS Network Restoration   87
    2.8.3 Distributed Self-Healing Mesh Network   88
    2.8.4. Spare Capacity Planning   90
    2.8.5 Impact of DCS Technology on SONET Self-Healing Networks   91
2.9 SONET Network Architectures Interworking   92
    2.9.1 Comparisons of Survivable SONET Architectures   92
    2.9.2 Two-Tier SONET Network Interworking Architecture   93
2.10 SONET Network Deployment and Standards Status   94
    2.10.1 SONET Network Deployment Status   94
    2.10.2 SONET Network Standards and Requirements   96
2.11 Summary and Remarks   97
References   97

Chapter 3 ATM Transport Networks

3.1 Introduction   99
3.2 ATM Technology Overview   100
    3.2.1 B-ISDN Reference Model   100
    3.2.2 ATM Cell Format and Encoding   101
    3.2.3 ATM Layered Structure   102
    3.2.4 ATM Cell Mapping and Multiplexing   103
    3.2.5 ATM Adaptation Layer (AAL)   106
    3.2.6 ATM Multiplexing and Switching   109
    3.2.7 ATM Protocol Processes   113
      3.2.7.1 Connection Establishment Process   114
      3.2.7.2 Connection Removal Process   115
      3.2.7.3 Link Parameter Update Process   115
      3.2.7.4 ATM Cell Transfer Sending Process   115
      3.2.7.5 ATM Cell Transfer Receiving Process   116
3.3 Impact of ATM Technology on Broadband Transport Infrastructure   116
    3.3.1 ATM Technology Characteristics   116
    3.3.2 Nonhierarchical ATM Multiplexing   117
    3.3.3 Flexible and Scalable ATM Switching   118
    3.3.4 Simpler Nodal Complexity   122
    3.3.5 Separation between Path and Capacity Assignments   124
    3.3.6 On-Demand OAM Feature   124
3.4 ATM Transport Network Architectures   125
    3.4.1 Broadband Transport Network Architecture Alternatives   125
      3.4.1.1 SONET Transport Networks   127

3.4.1.2 ATM VP Transport Networks   129
3.4.1.3 ATM VC Transport Network Architecture   130
3.4.2 Architecture Feasibility Studies   131
3.4.2.1 Case Study I: DS1 Transport   131
3.4.2.2 Case Study II: DS0 Transport   134
3.4.2.3 Case Study III: Multimedia Transport   139
3.4.3 VDT Transport   142
3.4.3.1 VDT Network Architecture   142
3.4.3.2 Video on Demand (VoD)   143
3.4.3.3 VDT Transport Network Platform   145
3.5 Broadband Transport Network Evolution   148
3.5.1 Network Evolution Strategies   148
3.5.2 Interworking between B-ISDN Transport and Other Networks   150
3.6 ATM Development and Deployment   152
3.7 Summary and Remarks   155
References   156

Chapter 4 ATM Signaling Networks

4.1 Introduction   159
4.2 Existing Signaling Networks   161
4.2.1 SS7 for Public Networks   161
4.2.2 Connection Setup for Packet-Switched Networks   162
4.3 Broadband Signaling Requirements and Design Impacts   162
4.3.1 Impact of Service on Signaling Evolution   162
4.3.1.1 Voice and Data Network Services   163
4.3.1.2 Intelligent Network (IN)   163
4.3.1.3 Mobility Management   163
4.3.1.4 Broadband Services   164
4.3.1.5 Multimedia Services   164
4.3.2 Impact of Traffic on Signaling Transport   165
4.4 Broadband Signaling Transport for Public Networks   166
4.4.1 ATM Role in Broadband Signaling   166
4.4.2 Broadband Signaling Protocols   168
4.4.2.1 Broadband Signaling Protocol Stack   168
4.4.2.2 Signaling AAL (SAAL)   169
4.4.2.3 Signaling Network Layer (MTP-3)   171
4.4.3 Broadband Signaling Transport Network Architectures   172
4.4.3.1 Quasi-associated Mode Only   172
4.4.3.2 Associated Mode Only   174
4.4.3.3 Hybrid Signaling Transport   176
4.4.3.4 Signaling Transport Architecture Comparisons   176
4.4.4 Signaling Network Architecture Analysis   178
4.4.4.1 Network Model   178
4.4.4.2 Cost and Reliability Analysis   180
4.4.5 ATM Signaling Network Survivability   181
4.4.5.1 Architecture Modeling   181
4.4.5.2 A Case Study for Survivable Signaling Network
         Architecture Analysis   184

4.4.6 Public Signaling Network Evolution   188
    4.4.6.1 General Considerations   188
    4.4.6.2 Evolution Scenarios and Interworking Architectures   189
4.5 Signaling for Broadband Enterprise Networks   192
    4.5.1 Private Network-to-Network Interface (PNNI) Specifications   193
        4.5.1.1 PNNI Signaling Protocols   194
        4.5.1.2 ATM Addresses   195
    4.5.2 PNNI Routing Protocol   195
        4.5.2.1 Source Routing   196
        4.5.2.2 Link State Routing   196
        4.5.2.3 Hierarchical Routing   197
    4.5.3 PNNI Signaling   202
        4.5.3.1 PNNI Signaling Protocol Model   202
        4.5.3.2 PNNI Control Features   203
4.6 IP Layer Signaling: RSVP   204
    4.6.1 Resource Reservation Protocol (RSVP)   205
    4.6.2 RSVP vs. ATM Signaling   206
4.7 Summary and Challenges   207
References   208

Chapter 5   ATM Network Traffic Management

5.1 Introduction   211
5.2 ATM Network Traffic Management Reference Model   212
5.3 Service Class and Quality of Service (QoS)   213
    5.3.1 QoS and Network Performance (NP)   213
    5.3.2 ATM Service Class   216
    5.3.3 Mapping between AAL QoS and ATM QoS   217
5.4 ATM Layer QoS   219
    5.4.1. ATM Layer QoS Parameters   220
    5.4.2 ATM Layer QoS Negotiation   224
    5.4.3 ATM Layer QoS Measurement   225
5.5 Traffic Characteristics and Declaration   226
    5.5.1 Traffic Parameters and Declaration   226
    5.5.2 Peak Cell Rate (PCR)   228
5.6 OAM Flow for Traffic Management   229
    5.6.1 OAM Cell Format and Types   230
    5.6.2 OAM Flow at the ATM Layer   230
        5.6.2.1 OAM Flow for Virtual Path Connection   231
        5.6.2.2 OAM Flow for Virtual Channel Connection   233
    5.6.3 Performance Monitoring Cells for In-Service Measurement   234
5.7 ATM Network Traffic Control   234
    5.7.1 Connection Admission Control (CAC)   235
    5.7.2 Usage Parameter Control (UPC) and Network Parameter Control
        (NPC)   235
    5.7.3 Traffic Shaping   239
    5.7.4 Resource Management   242
    5.7.5 Feedback Control   243
5.8 ATM Network Congestion Control   243

5.8.1 Measure of Congestion (MOC)   244
5.8.2 Congestion Control Functions   245
5.8.3 Congestion Control Mechanisms   246
    5.8.3.1 Rate-Based Congestion Control   247
    5.8.3.2 Credit-Based Congestion Control   249
5.9 Summary and Remarks   250
References   251

Chapter 6 ATM Protection Switching

6.1. Introduction   253
6.2 ATM Role in Broadband Network Survivability   254
    6.2.1 The Changing Telecommunications Service Paradigm   254
    6.2.2 Impact of ATM Technology on Broadband Network Survivability 255
    6.2.3 Multilayer Survivable SONET-Based ATM Network
          Architectures   258
6.3 Key Design Issues in ATM Protection Switching   258
    6.3.1 Protection Switching Communications Mechanism   259
    6.3.2 Protection Switching Fragmentation   260
    6.3.3 Multilayer Network Architecture Model and Interworking   262
6.4 APS for SONET-Based ATM Networks   264
    6.4.1 SONET Layer Protection for ATM Services   265
    6.4.2 ATM Layer Protection Switching Using OAM Cells or User
          AALs   265
        6.4.2.1 ATM Protection Switching Mechanisms   265
        6.4.2.2 Implementation Example of ATM Protection
                Switching Mechanisms   269
    6.4.3 ATM Layer Protection Switching Using SVC Signaling
          Capability   271
        6.4.3.1 The Monitoring and Detection Mechanisms   272
        6.4.3.2 The Recovery Mechanism   272
        6.4.3.3 Requirements for ATM Switches   274
        6.4.3.4 The Performance Evaluation   274
    6.4.4 Interworking between SONET and ATM Layer Protection
          Switching   276
6.5. ATM Self-Healing Rings   279
    6.5.1 Potential Role of ATM Technology in Self-Healing Rings   279
    6.5.2 SONET-Based ATM Self-Healing Ring Architectures   280
        6.5.2.1 A SONET-Based ATM VP Ring Architecture   281
        6.5.2.2 ATM/VP ADM Functional Diagram   283
        6.5.2.3 Self-Healing Control Schemes   284
        6.5.2.4 Spare Ring Capacity Engineering   285
    6.5.3 ATM Self-Healing Rings with Switching Capability   285
    6.5.4. Broadband Self-Healing Ring Evolution   287
    6.5.5 Applications to Mobility Management   289
        6.5.5.1 PCS Hand-off Feature   289
        6.5.5.2 Anchor Routing   290
        6.5.5.3 Ring Architecture for Reliable and Fast Anchor
                Rerouting   292

6.6 Summary and Remarks 293
References 293

Chapter 7 ATM Self-Healing Mesh Networks

7.1 Introduction 297
7.2 Potential Role of ATM in Self-Healing Networks 298
  7.2.1 Network Restoration Using Virtual Paths 298
  7.2.2 Digital Path Restoration vs. VP Restoration 298
7.3 ATM Self-Healing Network Architectures 300
  7.3.1 Class of ATM Self-Healing Network Protocols 300
    7.3.1.1 Centralized Restoration vs. Distributed Restoration 301
    7.3.1.2 Line Restoration vs. Path Restoration 301
    7.3.1.3 Guided Restoration vs. Unguided Restoration 302
    7.3.1.4 Coordinated Bidirectional Restoration vs.
      Uncoordinated Bidirectional Restoration 303
  7.3.2 Distributed Self-Healing Protocol 303
  7.3.3. Coordinated Bidirectional (Double-Search) Restoration 305
    7.3.3.1 Double-Search Restoration Algorithm 305
    7.3.3.2 Case Study 308
  7.3.4 Triggering Methods for Self-Healing Networks 309
7.4 Reliable Control Message Transfer for Restoration 311
  7.4.1 Restoration Message Structure 311
  7.4.2 Transfer Mechanism for Control Messages 312
  7.4.3 Reliable Restoration Message Transfer Protocol 313
  7.4.4 Error-Recovery Mechanism in ATM Adaptation Layer 314
7.5 Self-Healing Switching Systems 316
7.6 Backup VP Assignment and Spare Capacity Design and Analysis 319
  7.6.1 A Spare Capacity Design Algorithm for ATM Self-Healing
    Networks 319
  7.6.2 Spare Capacity Analysis for Path Restoration vs. Line Restoration 320
  7.6.3 Impact of VP Assignment on Network Restoration Ratio 322
  7.6.4 Spare Capacity Analysis for VP vs. SONET/STS Restorations 325
7.7 ATM VP Trace Methods 326
  7.7.1 VP Trace Function 326
  7.7.2 Alternative VP Trace Methods and Evaluation 327
7.8 Multilayer B-ISDN Self-Healing Networks 330
7.9 Summary and Remarks 331
References 332

Acronyms 335
Index 341
About the Authors 349

Preface

A trend toward merging information networking and telecommunication services has created the demand for higher bandwidth from telecommunication networks, looking to exchange ever larger volumes of data in a very short time interval. The creation and demand for World Wide Web services has also accelerated the need for these high-capacity, high-speed information and control transport networks. Due to its special characteristics, ATM (Asynchronous Transfer Mode) is considered as the wide-area transport network technology of choice for broadband communications. This book explains why.

Both the telecommunications community (traditionally supported by ITU-T and the ATM Forum) and the Internet community [traditionally supported by the Internet Engineering Task Force (IETF)] are attracted to ATM due to its technological characteristics. These include its high speed, flexibility of connection establishment and bandwidth allocation, bandwidth on demand, service integration, and technology scalability. It is the use of such powerful transport technology, however, that points up a difference between the traditional telecommunications community and the Internet community. For example, the traditional telecommunications community views ATM as an intelligent transport platform to support all services without adding any intelligence on the service layers [i.e., ATM Adaptation Layer (AAL)]. This approach is sometimes referred to as the ATM-oriented networking model. In contrast, an emerging view within the Internet community is that the ATM network may only be considered as a high-speed provisioned transport method. Intelligent control and signaling functions are implemented at a higher layer [e.g., Internet Protocol (IP) layer]. This approach is sometimes referred to as the IP-oriented networking model. The relative merits of these two views for supporting integrated broadband services have been debated for some time in research and engineering communities. Whether one view will prevail, or whether there will be a future meeting point for these two integrated transport networking views remains to be seen. Perhaps the final answer will be determined by the market instead of by technology.

The purpose of this book is to provide architecture design information on ATM-over-SONET transport networks that may be used in the ATM-oriented networking model as well as the IP-oriented networking model. For example, the ATM Permanent Virtual Connection (PVC)-based network restoration schemes may be applicable to both the ATM-over-SONET and IP-over-ATM networks, if the economics can be justified. This book will be particularly useful for those who are searching for an integrated multilayer interworking transport and control solution for ATM-over-SONET networks and/or IP-over-ATM-over-SONET networks.

The book focuses on how to use unique ATM characteristics to design a cost-effective SONET/ATM broadband transport network to meet Quality of Service (QoS) requirements under both normal and stress (i.e., congestion and/or failures) network conditions. These technical insights and this information are discussed from a network architecture perspective. In particular, technical insights provided by this book will help network planners and engineers address a crucial multilayer networking problem for ATM-over-SONET networks or future IP-over-ATM-over-SONET networks. In such a multilayer networking environment, some network functions, such as network restoration and QoS provisioning, can be implemented either at the SONET or ATM (or IP in the future) Layer. Thus, which layer functionality may or may not be needed, and how each layer should interwork with the other layers become crucial and challenging technical issues to broadband network engineers and planners.

This book covers all key network components that constitute a broadband ATM-over-SONET network supporting required QoS under both normal and stress network conditions. These key network components include information transport, signaling and control transport, traffic management and control to prevent the network from congestion and minimize the impact once congestion occurs, and network restoration after a network component fails. It also relates the status of the standardization efforts. The unique feature of this book is its end-to-end networking approach under a variety of network scenarios including normal network conditions, network congestion condition, and network failure conditions. To show the applicability of the described architectures, this book discusses several case studies for these broadband network architecture analyses. Although the results of these case studies may provide some insights on architecture application feasibility, the true worth of these case studies is in the descriptions of methodologies used for architecture analysis.

The goal of this book is to serve as a reference for telecommunications network researchers, planners, and engineers who are concerned about how to adapt their transport networks to meet emerging broadband service offerings and requirements, and how to evolve their transport networks cost-effectively from the present infrastructure to the future Broadband Integrated Service Digital Network (B-ISDN) infrastructure. This book can also be used as a textbook for a three-day or five-day seminar. The book may also be useful to researchers and engineers who are working on IP-over-ATM networking areas by providing them with technical insights into the network capabilities and functions that can be realized at the ATM Layer. These insights may help in planning cost-effective next-generation IP-over-ATM networks to best utilize potential PVC-based ATM capabilities to achieve the best network performance at affordable costs.

The book is organized into three parts. The first part consists of Chapters 1 and 2, which provide the background information and an overview of the subject matter in the book. Chapter 1 reviews broadband services, transport technologies, and control techniques for network integrity. Additionally, this chapter provides an overview of the rest of book. Chapter 2 reviews existing SONET/SDH network technology and architectures that have been implemented around the world. This chapter serves as background material for evolving into the broadband ATM world. *Fiber Network Service Survivability,* by Tsong-Ho Wu, is a more detailed reference for SONET transport and network integrity.

Chapters 3 and 4 constitute the second part of the book, which focuses on how to use unique ATM characteristics to provide transport and signaling capabilities that are very different from those provided by existing SONET/SDH networks. Specifically, Chapter 3 reviews high-speed transport capability and architectures that ATM can offer to support emerging broadband services. Many discussions in this chapter are primarily from the public carrier perspective, due to the authors' background. Chapter 4 reviews how ATM can be used to provide signaling capability on ATM networks, including both public and private networks. Signaling evolution issues from existing SS7 networks are also covered in this chapter.

The focus of the third part of the book, Chapters 5 through 7, is on control and architecture issues that may allow ATM to ensure network integrity. ATM network integrity means that the ATM transport network will function properly when the network is under normal conditions as well as stress conditions (i.e., congestion and/or failure). In Chapter 5, the focus is on traffic management issues and control schemes for preventing the network from congestion and minimizing effects when congestion occurs. Chapter 6 reviews control schemes and architectures for point-to-point ATM protection switching systems and ATM Self-Healing Rings, and Chapter 7 reviews control schemes and architectures for ATM self-healing mesh networks.

The following figure summarizes the relationships among the seven chapters. The arrows indicate prerequisites for the chapters. For example, reading Chapter 4 may require some basic understanding of Chapter 3.

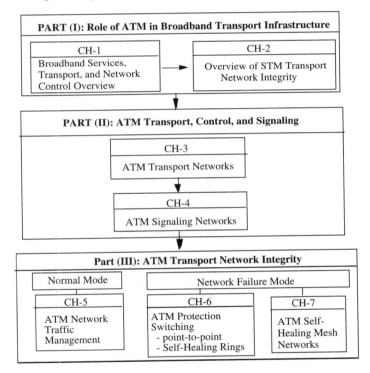

The material presented in this book is taken from public literature; most of it, however, is from our own research. Some information has come from our colleagues to ensure subject accuracy. This book does not reflect any opinion, policy, or position of Bellcore and Nippon Telephone and Telegraph (NTT). All ideas expressed herein are strictly those of the authors, who assume full responsibility for them.

Tsong-Ho Wu   Noriaki Yoshikai
Red Bank, New Jersey, USA   Kanagawa, JAPAN

Acknowledgments

The motivation to write a book can come from unusual sources. One motivation was that Arthur and Mae, children of the first author of this book, complained that their then 2-year-old sister's name, Sandy, did not appear in the book authored by their father (*Fiber Network Service Survivability*, by T.-H. Wu, published in 1992). Thus, to be fair, we decided to write another book in order to honor Sandy and the family of the second author (Yoshikai) as well. During the period of writing, our families, Shu-Jen, Yumiko, Arthur, Mae, Sandy, Mariko, and Hiroaki, have been very patient and supportive. Without our families' full support and understanding, it would have been virtually impossible to finish this book.

In addition to thanking our families, we would like to thank our Bellcore and NTT management, Stu D. Personick, Nim K. Cheung, Walter D. Sincoskie, and Yuji Inoue for their understanding and support of our writing efforts. We would also like to thank our colleagues, Deh-phone Hsing, Latha Kant, Jing-Jou Yen, Ken-ichi Sato, Ryutaro Kawamura, Yukari Tsuji, and Mitsuhiro Yuito, for their reviews and comments on the manuscript; and our friends Kathy Hintz and Rocco Tomazic for their comments on the Preface.

We would like to express our deep gratitude to Carroll Robinson and Karen Wachs for their efforts on the initial editing of the book manuscript; Bettina Burch, the production editor, for her thorough final copyediting of the manuscript and patience in helping prepare the camera-ready copy; and Shu-Jen, Arthur, and Mae for their help on the typesetting of the final version of the manuscript. Finally, we would like to thank Dr. Russell T. Hsing and Dr. Zvi Ruder for offering us the opportunity to publish this book.

Chapter 1

Broadband Services, Transport, and Network Control

1.1 Introduction

The merging information networking services demand a large bandwidth from telecommunication networks for large volumes of data to be exchanged within a very short interval. Transport technologies to meet these broadband communications needs include Synchronous Optical NETwork (SONET) [1] (or Synchronous Digital Hierarchy (SDH) [2]) and Asynchronous Transfer Mode (ATM) [3–5]. SONET and ATM have been adopted by the American National Standards Institute—Telecommunications (ANSI/T1) as Broadband Integrated Digital Services Network (B-ISDN) transmission and switching technologies, respectively. In the Telecommunication Standardization Sector of the International Telecommunication Union (ITU-T), SDH and ATM are adopted as B-ISDN transmission and switching technologies, respectively. SONET and SDH are similar, but not identical, digital hierarchies. They have similar sets of overheads and functions but differ in their use of overhead structures.

SONET comprises optical interfaces, line rates, and format specifications for high-speed optical signal transmission. ATM is a high-speed, integrated multiplexing and switching technology that transmits information across telecommunications and computer networks using the data transmission rate required by the customer through fixed cells that are connection-oriented. In North American standards (ANSI/T1), ATM cells are transported on the SONET transmission system; for ITU-T, ATM cells may be transported on the SDH system or a cell-based transmission system. The term *ATM transport,* which is used throughout this book, refers to the use of SONET/SDH transmission and ATM switching and multiplexing techniques to transport end-user traffic from a source to a destination within a network.

Physical SONET/SDH interfaces (155.52 and 622.08 Mbps) for the User-Network Interface (UNI) provide integrated support for high-speed information transfer and various communication modes, such as circuit and packet modes, and a variety of bit-rate communications (constant, variable, and burst). The high-speed capability of ATM is due largely to its short, fixed-length cell format, simple header overhead functions, and a connection-oriented nature. The elimination of header overhead functions such as error correction and flow control allows very simple, fast processing and low queuing delay in ATM nodes. The short and fixed-length cell format of ATM is inherently suitable for fast switching and potential hardware implementations.

The standards, product development, and commercial deployment of ATM have been maturing rapidly. ITU-T and ANSI/T1 have made significant progress for the basic set of ATM network capabilities; in addition, the contributions of the ATM Forum by late 1996 include 68 approved implementation specifications and 23 proposals for new work items.

The third generation of ATM switches and ATM access multiplexers have become commercially available in the market. A combined worldwide revenue of $557M in 1996 is projected to increase to $2.8B by the year 2000 [6]. ATM networks are being deployed by major public carriers, entertainment, manufacturing, education, military/defense, and law enforcement. ATM services have been available from many major public carriers. Application examples that have been explored in these ATM networks include graphics processing on high-end platforms, the movement of large image-based files, real-time multimedia services, and applications across the Local-Area Network (LAN)/Wide-Area Network (WAN) boundary, especially with the Internet.

Both switching and multiplexing functions of a broadband transport network can be performed at the SONET (physical) layer using Synchronous Transfer Mode (STM), or at the ATM Layer, or a combination of both. Each functional layer has different characteristics with respect to multiplexing, bandwidth management, network control, operations, and maintenance. In general, ATM technology used in ATM Layer transport is more flexible and efficient than STM technology used in SONET layer transport with regard to bandwidth allocation and transmission efficiency. However, these advantages come at the expense of the much more complex control systems that are needed, particularly when the network must be designed to operate under stress (i.e., network congestion and/or failures). Some of the broadband network equipment designed to transport broadband service traffic at the SONET layer [e.g., SONET Digital Cross-connect Systems (DCSs)] and the ATM layer [e.g., ATM Virtual Path Cross-connect Systems (VPXs)] could be commercially available in a similar time frame. Thus, understanding the relative merits of STM and ATM transport technologies will help network planners and engineers design a cost-effective transport network and plan for smooth evolution from the present transport network infrastructure to the target B-ISDN transport network infrastructure. Providing technical insights for differentiating these two broadband transport network technology alternatives forms the first part of this book.

In addition to broadband service transport under normal network operations, the network's capability of functioning under stress has become an ever more important advantage over its narrowband counterparts due to the significant network resources associated with network connections and a highly competitive business environment. Network reliability, in addition to network costs, is perceived as one of the crucial factors that will allow network providers not only to sustain the existing customer base, but also expand the emerging customer base to survive in a highly competitive business environment. Network reliability is a measure of the service integrity perceived by customers and is represented by Quality of Service (QoS) at the network layer. Thus, providing an adequate level of network reliability with minimum capital investment has become both a goal and a challenge for network providers.

The challenge of preserving QoS under network stress with minimum cost depends on the efficient use of broadband transport network technologies at both the SONET and ATM layers. Unlike SONET/STM transport, in which the QoS is affected only when the network components fail, the service quality in the SONET/ATM transport network can be affected even when the network components do not fail (i.e., network congestion caused by ATM queuing characteristics). Thus, efficient use of the multilayer broadband transport network structure is a key factor in dealing with the challenge presented here. This forms the second part of this book.

This chapter reviews network architectures and technologies that can be used to design a cost-effective, reliable SONET/ATM network. Details of these network architectures and associated technologies will be discussed in Chapters 2 through 7. In the remainder of this chapter, we first review emerging broadband services and associated QoS requirements. We then examine today's telecommunications network infrastructure and the potential problems and limitations of using it to support emerging broadband services. With the potential technology limitations of today's networks, ATM broadband transport network technology is reviewed as a possible solution to overcome these potential limitations and problems. Network integrity issues for SONET/ATM networks, such as congestion control and network restoration, will also be discussed. Finally, standards and development progress for ATM technology are briefly reviewed, and the challenge of building a National Information Highway using ATM technology is discussed in an effort to stimulate future research directions.

1.2 Business Drivers for ATM

ATM is growing and maturing rapidly. It is being delivered today in products offered by many vendors and implemented by many public carriers, private corporations, and government agencies. These ATM users are looking to gain business advantages that include

1. Enabling high-bandwidth applications including desktop video, digital libraries, and real-time image transfer
2. Coexistence of different types of traffic on a single network platform to reduce both transport and operational costs
3. Long-term network scalability and architectural stability
4. Equal implementations on local and wide area networks with protocol independence

Business drivers for ATM technology may be different in the computer/networking industry and telecommunication industry, although it is expected that such a service class distinction will disappear soon, due to a new Telecommunications Act passed by the U.S. Congress in February 1996.

Potential key applications for ATM technology include client–server applications, graphics processing on high-end platforms, the movement of large image-based files, real-time multimedia services, LAN replacement, disaster recovery, and applications across the LAN/WAN boundary at high speeds. Multimedia traffic types, including Computer-Added Design (CAD) images and full-motion video, are the primary forces driving ATM deployment. Although increased data speeds may be supported by several emerging technologies, the ability to handle multiple simultaneous multimedia video sessions and transcend the LAN/WAN boundary at high speeds are features that can be achieved only with ATM technology. We discuss these business drivers in subsequent subsections.

1.2.1 Business Drivers for the Computer and Networking Industries

There are three possible sectors that might use ATM technology in the computer and networking industry: ATM for desktop, ATM for LANs, and ATM for WANs. In general, ATM is winning its biggest place as a wide-area networking solution, but there are serious challenges from existing and emerging LAN switching products (e.g., Fast Ethernet and Gigabit Ethernet) in the LAN and desktop environments.

1.2.1.1 ATM for Desktop

Compared with existing and emerging technologies, ATM is not currently perceived as an attractive option in the desktop environment due to its cost. In addition to costs, most desktop applications today do not include the real-time multimedia for which ATM may be particularly suited. The challenge now is how to bring ATM to the desktop. The problem is that potential cost savings from eliminating Private Branch Exchanges (PBXs) must be offset by the cost of upgrading every desktop with a new ATM network interface card. ATM25, with an interface of 25 Mbps, is the most affordable option today, but it is still too expensive compared with existing technologies. However, an approach called "Cells-In-Frames (CIF)" has been proposed [7] that may bring ATM to the desktop at no higher cost than the current upgrade to switched Ethernet, a switching product that is all the rage today. It still remains to be seen whether CIF can compete with switched Ethernet systems in cost and performance.

CIF allows ATM to run over the existing Ethernet. CIF would reuse only the existing physical media, such as cabling and the Ethernet adapter, while ignoring the higher-layer Ethernet protocols. The ATM's QoS, flow control, and signaling protocols would be added to the workstation through software, thus eliminating the need for any hardware upgrades to the desktop. More details on CIF can be found in [7].

PC-ATM Bus is another option [8–9]. When handling a significant amount of multimedia information at the same time, the primary performance bottleneck may be on the system bus, not the processor. To dissolve this bottleneck, Nippon Telegraph and Telephone (NTT) has proposed that the media transmission path should be separated from the system control bus [8]. The PC-ATM bus is used as a media transmission bus that is independent of the existing system bus (e.g., Ethernet) to achieve real-time transmission of multimedia communications. The first specification of PC-ATM Bus was approved by the ATM Forum [9], and the detailed requirements are under discussion.

1.2.1.2 ATM for LANs

For local-area networking applications, much debate surrounds the issue of whether ATM can be equally attractive for LANs and associated applications as for WANs in cost and performance. Although client–server applications may be most suitable for ATM because their distributed processing requirement matches the bandwidth-on-demand characteristics of ATM technology, the higher cost of ATM compared with existing LAN products, which are being enhanced to handle speeds in the Mbps–Gbps range (e.g., Switched Fast Ethernet and Gigabit Ethernet) has made the ATM–LAN option unattractive in the near term. However, when the demand reaches Gbps and requires integration of voice, data, and video in a physical LAN, ATM may be the only technology that can offer the desired QoS and the needed scalable network capability.

1.2.1.3 ATM for WANs

ATM is generally considered to be a transport technology choice for wide-area networking applications. Although ATM is still more expensive than existing wide-area networking products [e.g., Fiber Distributed Data Interface (FDDI)], current applications have already demanded a network capability that is beyond the reach of existing products such as FDDI. For example, working with high-quality movie frames results in image file sizes that typically range from 20 to 40 MB and can be as large as 100 MB. Considering that a movie uses 24 frames/s, the result is data more than a shared 100 Mbps FDDI can handle effectively. This is where a scalable ATM technology may become very useful.

ATM could also play a crucial role with the Internet. Given its spectacular growth rate, the Internet could become one of the potential "killer applications" that will drive ATM acceptance. With the explosive growth rate in Internet and Web sites, the Internet may not be able to continue to support such a strong demand if the underlying infrastructure is not upgraded to ATM with appropriate flow control. The binary flow control mechanism contained in the Transport Control Protocol (TCP) exacerbates the problem. The Resource Reservation Protocol (RSVP) proposed by the Internet Engineering Task Force (IETF) is intended to provide QoS necessary for multimedia, but it does not address the capability of the network to meet these promises. Only ATM, armed with Available Bit Rate (ABR) design and the new explicit rate flow control, may accommodate the necessary tasks [7]. The future role of ATM on the Internet depends on how well ATM can work with the Internet Protocol (IP), which presents many technical and business challenges for network engineers and managers. Interested readers may refer to Comer and Mockapetris [10] for different approaches that will allow ATM and IP to work with each other.

1.2.2 Business Drivers for the Telecommunications Industry

ATM drivers for the telecommunications industry reside primarily in emerging broadband services. Broadband services generally refer to those services that require at least Mbps bandwidth to meet quality of services. Broadband services can be divided into three major classes (see Table 1-1): (1) high-speed data services, including Frame Relay (FR) services; Switched Multimegabit Data Service (SMDS); ATM cell relay services; gigabit data service; and circuit emulation services; (2) video and image distribution services (i.e., video dial tone services), including Video on Demand (VoD); interactive TV; multimedia applications; Cable Television (CATV); home shopping services; and so forth; and (3) enterprise information networking services such as LAN interconnection and LAN emulation. Initially, broadband ATM networks may support these emerging services through provisioned Virtual Channel/Virtual Path Connections (VCCs/VPCs), which are referred to collectively as Permanent Virtual Connections (PVCs). PVC services do not require call control, signaling, or processing functions. As the network evolves and signaling standards are developed, Switched Virtual Connection (SVC) service capabilities can be added if needed.

1.2.2.1 Frame Relay Service

Frame Relay (FR) service is a connection-oriented data service offering LAN–LAN communication and other data applications at access rates of between 56/64 Kbps and 1.544 Mbps. As with the X.25 network, FR is a standard interface protocol (ITU-T Rec. I.233 — "Frame Mode Bearer Service") that specifies an interface between a LAN or data terminal and public or private WANs. The term *relay* implies that the Layer 2 (the link layer) data frame is not terminated or processed at the endpoints of each link in the network, but is relayed to the destination, as is the case in a LAN. Table 1-2 compares FR technology with other existing narrowband data transport technologies.

Table 1-1: Telecommunications business drivers for ATM.

Service Class	Services	Service Availability
High-speed data services	Frame Relay service	1993
	Cell relay service	
	SMDS	1995
	Circuit emulation	1994
	service	1995
	High-performance	
	applications	trial
Video and image distribution	One-way video	now
	Interactive video	trial
	(e.g., VoD)	
Enterprise information networking services	LAN interconnection	1995
	LAN emulation	
		1995

FR is similar to the X.25 protocol, but transmission delay is reduced because neither error correction (packet retransmission) nor congestion control at intermediate nodes is required. So far, equipment vendors have offered FR equipment that operates at speeds ranging from 56/64 Kbps to 1.544 Mbps. However, FR has the potential to operate at speeds of 45 Mbps. The primary application of FR is to interface high-speed LANs to WANs, where very large LAN packets must be quickly injected into WANs. Other applications include image transfer and remote mainframe access. FR service, first offered commercially in 1993, has become one of today's most popular public telecommunications services. The worldwide revenue of FR switches is estimated to be $927M in 1996 with a projected increase to $1.5B by the year 2000 [6].

Initial exchange and exchange access FR services are PVC-based and offered over service-specific access interfaces such as Digital Signal Level 1 (DS1). If FR service is carried on the ATM network, the ATM Adaptation Layer (AAL) Type 5 has been recommended in ITU-T to accommodate FR service. Details of ATM service classes will be discussed in Chapter 3.

1.2.2.2 ATM Cell Relay Service

ATM Cell Relay Service (CRS) is a connection-oriented communications service providing users with high-speed, low-delay networking capabilities for data-intensive business computing and real-time mixed media communications applications. ATM CRS offers a flexible use of bandwidth because it can accommodate both constant and variable rates of information traffic. Initial CRS may be supported on a PVC basis, with potential access rates of 1.544, 44.736, and 155.52 Mbps. Lower-speed access rates (e.g., DS1) and SVC capabilities are anticipated as this service evolves. The CRS has been offered by most Local Exchange Carriers (LECs) since 1994.

Table 1-2: Relative comparison between FR technology and existing narrowband technologies.

Network attributes	Circuit switched	Leased circuits	X.25 packet switched	Frame relay
High speed transfer	no	yes	no	yes
Bandwidth on demand	no	no	yes	yes
Point-to-multipoint connections	no	yes	yes	yes
Network flexibility	yes	no	yes	yes
Cost flexibility	yes	no	yes	yes

ATM Cell Relay features can be found in the Bellcore Special Report (SR-3330 Issue 1) entitled "Cell Relay Service Core Features" [11]. This document presents a fundamental set of features for both exchange and exchange access PVC-based CRS, where PVC allows transmission across preestablished network connections. Bellcore SR-3330 includes a description of the features that end users can expect from a carrier providing ATM/Cell Relay as well as features that any carrier can expect to be offered by another carrier providing exchange access service. This core set of features will be instrumental in supporting interoperability among network and service providers.

1.2.2.3 Switched Multimegabit Data Service (SMDS)

Switched Multimegabit Data Service (SMDS) [12] is a high-speed, connectionless public switched data service that provides interconnection among LANs, host computers, and high-speed workstations. Customer access for SMDS can be over DS1 or Digital Signal Level 3 (DS3) service-specific access paths. SMDS has been made available to customers by several LECs since 1992.

SMDS adopts the philosophy that the network should deliver only packets that it can determine to be correct. End-to-end error recovery will be done by the customer on-site equipment, for example, by TCP (Transmission Control Protocol), that is attached to the SMDS network. Therefore, providing error recovery functionality in the SMDS network is not necessary, because it would be redundant with the customer's existing higher-layer protocol functionality. The TCP/IP protocol used in the Internet is a connectionless protocol, and its error and congestion control are performed at the transport layer.

1.2.2.4 Circuit Emulation Service

Circuit emulation service provides the transport of Constant Bit-Rate (CBR) signals, such as DS1 and DS3, over ATM. This service is facilitated by interworking functions between today's DS1/DS3 access interfaces and the ATM network. Circuit emulation service may also include both the CBR and Variable-Bit-Rate (VBR) video services that have been discussed in both the ITU-T and the ATM Forum. Circuit emulation service has been offered by some public carriers since 1995.

1.2.2.5 High-Performance Applications

Future networks capable of supporting data transfers between user interfaces at rates in excess of 1 Gbps are referred to as "gigabit networks." For an application to require an underlying high-performance data network, it must be geographically distributed, demand the transmission of large amounts of data, and require low latency. A useful way to characterize potential high-performance applications is to locate them within a delay sensitivity/data volume chart, as shown in Figure 1-1, that characterizes potential high-performance applications and those applications requiring gigabit networks [13].

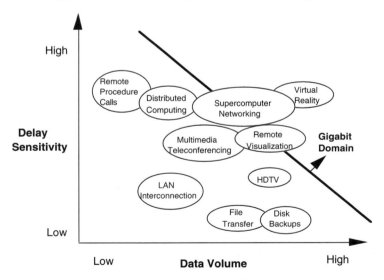

Figure 1-1: High-performance applications [13] (©1992 IEEE).

In general, applications that require large data transfers and low latency will demand high-performance gigabit networks. Some applications likely to require gigabit-per-second communications include supercomputer networking, remote visualization, and virtual reality.

1.2.2.6 Video Dial Tone (VDT) Services

Video Dial Tone (VDT) is the term used by the Federal Communications Commission (FCC) in referring to their concept of a common carrier network service that provides both transport and gateway functionalities that would allow and/or enable end users access to a variety of video information. These VDT network services can be grouped into five classes (1) distribution services; (2) interactive distribution services; (3) retrieval services; (4) interactive services; and (5) conversational services. Table 1-3 shows some applications associated with each VDT service class.

Table 1-3: VDT service class.

Service Class	Applications	Attributes
Distribution	Broadcast Video Subscription TV Interactive TV Pay-Per-View Targeted Advertising	Multicast one-way video
Interactive Distribution	Music Video Juke Box Interactive TV (e.g., game show) Tele-education (live, interactive)	Multicast one-way video Signaling Data path
Retrieval	Movies on Demand Music Video Pay-Per-View Video Mail Box Past TV Infomercials	Point-to-point one-way video Signaling
Interactive	Transaction Services Interactive Video Database Interactive Video Games Tele-education (record/multimedia) Telecommuting	Point-to-point one-way video Signaling
Conversational	Telecommuting Video Telephone	Point-to-point two-way video Signaling Data path as required

VDT services range from basic one-way video services, such as cable TV and pay-per-view, to advanced interactive video services, such as Video on Demand (VoD) and interactive games. VoD services allow subscribers to order movies whenever they want and control movie processing in a VCR-like environment (e.g., pause, rewind, stop, forward, and play). VoD has been considered as one of the "killer applications" that would drive ATM deployment in telecommunications networks. However, due to technological immaturity, high system component costs, and uncertainty of user demands, VoD must remain for now in the technology or service trial stage where most LECs reside today.

1.2.2.7 LAN Interconnection

With large-scale deployment of LANs, the need for LAN interconnection is increasing. Internetworking technologies that may be used to support the necessary QoS for LAN interconnection include FDDI, FR, SMDS, and ATM. FDDI is a high-speed metropolitan-area network running at 100 Mbps with a span up to 100 km. As discussed in Section 1.2.1.3, existing demand for distributing full-motion video may have already surpassed the capability of FDDI, and ATM may be the only viable option to support these demands in terms of scalability and speeds. FR and SMDS are also expected to be carried on the ATM network eventually, following network and operations system consolidation.

1.2.2.8 LAN Emulation (LANE)

To use the vast base of existing application software in computer communications, it is necessary to define ways to emulate services of existing LANs across an ATM network and support them via a higher layer in end systems. At present, there are two methods that may be used to support LAN emulation.

- IP over ATM: This can send an IP packet over ATM LAN, and it has been specified in IETF Request for Proposal (RFC) 1577, "Classical IP and Address Resolution Protocol (ARP) over ATM" [14].

- LAN emulation (LANE): This can send an Ethernet frame and Token-Ring frame over ATM LAN. This means that all higher-layer protocols supported by Ethernet and Token-Ring, such as TCP/IP, AppleTalk, Decnet, System Network Architecture (SNA) can be adopted to ATM LAN by using this mechanism. The specification for LAN emulation has been released by the ATM Forum [15].

1.3 Existing Telecommunications Network Infrastructure

1.3.1 Today's Narrowband Network Infrastructure

In general, there are two major types of network switching techniques, packet switching and circuit switching, that constitute the infrastructure of existing telecommunications networks. The message switching system may be viewed as a special case of the packet switching system. The traditional circuit switching concept is designed to handle and transport stream-type traffic such as voice and video. A circuit-switched connection is set up with a fixed bandwidth for the duration of a connection, which provides fixed throughput and constant delay. In contrast, packet switching is designed primarily to carry data traffic efficiently. Its characteristics include access with buffering, statistical multiplexing, and variable throughput and delay, which is a consequence of the dynamic sharing of communication resources to improve utilization. Today, it is used exclusively for data applications. Examples of circuit-switched networks include telephony networks and private line networks. Examples of packet-switched networks include X.25 (with variable packet sizes) and ATM networks (of fixed cell size).

Today's telecommunications services are supported primarily by three separate switched networks using two major switching concepts. As depicted in Figure 1-2 [16], the circuit-switched network supports voice transport, which includes copper, radio, and/or fiber transmission systems using asynchronous or SONET transmission technology. The signaling system needed for switched voice services is supported by an SS7 (Signaling System no. 7) network that is a specialized, delay-sensitive packet-switched network using a three-transport hierarchy: Service Switching Point (SSP), Signaling Transfer Point (STP), and Service Control Point (SCP). This hierarchical signaling network architecture was designed in the early 1980s based on economic considerations. The network supporting access from the service transport system to the Operations Systems (OSs) is a packet-switched data network such as an X.25 network.

In this section, we discuss the technology and limitations of today's service transport networks. Potential limitations for SS7 signaling networks in the emerging broadband service environment will be discussed in Section 1.5.

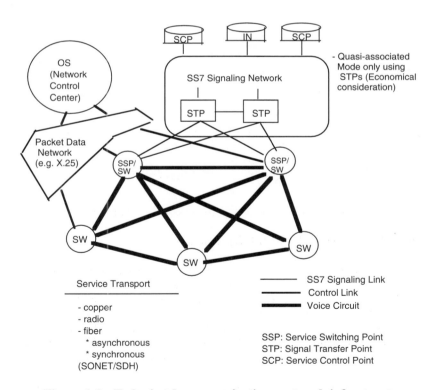

Figure 1-2: Today's telecommunications network infrastructure.

1.3.2 Synchronous Transfer Mode Service Transport Networks

The switching of the service connection for SONET transport networks uses Synchronous Transfer Mode (STM). STM is a time-division multiplexing and switching technology that uses a switching concept similar to that used by the traditional circuit switching system. In the following, we briefly explain the data format used for STM transport and its operational principle. This explanation may clarify why a transport technology that works well today may not be the best choice for emerging broadband services.

1.3.2.1 Channel Format for STM Transport

In SONET/STM, data channels are represented by time slots and identified through the relative time slot location within the frame, as depicted in Figure 1-3(a). STM switching is performed by Time Slot Interchange (TSI). One example of STM equipment is the Digital Cross-connect System (DCS) that cross-connects Virtual Tributary 1.5s (VT1.5s) of 1.728 Mbps and/or Synchronous Transport Signal (STS)-1s of 51.84 Mbps, depending on the type of DCSs (Wideband DCS or Broadband DCS).

Figure 1-3(b) shows an example of the SONET STS-1 frame that corresponds to a frame in Figure 1-3(a). In this example, each VT1.5 path carried on the payload of this STS-1 frame represents a channel (time slot) in Figure 1-3(a). The STS-1 frame is divided into two portions: transport overhead and information payload, where the transport overhead is further divided into line and section overheads. The STS frame is transmitted every 125 ms. The first three columns of the STS-1 frame contain transport overhead, and the remaining 87 columns and 9 rows (a total of 783 bytes) carry the STS-1 Synchronous Payload Envelope (SPE). The SPE contains a 9-byte path overhead that is used for end-to-end service performance monitoring. Optical Carrier Level 1 (OC-1) is the lowest-level optical signal used at equipment and network interfaces. OC-1 is obtained from STS-1 after scrambling and electrical-to-optical conversion.

The following section uses simple examples to explain the STM switching concept used in SONET transport networks.

1.3.2.2 SONET/STM System Configuration and Operations

One example of SONET equipment that performs STS path switching is SONET DCS. Figure 1-4 depicts a simplified SONET DCS system configuration and an operation for facility grooming. In this figure, an incoming OC-48 optical signal is demultiplexed to 48 STS-1s, which are cross-connected to the appropriate output ports destined for appropriate destinations. The STS-1 path cross-connection is performed through the TSI switching matrix. The TSI switching matrix within the DCS interfaces STS-1s/STS-Nc's and cross-connects STS-1s/STS-Nc's. The SONET DCS is primarily deployed at the hub for facility grooming and test access. For example, in the figure, STS1#7 (carried by an OC-48 fiber system between CO1 and the hub) and STS1#52 (carried by an OC-12 fiber system between CO2 and the hub) are destined to CO3. STS1#7 and STS1#52 terminate at different input ports of the DCS and are cross-connected to two output ports that connect to the same fiber system for CO3.

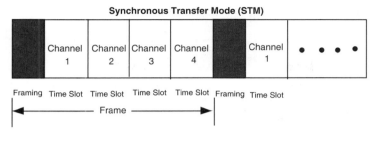

(a) STM (Framing) Concept

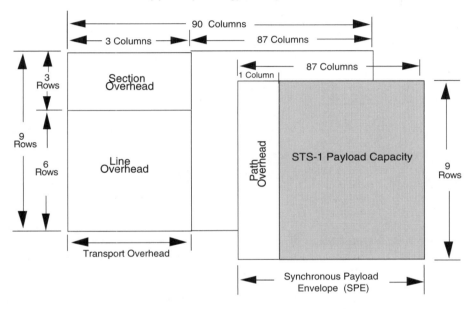

(b) An Example of SONET STS-1 Frame

Figure 1-3: STM concept and SONET STS-1 example.

Conceptually, a TSI can be viewed as a buffer that reads from a single input and writes onto a single output. The input is framed into m fixed-length time slots. The number contained in each input time slot is the output time slot for information delivery. The information in each input time slot is read sequentially into consecutive slots (cyclically) of a buffer of m slots. The output is framed into n time slots, and information from the appropriate slot in the buffer is transmitted on the output slot. Hence, over the duration of an output frame, the content of the buffer is read out in a random manner according to a read-out sequence as shown in Figure 1-5. This read-out sequence is uniquely determined by the connection pattern. In this manner, the information in each slot of the input frame is rearranged into the appropriate slot in the output frame, achieving the function of time slot interchanging. Since the TSI switching matrix performs physical STS path switching from one input port to one output port, TSI rerouting involves the physical switch reconfiguration.

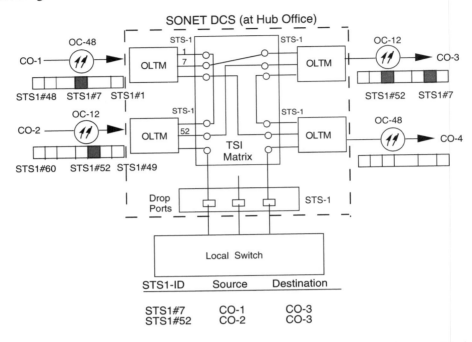

Figure 1-4: An example of SONET/STM system configuration and operation.

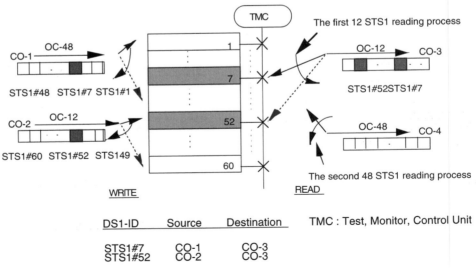

Figure 1-5: TSI switching concept.

1.3.3 Potential Limitations of SONET/STM Transport Networks

SONET/STM networks were originally designed to support private line services and provide a high-speed transport platform for switched services. The network connections supporting these services are either semipermanent or permanent; they are established and torn down through the service provisioning process. The network works well for these aforementioned services because necessary services do not require dynamic bandwidth control capability. However, due to the introduction of new broadband services such as frame relay services, SMDS, and cell relay services that introduce bursty traffic with a variety of bit rates, the SONET/STM network infrastructure may be inefficient enough to be unable to accommodate such bandwidth-on-demand and multirate transport requirements. This inability to meet these requirements is due primarily to the following characteristics of inherent STM technology:

1. The capacity of the connection (i.e., bps) is constant for the entire duration of the connection, even if the information flow in one or both directions is bursty; i.e., a fixed capacity may lead to waste of capacity.

2. In STM it is difficult to switch connections of different bit rates, even if various bit rates are exact multiples of a basic rate.

ATM technology may alleviate these bandwidth inefficiency and multirate transport concerns because of its inherent characteristics. For most broadband services, the stream of ATM cells in either direction of a connection is nonperiodic and its statistical character depends heavily on the nature of the information source. Where a source produces continuous output, the ATM cell stream will be quasi-periodic; however, for other types of sources, the flow of cells in both directions may occur in bursts with idle periods of unpredictable duration in between. Due to the principle of ATM multiplexing, the cells originating from different sources with highly different information rates may easily be combined and switched. When an information source is temporarily not producing any output, no ATM cells are generated and, consequently, no capacity is wasted. Furthermore, if a temporary peak occurs in the output of an information source, a properly dimensioned broadband network can cope with the resulting increase in ATM cells.

1.4 Broadband Integrated Service Network Infrastructure

In the emerging telecommunication network infrastructure, both today's narrowband services and emerging broadband services will be supported by a B-ISDN network where ATM has been adopted by ITU-T as the core technology to support transport. ATM is a high-speed, integrated multiplexing and switching technology that transmits information through uniform (or fixed) cells in the connection-oriented manner. New characteristics of ATM, compared with STM, include nonhierarchical path and multiplexing structure, higher Operations and Maintenance (OAM) bandwidth, separation between path assignment and capacity assignment, and bandwidth on demand. This section discusses these characteristics.

To maximize savings in the B-ISDN environment, ATM transport capability, services, control, signaling, and OAM messages are carried by the same physical network, but they are logically separated from each other by using different Virtual Channels (VCs) or Virtual Paths (VPs) with different QoS requirements. ITU-T Rec. 311 specifies an integrated ATM transport network model that includes one service intelligent layer and two layered transport networks [17]. The two-layered transport network comprises a service transport network and a control transport network that correspond to the user plan and the control plan, respectively. These two networks are coordinated by layer management and plane management systems.

1.4.1 B-ISDN Transport Network Reference Model

Figure 1-6 shows the B-ISDN signal transport protocol reference model defined in ITU-T Rec. I.321. Three layers are defined in the model: Physical, ATM, and ATM Adaptation Layer (AAL), where the Physical and ATM layers form the ATM transport platform.

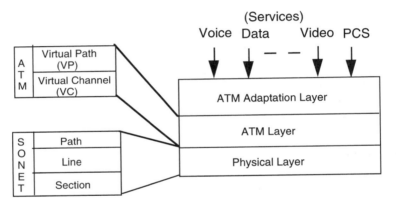

Figure 1-6: B-ISDN reference model.

In North America, the Physical Layer uses SONET standards, and in ITU-T, it uses the SDH or cell-based transmission system. The AAL layer is a service-dependent layer. SONET defines a progressive hierarchy of optical signal and line rates. The basic SONET building block is the STS-1 (Synchronous Signal Transport Signal at Level 1) signal operating at 51.84 Mbps. Higher rate signals (STS-N) are multiples (N) of this basic rate. Values of N currently used in the standards are 1, 3, 12, 24, 48, and 192. The function of the SONET layer is to carry ATM cells in a high-speed and transparent manner, and to provide protection switching capability to ATM cells whenever needed. SONET offers opportunities to implement new network architectures cost-effectively [18], (e.g., SONET self-healing rings), which have been widely deployed by public network carriers around the world since 1991.

The major function of the ATM layer is to provide fast multiplexing and routing for data transfer based on the header information. The ATM layer is divided into two sublayers: Virtual Path (VP) and Virtual Channel (VC). VC is a generic term used to describe unidirectional communication capability for the transport of ATM cells. The VC identifies an end-to-end connection that can be set via either provisioning (permanent VC) or near real-time call setup (switched VC).

The VP accommodates a set of different VCs having the same source and destination. The VPs are managed by network systems, whereas VCs can be managed, end-to-end, by users with ATM terminals. For example, a business customer may be provisioned with a VP to another customer location to provide the equivalence of leased circuits, and another VP to the serving central office for switched services. Each VP may include several VCs for WAN and video conferencing traffic.

The VC and VP are identified through a Virtual Channel Identifier (VCI) and a Virtual Path Identifier (VPI), respectively. For large networks, VCIs and VPIs are assigned on a per link basis (i.e., local significance). Thus, translation of VPI (or VCI) is necessary at intermediate ATM switches on an end-to-end VP (or VC). The basic switching operation principle for ATM cells will be discussed in Section 1.4.5.

The SONET physical link (e.g., an OC-48 optical system) provides a bit stream to carry digital signals. This bit stream may include multiple digital paths, such as SONET STS-3c, STS-12c, or STS-48c. Each digital path may carry ATM cells via multiple VPs, each having multiple VCs. The switching method used for SONET's STS paths is STM using a hierarchical TSI concept; the switching method for VPs/VCs uses a nonhierarchical ATM switching concept. In addition, STM performs network rerouting through physical network reconfiguration, and ATM performs network rerouting using logical network reconfiguration through update of the routing table.

1.4.2 Physical Path vs. Virtual Path

The switching principles used for SONET's STS Paths (STM) and ATM VPs/VCs (ATM) are completely different due to different characteristics of corresponding path structures. The STS path uses a physical path structure, and the VP/VC uses a logical path structure; these are depicted in Figure 1-7. The physical path concept used in the SONET STM system has a hierarchical structure with a fixed capacity for each physical path. For example, the VT1.5 and STS-1 have capacities of 1.728 and 51.84 Mbps, respectively. To transport optical signals over fiber, VT1.5s are multiplexed to an STS-1 and then to STS-12, STS-48 with other multiplexed streams for optical transport. Thus, a SONET transport node may equip a variety of switching equipment needed for each hierarchy of signals. In contrast, the VP transport system is physically nonhierarchical [see Figure 1-7(b)], and its capacity can be varied in a range of zero (for protection) up to the line rate or STS-Nc, depending on applications. This nonhierarchical multiplexing structure provides a natural grooming characteristic that may simplify the nodal system design [19–22].

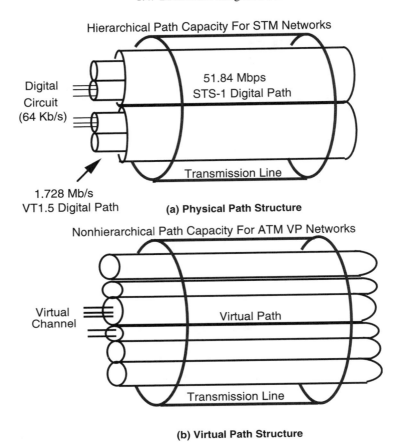

Figure 1-7: Physical path vs. virtual path.

1.4.3 Channel Format for ATM

In contrast with the time slot channel format used on STM networks, ATM channels are represented by a set of fixed-size cells [see Figure 1-8(a)] and are identified through the channel indicator in the cell header [see Figure 1-8(b)]. Thus, ATM switching is performed on a cell-by-cell basis based on routing information in the cell header.

The ATM cell is divided into two parts: the header and the payload. The header has 5 bytes and the payload 48 bytes. As already mentioned, the major function of the ATM layer is to provide fast multiplexing and routing for data transfer based on information included in the 5-byte header. This 5-byte header includes information not only for routing, but also fields used to (1) indicate the type of information contained in the cell payload (e.g., user data or network operations information), (2) assist in controlling the flow of traffic at the User–Network Interface (UNI), (3) establish priority for the cell, and (4) facilitate header error control and cell delineation functions.

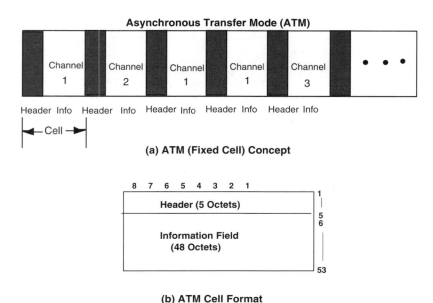

(a) ATM (Fixed Cell) Concept

(b) ATM Cell Format

Figure 1-8: ATM concept and cell formats.

One key feature of this ATM layer is that ATM cells can be independently labeled and transmitted on demand. This allows a facility bandwidth to be allocated as needed, without the fixed hierarchical channel rates required for STM networks. Since ATM cells may be sent either periodically or randomly (i.e., in bursts), both constant- and variable-bit-rate services are supported at a broad range of bit rates. The connections supported at the VP layer are either permanent or semipermanent connections that do not require call control, bandwidth management, or processing capabilities. The connections at the VC layer may be permanent, semipermanent, or switched connections (i.e., SVCs). SVCs require the signaling system to support its establishment, tear-down, and capacity management.

ATM cells are transported through the SONET STS-3c (STM-1) or STS-12c (STM-4) path. (The "c" stands for "concatenated"; see Section 2.2.1.) Other ATM cell mappings, such as ATM on DS1, can be found in The ATM Forum [23].

1.4.4 ATM Adaptation Layer (AAL)

The ATM Adaptation layer (AAL) is a protocol layer that performs the function of adapting services onto the ATM layer protocol. It represents a link between particular functional requirements of a service and the generic, service-independent nature of ATM transport. Depending on the service, the AAL can be used by end customers only [i.e., Customer Premise Equipment (CPE) having the ATM capability], or it can be terminated in the network (e.g., SMDS server at the ATM node for connectionless routing).

The ATM Forum has defined ATM service classes (i.e., Classes A, B, C, D) based on three parameters: delay, bit rates, and connection modes. Class A deals with connection-oriented services with constant bit rates (CBRs); their timing at the source and receiver are related. One example of Class A service is voice service. Class B represents connection-oriented services with variable bit rates (VBRs), and related source and receiver timing. These services are real-time. An example of Class B service is VBR video. Class C deals with bursty connection-oriented services with variable bit rates that do not require a timing relationship between the source and the receiver. An example of Class C service is connection-oriented data services such as X.25 and file transfer. Class D is essentially the same as Class C except that the services are connectionless. An example of Class D service is SMDS.

In addition to these four AAL service models, other AAL classes are under consideration in standards groups to accommodate newly created ATM service classes such as Available Bit Rate (ABR) and Unspecified Bit Rate (UBR). The ABR, primarily used in LAN and TCP/IP environment, is similar to the CBR, except that it provides variable data rates based on whatever is available through its use of the end-to-end flow control system. In contrast, a UBR user does not specify the required bit rate, and cells are transported by the network whenever the network bandwidth is available and no flow control is used for the UBR. We will discuss these ATM service classes in more detail in Chapter 3. Some specifications for CBR, VBR, ABR, and UBR can be found in The ATM Forum [24].

For near-term applications, three types of AAL has been identified: AAL Type 1, Type 3/4, and Type 5. AAL Type 1 has an available cell payload of 47 bytes for data and is used to carry Constant-Bit-Rate (CBR) applications, such as the transparent transport of a synchronous DS1 through the asynchronous ATM network. AAL Type 3/4 has an available cell payload of 44 bytes and is designed for error-free transmission of Variable-Bit-Rate (VBR) information. For example, this AAL type is used for connectionless SMDS. AAL Type 5 is used for supporting VBR data transfer with minimal overhead (i.e., an available cell payload of 48 bytes for data). It is primarily used to transport Frame Relay Service and user–network signaling messages over ATM. In addition, a null AAL can be used to provide the basic capabilities of ATM switching and transport directly, as in Cell Relay Service. End customers may use proprietary AALs for special applications as well.

1.4.5 ATM Switching Operations

In this section, we will show an example of ATM switching operations. Figure 1-9 illustrates a simplified ATM cell switching system configuration and a switching principle. The ATM cell entering input port #3 of the switch arrives with VPI = 9 and VCI = 4 (e.g., the first ATM cell in the figure). The call processor has been alerted through the routing table that the cell must leave the ATM switch with VPI = 4 and VCI = 5 on output port #4. The call processor directs the virtual channel identifier converter to remove the "9" VPI and "4" VCI and replaces them with "4" and "5," respectively. If the switch is a large multistage switch, the call processor further directs the virtual channel identifier converter to create a tag that travels with the cell, identifying the internal multistage routing within the ATM switch matrix. Note that ATM cells belonging to the same VP must have the same VPIs in the input port (e.g., VPI = 9 in input port #3) or the output port (e.g., VPI = 4 in output port #4).

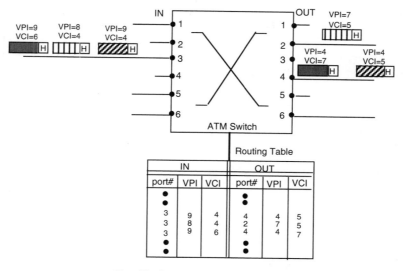

Note: The first and the third cells are in the same VP.

Figure 1-9: An example of ATM cell switching principle.

1.4.6 Comparison between STM and ATM

Table 1-4 summarizes the differences between the STM and ATM switching concepts [16]. In general, the STM system switches signals on a physical and hierarchical basis; the ATM switches signals on a logical and nonhierarchical basis, due to their corresponding path multiplexing structures. Thus, ATM may be more flexible than its STM counterpart in terms of bandwidth on demand, bandwidth allocation, and transmission system efficiency, but a relatively complex control system is required. As STM physically switches the signals, its rerouting requires physical switch reconfiguration that may make network rerouting slower than its ATM counterpart, which only requires the update of the routing table for large-scale networks (e.g., large-scale mesh networks). We will discuss network rerouting subjects in detail in Chapters 6 and 7.

Table 1-4: Comparison between STM and ATM.

System Parameters	STM	ATM
Switching unit	time slot (STS paths)	fixed-size cell (VPs/VCs)
Path structure	physical and hierarchical	logical and non-hierarchical
Path capacity hierarchy	limited	more
Switching system complexity	simpler	complex
Switching nodal system	complex (hierarchical)	simpler (non-hierarchical)
Transmission system efficiency (in average)	lower	higher
Method of rerouting	physical network reconfiguration	logical network reconfiguration
Bandwidth on demand	difficult	easier

1.4.7 End-to-End Broadband Transport Path

Figure 1-10 depicts a relationship between the SONET transmission system and STM and ATM switching systems. The end-to-end ATM connection is established and transported through a set of transmission links that are terminated at nodes for processing and switching (called the transfer mode). The nodal transfer mode may be either existing STM or emerging ATM, depending on the types of services and QoS being supported.

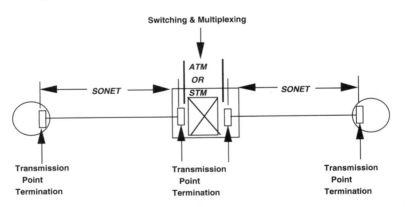

Figure 1-10: Relationship between SONET transmission and STM and ATM switching.

Figure 1-11 depicts two examples of SONET/ATM network configurations that reflect the concept that is depicted in Figure 1-10. In the example shown in Figure 1-11(a), the primary transport function is performed at the SONET layer using STM technology and the end-to-end path is terminated at the ATM layer. In Figure 1-11(b), the transport function is performed at the ATM layer using ATM Virtual Path Cross-Connect Systems (VPXs) supporting the end-to-end ATM VC connection.

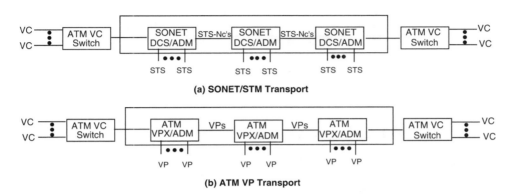

(a) SONET/STM Transport

(b) ATM VP Transport

Figure 1-11: Examples of SONET/ATM networks.

1.5 Broadband Signaling Transport Networks

Today's switched voice services in public telephony networks are supported primarily by the SS7 signaling network, a delay-sensitive packet data network. As public carriers plan to offer new broadband services and consolidate different types of services into a single ATM network platform, how today's SS7 signaling networks will evolve to meet future broadband signaling needs is an important and timely issue, especially as many public carriers plan to offer switched broadband services in 1997–98. The identification of an appropriate target broadband signaling transport network architecture is necessary to ensure smooth and cost-effective signaling network evolution. This section briefly reviews the progress made to date on broadband signaling transport. Details will be discussed in Chapter 4.

1.5.1 Today's SS7 Signaling Network Architecture

The SS7 architecture comprises three major components: SSP (Service Switching Point), STP (Signal Transfer Point), and SCP (Service Control Point). All nodes in the network that have CCS (Common Channel Signaling) capability are SSPs that are interconnected by signaling links. Nodes that serve as intermediate signaling message transport switches are called STPs. SCPs are the SSPs that provide database access to support transaction-based services such as 1-800 service, Personal Communication Service (PCS), and others. Signaling links are the transmission facilities that convey the signaling messages between two SSPs.

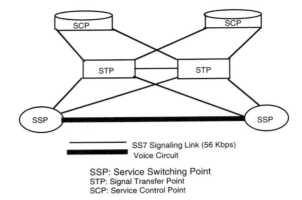

SS7 Signaling Link (56 Kbps)
Voice Circuit

SSP: Service Switching Point
STP: Signal Transfer Point
SCP: Service Control Point

Figure 1-12: SS7 network configuration (quasi-associated mode).

The STP is engineered on a paired basis to enhance reliability. The SS7 network is physically separated from the Plain Old Telephone Service (POTS) transport network because the SS7 network is a packet-switched network and the POTS network is a circuit-switched network. There are two possible signaling modes defined in SS7 standards: the associated mode and the quasi-associated mode. The associated signaling mode is referred to as point-to-point signaling transport between an SSP pair or an SSP–SCP pair, but for the quasi-associated mode, the SSP pair or the SSP–SCP pair communications must be through the STP. Although the SS7 network architecture allows for supporting both the associated and quasi-associated signaling modes, today's SS7 networks in North America implement only the quasi-associated mode (see Figure 1-12) for strictly economic reasons. However, as will be discussed later in this section, the design assumptions used to optimize the SS7 network in early 1980s may not fully apply to the new ATM network environment (e.g., the service and signaling messages may be carried on the same physical ATM network with overlay end-to-end connections). Note that in European countries the associated signaling mode is used in SS7 networks; Japan's NTT, though, has implemented both the associated and the quasi-associated signaling modes in its SS7 network.

1.5.2 Potential Limitations of SS7 Network Architecture

A number of studies regarding the impact of the introduction of new services on today's SS7 signaling networks have been conducted and reported [25–29]. These studies evaluated the signaling network capacity and delay impacts due to the penetration of PCS and Advanced Intelligent Network (AIN) services. The results in [25] have suggested, from a delay perspective, that use of today's 56 Kbps links may lead to unacceptable network response times, particularly for services requiring long network paths and extensive user–network or SCP–SCP interactions. Higher-speed signaling links are needed to alleviate this signaling delay performance problem. It has been reported that the total network response time improves by approximately 30 percent when 56 Kbps links are replaced by 1.5 Mbps links under the same load and network configuration. In addition, the results in [25] suggest that a significant increase in the speed of SCP A-links will be needed, and the number of SCPs increases significantly due to the capacity limitation of the incoming SCP A-linksets (which, therefore, limit the traffic entering the

SCP). These potential limitations would eventually result in increased network complexity and costs due to additional investments in new SCP, reconfiguration of network components (nodes and links), and the additional resources needed to maintain and administer the network.

There is no impact analysis presently available for broadband services on the SS7 network capacity. However, it is expected that the network capacity requirement for broadband signaling will be much higher than its narrowband counterpart due to its more intelligent nature and the need to support both point-to-point and point-to-multipoint connections. It has also been suggested that existing SS7 network physical interfaces may not be able to support stringent end-to-end signaling time goals (20–100 ms), which are requested by potential users of the broadband SVC (Switched Virtual Connection) services.

Luetchford *et al.* [26] have identified the enhancements needed to meet the signaling requirements of network services, intelligent networks, mobility management, mobility services, broadband services, and multimedia services. These broadband signaling requirements include

- Increased signaling link speeds and processing capabilities
- Increased service functionality, such as version identification, mediation, billing, mobility management, quality of service, traffic descriptors, and message flow control; (Note that it requires bandwidth-on-demand capability)
- Separate call control from connection control
- Reduce operations costs (including provisioning) for services and signaling; (Note that it requires flexibility of connection establishment)

These broadband signaling requirements demand a signaling network infrastructure that is much faster, more flexible, and more scalable than the existing dedicated SS7 signaling network. ATM by nature is a viable choice of signaling transport to meet these new broadband signaling requirements.

1.5.3 Role of ATM Technology in Broadband Signaling

Several approaches have been proposed [25–27] to alleviate the potential limitations of today's SS7 networks. The ATM signaling network platform, due to its flexible connection and bandwidth management capabilities, may be better suited to accommodate signaling traffic growth and the stringent delay requirement. In general, in a comparison with the SS7 network, the ATM signaling network has more flexibility in establishing connections and allocating needed bandwidth on demand. The bandwidth allocation for each ATM signaling connection can be 173 cells per second (approximately 66 Kbps) or multiples of that, up to 23 (i.e., approximately 1.5 Mbps) [29], depending on the application or service being supported. This bandwidth-on-demand feature makes the ATM network attractive for supporting services with unpredictable or unexpected traffic patterns, such as PCS and new broadband multimedia services. In the next subsection, some potential long-term broadband signaling transport network architectures based on the progress of ITU-T and the ATM Forum will be discussed.

1.5.4 Potential ATM Signaling Network Architectures

Signaling is a family of protocols used for call and connection setup. ATM signaling is a set of protocols used for call/connection setup over ATM interfaces. Two ATM signaling network architectures having very different design philosophies have been specified by both the North American and international standards groups. One is designed for public networks and defined by ITU-T, and the other is designed for enterprise networks and specified by the ATM Forum. The latter signaling architecture is called Private Network-to-Network Interface or Private Network Node Interface (P-NNI) [30]. The different signaling network design philosophies between public and enterprise networks are due primarily to the different nature of public and enterprise networks. Major differences between the public networks and enterprise networks are in network size, stability frequency, nodal complexity, and intelligent residence. In public networks, such as public telecommunications networks operated by LECs, an interoffice network is generally on the order of up to several hundred nodes. A careful, long planning process for node additions and deletions is required. In contrast, an enterprise network could easily extend to thousands, even tens of thousands, of nodes, and frequent node deletion and addition is anticipated. The network transport, control, and management capabilities in the public network node are much more complex, reliable, and expensive than its enterprise counterpart. This implies that intelligence in the public networks is designed primarily in the network nodes, and that in enterprise networks these intelligent capabilities reside in customer premise equipment (e.g., LAN environment).

The ATM signaling network architecture specified by ITU-T for public networks has evolved from Narrowband ISDN (N-ISDN) signaling design philosophy, and the Private Network Node Interface (PNNI) [30] specified by the ATM Forum for enterprise networks has evolved mostly from TCP/IP design philosophy. These two emerging broadband signaling network architectures create a challenging task for network providers, who need these signaling capabilities to offer broadband switched services in terms of the implementation time frame, QoS support, and capital investment. For example, due to slow progress in ITU-T on the NNI signaling network architecture and protocols, some network planners and engineers have been trying to use a simplified PNNI (i.e., source routing with nonhierarchical routing structure) for public network signaling, as the PNNI version 1.0 was approved by the ATM Forum in early 1996 and the switch equipment with PNNI capability is available in early 1997. However, the primary concern in using this approach is whether or not the PNNI can support the stringent QoS and reliability needed for public networks. This challenge remains.

1.5.4.1 Enterprise ATM Signaling Approach

Enterprise ATM network routing and signaling based on a TCP/IP-like structure and hierarchical routing philosophy have been specified in PNNI [30] by the ATM Forum. PNNI is important to ATM enterprise networks because not only does it contain signaling for SVCs and dynamic routing capabilities, it also allows the ATM enterprise network to be scaled to a large network. An earlier version, known as Interim Interswitch Protocol (IISP), allowed users to establish call routing tables manually so as to enable SVC interoperability; however, QoS support was not provided.

PNNI is a hierarchical, link-state routing protocol similar to Open Shortest Path First (OSPF). The protocol performs two roles. The first is to distribute topology information between switches and clusters of switches used to compute routing paths from the source node (called "source routing") through the network. This routing protocol uses a hierarchical mechanism to ensure network scalability and has the ability to automatically configure itself in networks in which the address structure reflects topology. In addition, the path must support a connection's requested bandwidth and QoS based on parameters such as maximum cell rate, available bit rate, cell transfer delay, and cell loss ratio. Note that the routing path for signaling and the routing path for service data are the same under the PNNI routing protocol, because the service transport path is established by signaling path tracing.

The second function is to use the signaling protocol based on Q.2931 to establish point-to-point and point-to-multipoint connections across the ATM network and to enable dynamic alternative rerouting in the event of a link failure. If a connection goes down, the PNNI can reestablish the connection over a different route without manual intervention. This signaling protocol is based on the ATM Forum's UNI signaling with additional features that support source routing, crankback, and alternative routing of call setup requests when there has been a connection setup failure.

The ATM addressing structures defined in the ATM Forum not only can be used in enterprise ATM address structure, but also are applicable to public networks if the public E.164 address is included. Thus, theoretically, PNNI can be applied to both enterprise and public networks. This addressing option may allow public network providers to run PNNI in its edge switches to integrate with private networks.

1.5.4.2 Public ATM Signaling Approach

Two major aspects of changing requirements from SS7 signaling networks to future broadband signaling networks are (1) the evolution of the signaling user parts and (2) the evolution of signaling transport in the broadband environment. The evolution of signaling user parts [i.e., B-ISDN User Part (B-ISUP)] has occurred quickly over the past few years, while at the same time work on broadband signaling transport is underway in ITU-T (e.g., ITU-T Recs. I.311 and Q.2010) [17,31]. The technical aspects of broadband signaling transport that remain to be addressed include signaling transport architectures and protocols. These architectures and protocols may be used in the ATM environment to provide reliable signaling transport, while also making efficient use of the ATM broadband capabilities in support of new, vastly expanded signaling applications. Some potential benefits of using an ATM transport network to carry the signaling/control messages include possible simplification of the existing signaling transport protocols, shorter control and signaling message delays, and reliability enhancement via the possible self-healing capability at the VP level [17].

A number of suggestions have been made for possible broadband signaling transport architectures, ranging from the retention of Signal Transfer Points (STPs) to the adoption of a fully distributed signaling transport architecture supporting the associated signaling mode only. Wu *et al.* [27] discusses three possible ATM signaling transport network architectures and presents some preliminary results for architecture comparison. These possible architectures are classified based on the signaling message routing principle: (1)

the quasi-associated mode only, (2) the associated mode only, and (3) a hybrid of the quasi-associated mode and the associated mode. For architecture options (1) and (3), the Message Transfer Part Levels 1 and 2 (MTP-1 and MTP-2) functions provided by the current STP will be replaced by ATM and AAL layer functions in the ATM-based signaling transport network. The first architecture evolving from today's SS7 networks uses ATM technology as a one-to-one replacement of the dedicated data link. The second architecture tends to use ATM's unique characteristics to meet the broadband signaling requirements. The third architecture is based on a design assumption that the quasi-associated mode will be used in the low signaling traffic environment, whereas the associated mode will be used in a highly connected mesh network [17]. Details of these architectures and analysis will be presented in Chapter 4.

1.5.4.3 Signaling Network Interconnection

A crucial capability for public service providers is the Broadband InterCarrier Interface (B-ICI) specification, of which Version 2.0 [32] was approved by the ATM Forum in April 1996. B-ICI enables the interconnection of carrier networks for ATM services. With most carrier interconnections today based on PVCs only, the B-ICI, in conjunction with a set of Bellcore generic requirements GR-1115-CORE [33], will enable SVC-based interconnected services. Although B-ICI is designed to interconnect carrier networks, it has been suggested by some researchers and engineers that it may be considered as part of the Network Node Interface (NNI) signaling protocol for public networks. Further study is needed for this proposal.

Among the great challenges for a network of networks is to integrate public and private networks. The PNNI will be especially useful in building large private networks. One working item in the ATM Forum is a joint effort of the B-ICI and the PNNI working groups to enable public–private network interworking. Items to be studied include whether a public service provider could run PNNI on the edge switches that interface with private networks, and how to reconcile different addressing schemes. Public networks generally use an international addressing standard known as E.164, and the PNNI uses ATM addressing.

1.6 ATM Traffic Management

1.6.1 ATM Traffic Management Objectives and Functions

Network Traffic Management (NTM) uses network resources effectively during network congestion/failure periods to provide the required QoS to customers. To accomplish this, three sets of NTM functions are needed: measure of congestion (MOC), surveillance, and control functions [24,34,35]:

1. *Measures of Congestion (MOCs)* are defined at the ATM level based on measures such as cell loss, buffer fill, utilization, and other criteria.

2. *NTM surveillance functions* are used to gather network usage and traffic performance data to detect overloads as indicated by MOCs.

3.   *NTM control functions* are used to regulate or reroute traffic flow to improve traffic performance during overloads and failures in the network.

The key issues associated with ATM network traffic management include

- How QoS is defined at the ATM layer
- How users define their particular traffic characteristics that a network can recognize and use them to monitor traffic
- How the network measures traffic to determine if the call can be accepted or if congestion control should be triggered
- How the network avoids congestion whenever and wherever possible
- How the network reacts to network congestion to minimize effects

The following characteristics of high-speed ATM networks have made ATM traffic management design much more complex than those designs used for present low-speed packet- and circuit-switched networks:

- Various B-ISDN's VBR sources generate traffic at significantly different rates (few Kbps–hundreds of Mbps) with very different QoS requirements.
- Traffic characterization of various B-ISDN services is not well understood.
- A single source may generate multiple connections with different types of traffic patterns and characteristics.
- High-speed transmission speed results in a large number of cells remaining in the network.
- High-speed transmission speed limits the available time for message processing at immediate nodes.

ATM network traffic management includes proactive ATM network Traffic Control and reactive ATM network Congestion Control. ATM network Traffic Control is the set of actions taken by the network to avoid congested conditions. ATM network Congestion Control is a set of actions taken by the ATM network to minimize intensity, spread, and duration of congestion, where these actions are triggered by congestion in one or more network elements.

The objectives of ATM network Traffic Control and Congestion Control for B-ISDN are as follows:

1.   ATM network Traffic Control and Congestion Control should support a set of ATM layer QoS classes sufficient for all foreseeable B-ISDN services.

2.   ATM network Traffic Control and Congestion Control should not rely on AAL protocols that are B-ISDN service-specific, nor on higher-layer protocols that are application-specific. Protocol layers above the ATM layer may make use of information provided by the ATM layer to improve QoS performance at these layers.

3.   The design of an optimum set of ATM layer traffic controls and congestion controls should minimize network and end-system complexity while maximizing network utilization.

The ATM layer QoS is defined by a set of parameters such as delay, delay variation, and cell loss ratio. Other QoS parameters are being studied by ITU-T, the ATM Forum, and other industry forums [24,34–37]. A user requests a specific ATM layer QoS from QoS classes provided by a network. This is part of the Traffic Contract at connection establishment. It is a commitment for the network to meet the requested QoS as long as the user complies with the Traffic Contract. If the user violates the Traffic Contract, the network need not respect the agreed QoS.

The values of the traffic contract parameters can be specified either explicitly or implicitly. A parameter value is explicitly specified when it is specified by the user via Q.2931 access signaling messages for SVCs or when it is specified via a provisioning system for PVCs. A parameter value is implicitly specified when its value is assigned by the network operator using default rules, which, in turn, can depend on the information explicitly specified by the user. A default rule is the rule used by a network operator to assign a value to a traffic contract parameter that is not explicitly specified.

The Connection Traffic Descriptor, which constitutes a traffic contract, consists of all parameters needed to specify unambiguously the conforming cells of the ATM connection. For example, these parameters include peak cell rate, sustainable cell rate, burst tolerance, and cell delay variation. In general, traffic parameters in a Connection Traffic Descriptor should fulfill the following requirements: (1) they can be understood by the user or terminal equipment; (2) they are useful in resource allocation schemes to meet network performance requirements; and (3) they are enforceable by the network.

Network performance objectives at the ATM layer are intended to capture the network's ability to meet the requested ATM layer QoS. It is the role of the upper layers, including the AAL, to translate this ATM layer QoS to any specific application requested QoS. For example, according to video requirements specified in ITU-T SG13's Integrated Video Services baseline document [37], the acceptable Cell Loss Ratio (CLR) for a two-hour movie may not exceed the order of 10^{-8}, where one cell loss in a two-hour time period is equivalent to the cell loss ratio on the order of 10^{-8} for 1.5 and 4 Mbps video.

1.6.2 ATM Traffic Control Functions

The goal of ATM traffic control is to protect the ATM network from congestion. To meet this goal, the following functions are needed and may be used in appropriate combinations.

- Connection admission control (CAC)
- Usage/network parameter control (UPC/NPC) (traffic policing)
- Traffic shaping
- Feedback control

1.6.2.1 Connection Admission Control (CAC)

Connection Admission Control (CAC) is defined as the set of actions taken by the network during the call setup process (or during the call renegotiation process) to determine whether a virtual channel or virtual path connection request is to be accepted or rejected (or whether a request for bandwidth reallocation can be accommodated). The routing is part of connection admission control actions.

On the basis of CAC in an ATM network, a connection request is accepted only when sufficient resources are available to establish the connection through the network at its required QoS [and the Cell Delay Variation (CDV) tolerance, if any] and to maintain the agreed QoS of existing connections. This applies as well to renegotiation of connection parameters within a given call. The CAC makes use of the information derived from the traffic contract to determine (1) whether the connection can be accepted; (2) those traffic parameters needed by usage parameter control; and (3) routing and network resource allocation.

1.6.2.2 Usage/Network Parameter Control (UPC/NPC) (Traffic Policing)

Usage Parameter Control (UPC) and Network Parameter Control (NPC) are defined as a set of actions taken by the network to monitor and control traffic, in terms of traffic offered and validity of the ATM connection, at the user access and the network access, respectively. The main purpose is to protect network resources from malicious as well as unintentional misbehavior that can affect the QoS of other existing connections by detecting violations of negotiated parameters and taking appropriate actions.

For each cell arrival, the traffic policing mechanism determines whether the cell conforms with the Traffic Contract of the connection; thus, this mechanism is used to provide a formal definition of traffic conformance to the Traffic Contract. The Leaky Bucket method is the major principle used to implement a traffic policing function. The basic idea of Leaky Bucket method is that a cell must obtain a token from the token pool before entering the network. An arrival cell will consume one token and then immediately depart from the leaky bucket if there is at least one token available in the token pool. Tokens are generated at a constant rate and placed in the token pool. There is an upper bound on the number of tokens that can be waiting in the pool; thus, tokens arriving when the token pool is full are discarded. To police the peak rate of VPs only, it is sufficient to set the token generation rate to a VP bandwidth. For example, policing a 10 Mbps VP would require one token generated every 42.4 ms.

There are several types of enforcement actions that can be used with the leaky bucket method. Each has advantages and disadvantages in terms of network utilization, delay, and control system complexity. We will discuss these mechanisms in detail in Chapter 5.

1.6.2.3 Traffic Shaping

ATM network congestion is sometimes caused by the burst effects of ATM traffic. To minimize the possibility of network congestion, traffic shaping may be used to reduce the level of bursty traffic. Traffic shaping is a mechanism that alters the traffic characteristics of a stream of cells on a VCC or a VPC to achieve a desired modification of those traffic characteristics. Traffic shaping must maintain cell sequence integrity on an ATM connection. Examples of traffic shaping include peak cell rate reduction, burst length limiting, reduction of CDV by suitably spacing cells in time, and queue service schemes.

Traffic shaping may be used in conjunction with suitable UPC functions, provided the additional delay remains within the acceptable QoS negotiated at call setup. It may also be used within the customer equipment or the terminal to ensure that the traffic generated by the source or at the UNI conforms to the traffic contract.

The amount of bandwidth reserved for a connection falls between the average rate and the peak rate. For most VBR applications, cells are generated at the peak rate during the active period but no cells are generated during the silent period. The purpose of traffic shaping is to buffer cells before they enter the network so that the departure rate is less than the peak arrival rate of cells (but still greater than the average rate). For example, the equivalent bandwidth for a LAN interconnection may be decreased from 15.8 to 2.8 Mbps (see Figure 5-12, Chapter 5) when the traffic shaper is used. However, the use of the traffic shaper introduces delays that may not be appropriate for delay-sensitive services or applications, such as signaling. Thus, the trade-off analysis between the network delay and network effectiveness on congestion would help determine the application of traffic shaping systems.

1.6.2.4 Feedback Control

Feedback control is defined as the set of actions taken by the network and by users to regular the traffic submitted to ATM connections according to the state of network elements. The purpose of feedback control is to coordinate user traffic volume and the available network resource to avoid potential network congestion.

1.6.3 ATM Network Congestion Control

Network congestion is defined in ITU-T Rec. I.371 [34] as a state of network elements (e.g., switches, concentrators, cross-connects, and transmission links) in which the network cannot meet the negotiated network performance objectives for the already established connections and/or the new connection requests. Note that congestion should be distinguished from the state in which buffer overflow causes cell losses but still meets the negotiated QoS.

There are two possible causes of network congestion. The first is unpredictable statistical fluctuations of traffic flows in normal conditions. The second possible cause is when the network is under fault conditions. These fault conditions could be software faults and/or hardware failures. Software faults typically cause undesired rerouting that would exhaust some particular subset of network resources. Hardware failure can be overcome by using network restoration procedures that may require network resources to compete with existing unaffected connections in an ATM network.

The Explicit Forward Congestion Indication (EFCI) is a congestion mechanism that the ATM service user may employ to improve QoS performance. A network element in an impending-congested state or a congested state may set an explicit forward congestion indication in the cell header so that this indication may be examined by the destination CPE. For example, the end user's CPE may use this indication to implement protocols that adaptively lower the cell rate of the connection during congestion or impending congestion. A network element that is not in a congested state or an impending-congested state will not modify the value of this indication. An impending congestion state is that state during which a network equipment is operating at its engineered capacity level. Note that the mechanism by which a network element determines whether it is in an impending-congested state is an implementation issue and is, therefore, not subject to standardization.

When it is determined to be in a congested state, the network element may selectively discard cells with low Cell Loss Priority (CLP) (i.e., the CLP = 1 flow), while still meeting network performance objectives on both the high priority (i.e., CLP = 0) and aggregated flows (with CLP = 0).

There are three possible congestion control mechanisms that could be implemented in ATM networks. The first mechanism is Priority Control and Selective Cell Discard. In this scheme, the user may generate different priority traffic flows by using the Cell Loss Priority (CLP) bit. A congested network element may selectively discard cells with low priority if necessary to maintain to the extent possible the network performance for cells with high priority.

The second congestion control mechanism is the credit-based congestion control system that performs congestion control on a link-by-link basis based on credits allocated to the node. The third mechanism is the rate-based congestion control system that adjusts the access rate based on the end-to-end or segmented network status information. The ATM Forum has voted that the rate-based control mechanism should be used in ATM networks [24]. We will discuss these congestion control schemes fully in Chapter 5.

1.6.4 Traffic Management Options and Time Scale

Functions needed in ATM traffic management and the response time scale at which they are most effective were discussed in ITU-T Rec. I.371 [34]. Each class of control is applicable at a different response time scale, where the response defines how quickly the controls react. Congestion control functions can necessarily operate at response time scales greater than the propagation delay, whereas traffic control techniques are effective at cell transmission times due to their propagation long-term resource provisioning. For example, cell discarding can react on the order of the insertion time of a call. Similarly, feedback controls can react on the time scale of round-trip propagation times. Since traffic control and resource management functions are needed at different response time scales, no single network traffic management function is likely to be sufficient.

1.7 Broadband Transport Network Restoration

Network survivability is an issue of great concern to any part of the telecommunications industry that wants to deploy high-capacity broadband networks, because loss of broadband services in high-capacity networks due to disasters and catastrophic failures could be devastating and very likely would result in significant revenue loss as well. However, providing protection against broadband network failures could be very expensive due to the high costs associated with broadband transport equipment as well as the requirement of advanced control capability. Thus, how to reduce network protection costs while maintaining an acceptable level of survivability has become a crucial challenge for those network providers seeking to position themselves competitively in today's emerging telecommunications and information networking market. Technology advancement is certainly crucial for meeting this challenge, especially in the future B-ISDN environment [38–40]. This section briefly reviews the network restoration technologies that will be used to protect the broadband network from network failures. Detailed discussions on broadband network survivability will follow in Chapter 2 (for survivable SONET/SDH networks), Chapter 6, and Chapter 7 (for survivable SONET/ATM networks).

1.7.1 Network Restoration Concept and Methods

The function of network restoration is to reroute new and existing connections around the failure area if a network failure occurs. One network restoration objective is to provide a cost-effective, acceptable level of network survivability. ITU-T has defined three network protection schemes (i.e., protection switching, rerouting and self-healing) in ITU-T Rec. I.311. Protection switching is the establishment of a preassigned replacement connection by using equipment but no network management control function. An example of protection switching is an Automatic Protection Switching (APS) system.

Rerouting is the establishment of a replacement connection by the network management control connection. When a connection failure occurs, the replacement connection is routed depending on network resources available at that time. An example of rerouting is the centralized control DCS network restoration. Self-healing is the establishment of a replacement connection by a network without utilizing a network management control function. When a connection failure occurs, the replacement connection is found by the network elements and rerouted depending on network resources available at that time. Examples of self-healing include the distributed control DCS network restoration and SONET self-healing rings.

In addition to these network restoration techniques, network restoration of SONET/ATM networks can be performed at the physical (SONET) or ATM Layer [16,39]. For physical layer protection, service restoration is achieved through STM technology at the STS-1, STS-3Nc level (where N = 1, 4, or 16), or at the physical transmission link level. SONET networks restore services on physical paths (e.g., STS-3c) and/or links (e.g., SONET links) through physical network reconfiguration using Time Slot Interchange (TSI) technology. In contrast, ATM networks restore services carried on VPs/VCs through a logical network reconfiguration by modifying the VPI/VCI routing table.

For ATM layer protection, network restoration can be performed at the VC or VP (or group of VPs) level. SONET/STM network protection is through simple and cost-effective self-healing capability (such as SONET self-healing rings). When the network moves to ATM transport technology for accommodating bandwidth-on-demand requirements, the challenge comes from whether or not the SONET layer is sufficient to provide protection for new broadband services that may have very different delay and cell loss requirements. Determining which layer (SONET, VP, or VC) is an appropriate restoration layer depends on the SONET/ATM network architecture and other factors, such as costs and QoS. Note that unlike STM transport, in which the QoS is affected only when the network components fail, the QoS in ATM layer transport can be affected not only by network failure conditions, but also by network congestion conditions (due to ATM queuing characteristics).

1.7.2 ATM Technology Impact on Network Restoration

SONET network restoration systems, such as Automatic Protection Switching (APS) systems and SONET self-healing rings, have been widely commercially deployed around the world. The effects of emerging ATM technology on network restoration (compared with SONET network restoration) may be significant, due to the following factors:

* Separation of capacity allocation and physical route assignment for VPs/VCs that would reduce required spare capacity (e.g., the capacity of protection VPs/VCs can be zero in normal conditions).

* More OAM bandwidth with allocation on demand that would reduce delays for restoration message exchanges, and detection of system degradation (i.e., soft failure) in the ATM network faster than in its SONET counterpart.

* Nonhierarchical path multiplexing (for VPs) that would simplify the survivable network design and help reduce intranode processing delays and required spare capacity.

The layer of ATM network protection depends on whether the ATM/VC or VP transport network is considered. The ATM/VC transport network involves a large number of smaller ATM switches that terminate at VPs and VCs. The transmission path between two adjacent ATM switches is SONET/STS path. The alternative transport system is the VP transport network that deploys a smaller number of larger ATM switches in strategic locations and uses less expensive ATM/VPXs and/or ATM Add–Drop Multiplexers (ADMs) to bring remote customer traffic to the strategically located ATM switch for switching and processing. The latter approach is sometimes referred to as the "Virtual CO" approach, which has been implemented in the present SONET network infrastructure by many LECs. For each type of ATM transport network, three possible survivable network architectures (i.e., point-to-point systems, self-healing rings, and mesh self-healing networks) used in today's SONET networks may also be applicable to ATM transport networks with necessary modifications to accommodate unique ATM characteristics.

These survivable network architectures are generally divided into two categories: dedicated facility restoration and dynamic facility restoration. Dedicated facility restoration uses the dedicated protection facility for service restoration, and dynamic facility restoration uses the spare capacity within working facilities for service restoration. The former restoration category includes APS and rings, whereas the latter includes dynamic path rearrangeable mesh architecture and dual homing. There are trade-offs between the flexibility (thus, system complexity) and the additional spare capacity required for each restoration category. In general, the more sophisticated techniques require less spare capacity but slow down the restoration procedure. Details of these architectures and operations will be discussed in Chapters 2, 6, and 7.

1.7.3 Feature Comparisons

Table 1-5 summarizes a feature comparison between SONET/STM and ATM network restoration systems. In general, ATM technology used in ATM layer transport is more flexible and efficient than STM technology used in SONET layer transport, in both bandwidth allocation and transmission efficiency. However, these advantages are offset by the highly complex control systems that are necessary, especially when the network must be designed to function under stress (i.e., network congestion and/or network failures). The best use of a combination of these two-layer protection schemes in the same network is a challenging task for network planners and engineers who are seeking cost-effective solutions for SONET/ATM network protection.

Table 1-5. Comparison of SONET and ATM Layer Protection Technologies.

Protection Layer	ATM Layer	SONET Layer (STM)
Protection protocol complexity	moderate	simple (APS/rings) most-moderate (DCS mesh)
Restoration time	moderate	fast (APS/rings) slow (DCS mesh)
Spare capacity needed	least	moderate
Protection electronic equipment needed	less	more
Network management systems/OSs	developing	exist
Protection system availability	available soon	available now
Protection targets	node and link	node and link

1.7.4 Multilayer Network Restoration

Since the B-ISDN transport network is a multilayer transport network, network protection may be implemented at the SONET layer (SONET Line and Path layers), the ATM layer (VP and VC layers), and the Open System Interconnection (OSI) network layer (e.g., network IP rerouting). These network protection schemes are managed and controlled (triggered) by the network management entities [e.g., fault management for the "hard" network failures and performance management for "soft" failures (performance degradation)] at each layer, which are coordinated by the system management entities for across-layer activities. Thus, a key requirement for implementing a cost-effective network protection scheme in a B-ISDN transport network is to minimize the redundancy of network protection mechanisms and management activities across the layers, while achieving the needed restoration speed to support services. The multilayer transport network models for broadband signaling and network restoration will be discussed in Chapters 4 and 6, respectively.

1.8 Standards Progress and ATM Network Deployment

There are two major official standards groups working for ATM standards: ITU-T and ANSI T1S1. The ATM standards targeted by ITU-T and T1S1 are primarily designed for public networks. However, the slower standards development process undertaken by ITU-T and T1S1 is in conflict with the rapid technological advances needed by both the computer and networking industries (18-month product cycles have become normative). The ATM Forum was created in 1991 by four computer and communications companies to deal with this business-driven need for a faster standards development process. Since then the Forum has grown to include nearly 800 member companies, including 200 user companies. The Forum has stated that its goal is not to replace the standards groups, but to supplement their work with vendor-derived implementation agreements that would reference official standards whenever possible. In the following subsections, we review standards progress in ITU-T and T1S1 only. The implementation agreements specified by the ATM Forum will be discussed throughout the rest of the book whenever and wherever appropriate.

1.8.1 Standards References and Progress

In ITU-T and ANSI T1 standards, there exists no clear mapping between the OSI model and B-ISDN transport model. Figure 1-13 shows a possible correspondence between the OSI model and the B-ISDN protocol reference model. The SONET physical layer corresponds to Layer 1 of the OSI model. Depending on the functionalities implemented, the ATM Layer could either correspond to the Physical Layer (Layer 1) or the Data Link Layer (Layer 2) of the OSI model. The AAL Layer could correspond to Data Link Layer (Layer 2), the Network Layer (Layer 3), or the Transport Layer (Layer 4) in the OSI model, because AAL contains Service-Specific Coordination Functions (SSCFs) that may map up to the Transport Layer. For example, for the B-ISDN signaling protocol stack, the AAL corresponds to the Data Link Layer (Layer 2) in the OSI model. For services having CONS (Connection-Oriented Network Service), which is an SSCF, the AAL corresponds to the Network Layer (Layer 3) in the OSI model. For services having Connection-Oriented Transport Service (COTS) which is also an SSCF, the AAL layer has functions of the Transport Layer (Layer 4) in the OSI model.

Figure 1-14 shows recommendations related to B-ISDN in ITU-T study groups [41,42]. B-ISDN principles and services are covered by Study Group 1 (SG1) and SG13. UNI specifications including the Physical Layer, the ATM Layer, and the AAL Layer are covered by SG13; the signaling system is covered by SG11; ATM traffic management is covered in SG4; NNI (Network Node Interface) specifications are covered by SG13 and SG15; issues on interworking with other networks (FR, N-ISDN, etc.) are covered by SG13.

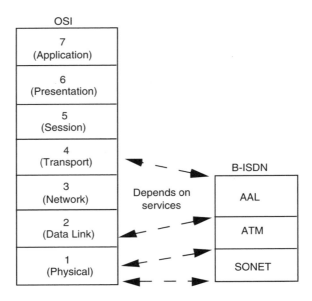

Figure 1-13: Possible mapping between B-ISDN model and OSI model.

B-ISDN detailed protocol specifications are being released in three steps in ITU-T. Release 1 (–1993) supports a minimum set of functions for point-to-point, CBR services. The first set of UNI/NNI signaling system specifications (recommendations) corresponding to Release 1 (under SG11) was released in late 1994, and these ensure compatibility with existing Q.931 (N-ISDN UNI signaling) and ISUP (N-ISDN NNI signaling). Release 2, which was completed in early 1996, supports a VBR capability for voice and video, point-to-multipoint connections, and a multigrade quality of service. LAN-to-LAN interconnections and B-ISDN interworking with existing networks will also be emphasized to catch up with emerging Customer Premises Network (CPN) development. Release 3, which has been almost completed, covers the enhanced connection configurations and broadcast type communications.

Figure 1-15 shows a relationship among several Bellcore B-ISDN/ATM documents, including the ATM broadband switching platform, the ATM operations platform and associated interface requirements.

1.8.2 ATM Network Deployment

ATM standards work is progressing in ITU-T and the ATM Forum. However, ATM network deployment in Japan and Europe's PTTs is proceeding much more quickly and is more widely disseminated than is the case for their U.S. counterpart. This may be due, in part, to a different operations environment.

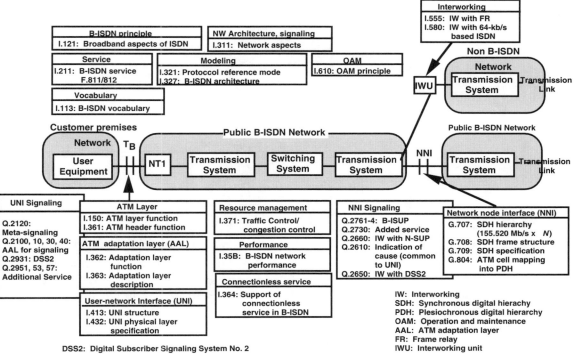

Figure 1-14:   ITU-T B-ISDN recommendations.

Telecommunications infrastructure work in Japan and Europe is more policy-driven, although cost remains a major concern. In contrast, the infrastructure development in the U.S. is more cost-driven, a situation in which American governmental policy is rarely involved.

In a policy-driven operations environment, the building of an ATM network infrastructure is based on the country's long-term planning and is often used to stimulate applications development (such as multimedia applications). For an economics-driven operations environment, network infrastructure work must be economically justifiable and primarily depends on the success of those services that require such an infrastructure. In the U.S., the ATM transport infrastructure is considered the transport of choice for Video on Demand (VoD) services and other advanced interactive video services by many network providers; however, both costs and uncertain market demand for these advanced interactive video services have made it difficult to justify initial ATM network deployment. Slow-down of these advanced interactive video service deployments results in slower ATM network deployment, because the ATM network may not be justifiable if it is used only to support emerging broadband data services (other than video services). More on the ATM deployment status in Japan and the United States will be discussed in Chapter 3.

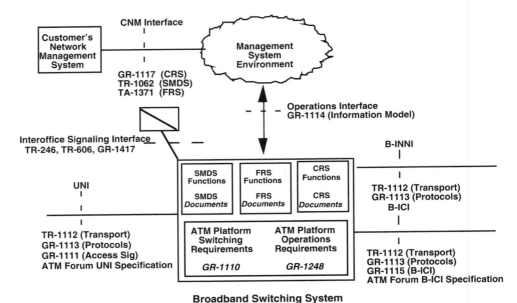

Broadband Switching System
Figure 1-15: Relationship of selected Bellcore B-ISDN/ATM documents.

1.9 Summary and Challenges

We have reviewed the business potential, transport technologies (including service, control, and signaling), traffic management, and network restoration needed to build a large-scale, robust ATM network. One of the major challenges for ATM networks to support a variety of services is how to evolve from today's SONET/STM network infrastructure and how to design an ATM network to meet the QoS for each B-ISDN service under both normal and stress conditions (i.e., network congestion and/or failures). Different services may have very different network performance requirements (e.g., video vs. data), and some services may have a multigrade QoS requirement within the same connection (e.g., multimedia applications). QoS implementation issues are crucial to B-ISDN deployment, particularly in the next-generation Internet, where ATM has been suggested as a viable network infrastructure for the Internet as it tries to provide QoS for a variety of services. The challenge here is to decide in which layer should the QoS for multimedia services be implemented. One group has suggested that the QoS may be implemented by a new IP layer signaling protocol, called Resource Reservation Protocol (RSVP), which is specified by Internet Engineering Task Force (IETF). The RSVP may use the ATM network as a high-speed transport platform. This approach bypasses all QoS and related models defined by the ATM Forum. The other group suggests that QoS provided at the ATM Layer (based on the model and parameters defined by the ATM Forum) may provide adequate QoS performance for these upper-layer applications. Technical assessment and analysis of trade-offs in these two approaches for an integrated service Internet require significant efforts if network providers are to make wise choices.

Broadband signaling remains a challenging issue due to two potentially competing sets of standards from enterprise and telecommunications public network perspectives. This raises the possibility that two sets of ATM switched network infrastructures may coexist for a period of time. A challenging question is how and when these two infrastructure sets might be merged as an integrated ATM network infrastructure, if possible. This challenge may require a change of traditional QoS perspectives, definition, and requirements from both the computer and telecommunications industries.

Emerging B-ISDN service survivability requirements will also significantly affect SONET/ATM transport network design. Existing SONET/STM network protection is through simple and cost-effective self-healing capability such as SONET self-healing rings. When the network moves to ATM transport technology for accommodating bandwidth-on-demand requirements, a natural and challenging question is whether the SONET layer can provide protection for new broadband services that may have very different delay and cell loss requirements. It is expected that very fast network restoration may be needed for high-speed B-ISDN networks if they are to meet the service and signaling/control requirements and minimize the impact of multilayer network restoration on ATM Layer network congestion. The complexity of designing an ATM network that would operate uninterrupted even when the network becomes congested or fails is much higher than that of any existing network due to inherent ATM high-speed transport characteristics and its service integration requirement. The area of B-ISDN transport network survivability is currently in the early research and development stage, and ITU-T and ANSI T1 began efforts on B-ISDN network protection studies in 1996.

References

[1]   R. Ballart, and Y.-C. Ching, "SONET: Now It's the Standard Optical Network," *IEEE Commun. Mag.*, pp. 8–15, March 1989.

[2]   ITU-T Recommendations G.803, "Architectures of Transport Networks Based on Synchronous Digital Hierarchy," 1992.

[3]   M. D. Prycker, *Asynchronous Transfer Mode: Solution for Broadband ISDN*, Ellis Horwood, 1995.

[4]   U. Black, *ATM: Foundation for Broadband Networks*, Prentice–Hall, Englewood Cliffs, New Jersey, 1995.

[5]   J.-Y. L. Boudec, "The Asynchronous Transfer Mode: A Tutorial," *Computer Networks and ISDN Systems* , Vol. 24, pp. 279–309, 1992.

[6]   *Broadband Networking News*, Phillips Business Information, February 20, 1996.

[7]   "THE ATM REPORT," Broadband, April 15, 1996.

[8]   K. Yoguchi, "Proposal for the PC-ATM Bus Architecture," The ATM Forum Contribution 96-0158, February 1996.

[9]   The ATM Forum, *PC-ATM Bus Specification, Release 1.0, Go-MVIP*, April 25, 1996 (http://www.mvip.org).

[10] D. Comer, and P. V. Mockapetris, "ATM and IP: Theory and Practice," Interop Tutorial Course Notes, April 1–5, 1996.

[11] Bellcore SR-3330, "Cell Relay Service Core Features," Issue 2, December 1996.

[12] Bellcore TR-TSV-000772, "Generic System Requirements in Support of Switched Multi-Megabit Data Service," Issue 1, May 1991.

[13] M. N. Ransom and D. R. Spears, "Applications of Public Gigabit Networks," *IEEE NETWORK*, Vol. 6, No. 2, pp. 30–40, March 1992.

[14] IETF RFC 1577, "Classic IP and Address Resolution Protocol over ATM," January 1994.

[15] The ATM Forum, *LAN Emulation over ATM, Version 1.0 (LANE 1.0)*, January 1995.

[16] T.-H. Wu, "Network Switching Concept," *The Electronics Handbook*, Chapt. 109, Sect. 15.1, CRC Press, Boca Raton, Florida, 1996.

[17] ITU-T Recommendation I.311, "B-ISDN General Network Aspects," Temporary Document 5G (XVIII), January 1993.

[18] T.-H. Wu, *Fiber Network Service Survivability* Artech House, May 1992.

[19] K. Sato, S. Ohta, and I. Tokizawa, "Broadband ATM Network Architecture Based on Virtual Paths," *IEEE Trans. on Commun.*, Vol. 38, No. 8, pp. 1212–1222, August 1990.

[20] K. Sato, H. Ueda, and N. Yoshikai, "The Role of Virtual Path Crossconnection," *IEEE Mag. Lightwave Telecommunications Systems*, Vol. 2, No. 3, pp. 44–54, August 1991.

[21] T. Aoyama, I. Tokizawa, and K. Sato, "Introduction Strategy and Technologies for ATM VP-Based Broadband Networks," *IEEE J. Selected Areas in Commun.*, Vol. 10, pp. 1434-1447, Dec. 1992.

[22] T.-H. Wu, J. Bartone, and V. Kaminisky, "A Feasibility Study of ATM Virtual Path Cross-Connect Systems in LATA Transport Networks," *Proc. IEEE GLOBECOM*, pp. 1421–1427, San Francisco, CA, November 1994.

[23] The ATM Forum, *ATM User-Network Interface (UNI) Signalling Specification, Version 4.0*, April 1996.

[24] The ATM Forum, *Traffic Management Specification Version 4.0,* April 1996.

[25] Bellcore SR-NWT-002897, "Alternatives for Signaling Link Evolution," Issue 1, February 1994.

[26] J. Luetchford, N. Yoshikai, and T.-H. Wu, "Network Common Channel Signaling Evolution," *Conference Records of International Switching Symposium'95*, Vol.2, pp. 234–238, April 1995.

[27] T.-H. Wu, N. Yoshikai, and H. Fujji, "ATM Signaling Transport Network Architectures and Analysis," *IEEE Commun. Mag.*, pp. 90–99, December 1995.

[28] E. H. Lipper, and M. P. Rumsewicz, "Teletraffic Considerations for Widespread Deployment of PCS," *IEEE Network*, pp. 40–49, September 1994.

[29] Bellcore GR-1111-CORE, "Broadband Access Signaling Generic Requirements," Issue 2, October 1996.

[30] The ATM Forum, *Private Network-Network Interface Specification Version 1.0 (PNNI 1.0)*, March 1996.

[31] ITU-T Draft Recommendation Q.2010, "Broadband Integrated Service Digital Network Overview Signaling Capability," Geneva, December 1993.

[32] The ATM-Forum, *Draft B-ICI Specification Document, Version 2/0*, April 10–14, 1995.

[33] Bellcore GR-1115-CORE, "Broadband Inter-Carrier Interface (B-ICI) Generic Requirements," Issue 2, December 1995.

[34] TU-T Rec. I.371, "Traffic Control and Congestion Control in B-ISDN," March 1993.

[35] Bellcore GR-477-CORE, "Network Traffic Management," Issue 2, December 1995.

[36] The Multimedia Communications Forum, "Multimedia Communications Quality of Service," ARCH/QoS/94-001, Rev. 2.1, June 1995.

[37] ITU-T Document, "Integrated Video Service (IVS) Baseline Document," Study Group 13, March 1994, Geneva.

[38] T.-H. Wu, J. C. McDonald, K. Sato, and T. P. Flanagan, (eds.), "Integrity of Public Telecommunications Networks," *IEEE J. Selected Areas in Commun.*, January 1994.

[39] T.-H. Wu, "Emerging Technologies for Fiber Network Survivability," *IEEE Commun. Mag.*, pp. 58–74, February 1995.

[40] C. A. Siller, Jr. and M. Shafi, *SONET/SDH: A Sourcebook of Synchronous Networking*, IEEE Press, New York, 1996.

[41] N. Yoshikai, "Broadband ISDN," NTT Review, Vol. 6, No. 5, pp. 57–59, May 1994.

[42] K. Asatani, "Standardization of Network Technologies and Services," *IEEE Commun. Mag.*, pp. 86–91, July 1994.

Chapter 2

SONET/SDH Transport and Network Integrity

2.1 Introduction

Increasing deployment of fiber facilities in telecommunications networks raises concerns about service efficiency and equipment interoperability on an end-to-end basis. These service efficiency and equipment interoperability concerns, along with the need for supporting broadband services, which require bandwidth beyond the DS3 level (i.e., 45 Mbps), led to the establishment of international fiber transmission network standards, called Synchronous Digital Hierarchy (SDH) [1,2]. SDH is derived from North America's standards, SONET (Synchronous Optical NETwork) [2,3], with provisioning that may accommodate any non–North American signal hierarchy. SONET comprises both an optical interface and specifications for rate and format of optical signal transmission. It can be used to support broadband and narrowband services.

SONET was initiated by Bellcore in February 1985 in response to MCI's proposal to interconnect multiowner, multimanufacturer, fiber-optic transmission terminals in the Interchange Carrier Compatibility Forum (ICCF) in 1984. In 1988, the American National Standards Institute (ANSI) approved Phase I of SONET as an American standard, and ITU-T approved it as an international standard. SONET Phase I specifies transmission rates, signal formats, optical interface parameters, and some payload mappings. Phase II of SONET, which defines the message set and protocols for using overhead channels for Operations, Administration, Maintenance, and Provisioning (OAM&P), was completed by U.S. standards groups (ANSI) in 1993. Phase II includes four major components: a protocol stack, a language, a message structure, and a common view of the data.

In a comparison with pre-SONET asynchronous fiber transmission systems, SONET offers several advantages that cannot be realized by its asynchronous counterpart. These advantages include an interoperability system, efficient multiplexing, higher network reliability, embedded operations and maintenance channels, and cost-effective realization of new network architectures (e.g., self-healing rings). SONET networks, including SONET self-healing rings, have been widely deployed around the world. They support an increase in speed to 10 Gbps. Figure 2-1 depicts a variety of SONET-based networks that have been or are expected to be deployed, and services that may be supported by these networks.

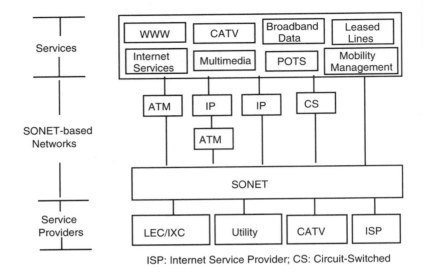

ISP: Internet Service Provider; CS: Circuit-Switched

Figure 2-1: Possible SONET-based networks and services being supported.

SONET deployment in North America can be loosely divided into two phases. The first phase consists of deploying point-to-point systems in both the interoffice and feeder portions of the telecommunication networks. Because of standard interfaces, the second phase of deployment interconnects these point-to-point SONET systems to provide multivendor, end-to-end optical networks through high-speed SONET Add–Drop Multiplexers (ADMs), and through SONET wideband and broadband Digital Cross-connect Systems (DCSs). More survivable network architectures, such as SONET self-healing rings, have been widely deployed since 1992. In 1994, SONET interfaces on digital switches, including Asynchronous Transfer Mode (ATM) switches, became commercially available. SONET was initially deployed only by Regional Bell Operating Companies (RBOCs) to reduce network operations costs and implement more survivable network architectures to support voice services as well as private line services. The technology has now been used by virtually all kinds of industry (such as CATV, utility companies, and government) to provide high-speed and reliable transport for their applications.

2.2 SONET/SDH Technology Overview

SONET comprises optical interface, rate, and format specifications for broadband optical signal transmissions. SONET is designed to transport a wide variety of signal types with a basic signal format containing fixed overhead to support various "in-band" operations features.

2.2.1 SONET Frame Structure

The basic building block of the SONET signal hierarchy is called the Synchronous Transport Signal Level 1 (STS-1). Its frame structure is depicted in Figure 2-2. The STS-1 has a bit rate of 51.84 Mbps and is divided into two portions: transport overhead and information payload, where the transport overhead is further divided into line and section overheads. The STS-1 frame consists of 90 columns and 9 rows of 8-bit bytes, as depicted in the figure. The transmission order of the bytes is row by row, from left to right, with one entire frame being transmitted every 125 ms. The first three columns contain transport overhead, and the remaining 87 columns and 9 rows (a total of 783 bytes) carry the STS-1 Synchronous Payload Envelope (SPE). The SPE contains a 9-byte path overhead that is used for end-to-end service performance monitoring. Optical Carrier Level-1 (OC-1) is the lowest-level optical signal used at equipment and network interfaces. OC-1 is obtained from STS-1 after scrambling and electrical-to-optical conversion.

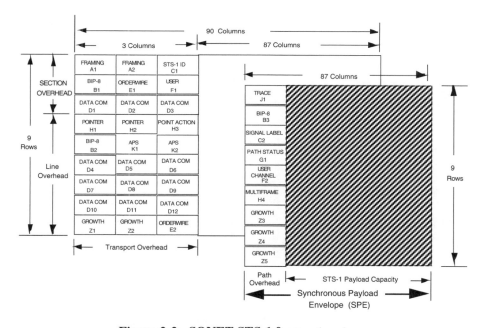

Figure 2-2: SONET STS-1 frame structure.

Broadband services requiring more than one STS-1 payload capacity are transported by concatenated STS-1s. For example, ATM services requiring 135 Mbps can be carried by three concatenated STS-1s, denoted by STS-3c, the transport overheads and payload envelopes of which are aligned. Figure 2-3 depicts an STS-3c frame overhead and information payload format. In this STS-3c, the first of three H1 and H2 bytes contains a valid SPE pointer, whereas the second and third H1 and H2 bytes contain a concatenation indicator that prevents the STS-3c signals from being demultiplexed. Concatenation specifies that these signals are considered as a single unit and transports them as such through the network. Compared with an STS-3 (nonconcatenated), the STS-3c can carry more information bits because only one set of path overhead (9 bytes) is required for the STS-3c. Within the STS-3 SPE, one path overhead is required for each STS-1. In ANSI T1S1, STS-3c and STS-12c formats carrying ATM cells have been specified. These high-speed SONET interface cards are commercially available.

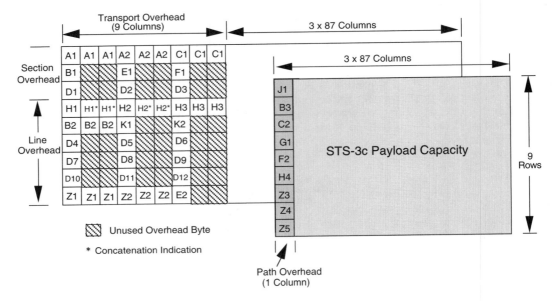

Figure 2-3: SONET STS-3c frame structure.

2.2.2 SONET Data Rates

SONET/SDH defines a progressive hierarchy of optical signal and line rates. The basic SONET building block is the STS-1 signal operating at 51.840 Mbps. Higher rate signals (STS-N) are multiples (N) of this basic rate. Values of N currently used in the standards are 1, 3, 12, 24, 48, and 192, as shown in Table 2-1. There are only three high-speed rates defined in SDH: STM-1, STM-4, and STM-16, which correspond to SONET's STS-3, STS-12, and STS-48, respectively, in terms of the transmission line rate.

Table 2-1: SONET/SDH transmission rates.

Level	Line Rate (Mbps)	Digital Hierarchy (ANSI)	SDH (ITU-T)
OC-1	51.84	STS-1	-
OC-3	155.52	STS-3	STM-1
OC-12	622.08	STS-12	STM-4
OC-24	1244.16	STS-24	-
OC-48	2488.32	STS-48	STM-16
OC-192	9953.28	STS-192	STM-64

2.2.2.1 Virtual Tributaries (VTs)

The STS-1 SPE can be used to carry one DS3 or a variety of sub-DS3 signals. The DS1 of 1.5 Mbps is a commonly used sub-DS3 signal that can be mapped into a SONET unit called Virtual Tributary 1.5 (VT1.5). Each STS-1 SPE can carry up to 28 VT1.5s. Other types of VT signals include VT2, VT3, and VT6, as shown in Table 2-2.

Table 2-2: SONET virtual tributary

VT Type	Data Rate (Mbps)
VT1.5	1.728
VT2	2.304
VT3	3.456
VT6	6.912

Before mapping VT formats to the STS-1 SPE, one more mapping, called VT group mapping, is needed. The VT group is a fixed-size container for VT signals (approximately 6.9 Mbps). Each VT group contains up to four VT1.5s, three VT2s, two VT3s, or one VT6. When mapping services (such as DS1) into an STS-1 SPE, the service is first mapped into the appropriate VT format, and one or more VTs of the same size are then combined to form a VT group. Seven VT groups is then inserted into the STS-1 SPE. Note that the STS-1 SPE can contain VT groups of different sizes (e.g., four VT1.5 groups and one VT6 group), but each VT group contains VTs of the same size.

The VT structure has two possible modes of operation, floating and locked, based on the method of VT group mapping to the STS-1 SPE. Floating VTs use a flexible mapping method to adjust locations of VT groups inside the STS-1 SPE. In contrast, locked VTs use a fixed VT group mapping method (e.g., fixed-stuffing). The floating mode minimizes delay for distributed VT switching (at least 1.5 Mbps), whereas the locked mode minimizes interface complexity in distributed DS0 (64 Kbps) switching. An STS-1 payload can be structured only as all floating or all-locked VTs. Details of VT floating and locked-operations modes can be found in References [3,4].

2.2.3 SONET Layer Model

Figure 2-4 depicts a SONET network layer model that may help clarify SONET system architecture. In SONET system, an end-to-end connection is carried on the STS path (e.g., STS-1, STS-3c, . . .). This end-to-end STS path is accommodated by a set of SONET links terminated at immediate SONET Line Terminating Equipment (LTE) (e.g., SONET ADM) along the path. Each SONET link may contain one or more regenerator sections that are terminated at SONET regenerators.

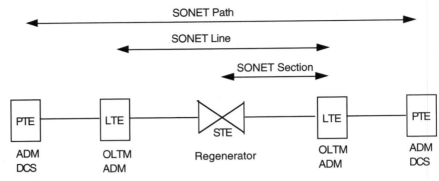

STE: Section Terminating Equipment
LTE: Line Terminating Equipment
PTE: Path Terminating Equipment
OLTM: Optical Line Terminating Multiplexer
ADM: Add-Drop Multiplexer
DCS: Digital Cross-Connect System

Figure 2-4: SONET layered structure.

The layered approach to overhead allows SONET equipment to be built cost-effectively according to the associated functionalities. Based on this approach, SONET-based Network Elements (NEs) are divided into three types: Section Terminating Equipment (STE), Line Terminating Equipment (LTE), and Path Terminating Equipment (PTE). The LTE can be any SONET equipment [such as Traffic Management (TM), an ADM, or a DCS] except regenerators. The STE, an NE that originates and/or terminates the SONET physical and section layers, interprets and processes the section overhead. The STE is either a regenerator or part of an LTE. The PTE, an NE that terminates and/or originates path overhead, can be any SONET equipment (such as TM or a DCS) that terminates SONET STS-1 payload. (The path overhead is included in the STS-1 payload.) The VT STE can be any SONET equipment that terminates SONET VTs, such as a VT multiplexer.

2.2.4 SONET Overhead Channels

SONET, which adopted a layered approach to overhead, allocates bandwidth to a layer based on the function addressed by that particular channel. As already discussed, SONET overhead is divided into three layers: section, line, and path layers. The overhead bytes and their relative positions in the STS-1 frame structure are illustrated in Figure 2-2. The line and section overheads (called transport overhead) are contained in the first three columns (27 bytes) of the STS-1 frame, and the path overhead is contained in the first column (9 bytes) of the SPE within the STS-1. Details of the SONET overhead communications can be found in Reference [5].

2.2.4.1 Section Overhead

The *section* overhead is the overhead necessary to verify reliable communication between network elements such as terminals and regenerators. A minimum amount of overhead is placed here to allow regenerators to remain cost-effective. The section overhead channels for an STS-1 include

- Two frames bytes (Bytes A1 and A2) that indicate the beginning of each STS-1 frame
- An STS-1 identification byte (Byte C1)
- An 8-bit Bit-Interleaved Parity (BIP-8) check for section error monitoring (Byte B1)
- An orderwire channel for craftsperson (network maintenance personnel) communications (Byte E1)
- A channel for unspecified network user (e.g., operator) applications (Byte F1)

Three bytes (Bytes D1, D2, and D3) for a section level 192-Kbps message-based data channel carry OAM&P information and other information between STEs.

2.2.4.2 Line Overhead

The *line* overhead is designed to verify reliable communication between more complicated network elements, such as terminals, DCSs, ADMs, and switches. The line overhead includes

- The STS-1 pointer bytes (Bytes H1, H2, and H3)
- An additional BIP-8 for line error monitoring (Byte B2)
- A two-byte Automatic Protection Switching (APS) message channel (Bytes K1 and K2)
- A 9-byte, 576-kbps, message-based Data Communications Channel (DCC) (Bytes D4 through D12)
- Bytes reserved for future growth (Bytes Z1 and Z2)
- A line orderwire channel (Byte E2)

Note that within the section and line layers of the SONET structure, DCCs are physical channels that can be used to transport data for any application. When DCCs are embedded in a signal format and dedicated to network operations and management, they are referred to as Embedded Operations Channels (EOCs). More details regarding DCCs can be found in References [3,5].

2.2.4.3 STS Path Overhead

The *path* overhead is used to communicate operations-type functions (e.g., performance checking) from the point where a service is mapped into the STS SPE to the point where it is delivered. The path overhead, which is part of the STS SPE, allows complete performance monitoring on an end-to-end basis. The path overhead includes

- A byte used to verify a continued connection (Byte J1)
- A path BIP-8 for end-to-end payload error monitoring (Byte B3)
- A signal label byte to identify the type of payload being carried (Byte C2)

- A path status to carry maintenance signals (Byte G1)
- A multiframe alignment byte (Byte H4) for identifying the location of tributary signals
- Three bytes for future growth (Bytes Z3, Z4, and Z5)

2.2.5. SONET Mapping and Multiplexing

Using SONET systems for physical transport, services with different rates would need to be mapped into appropriate SONET payloads. For example, the DS1 service must be mapped into a SONET VT1.5, which is then multiplexed with other VTs into one STS-1 for transport. A DS3 may be directly mapped into one STS-1. ATM services may be directly mapped into one STS-3c frame. These SONET paths are then multiplexed using byte-interleaving to form an STS-N multiplexed frame, which is then converted to an OC-N signal for optical transmission. Note that services may or may not be translated into ATM cells before being accommodated into SONET paths, depending on the applications and network economics. Figure 2-5 depicts an example of SONET mapping and multiplexing.

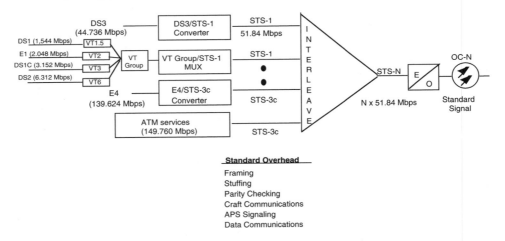

Figure 2-5: SONET mapping and multiplexing.

Higher-rate optical signals are formed by byte-interleaving an integral number of STS-1s. Figure 2-6 shows how an OC-N signal is formed. In the figure, services such as DS1 (along with path overhead) are first mapped into an SPE. Line overhead is then added to form an STS-1. A number of STS-1s are then byte-interleaved and multiplexed to form an STS-N. The frame is scrambled after section overhead (except framing and STS-N ID) is added. Framing and STS-N ID are then added into the section overhead of the scrambled STS-N, and finally, the STS-N signal is converted into optical signal OC-N.

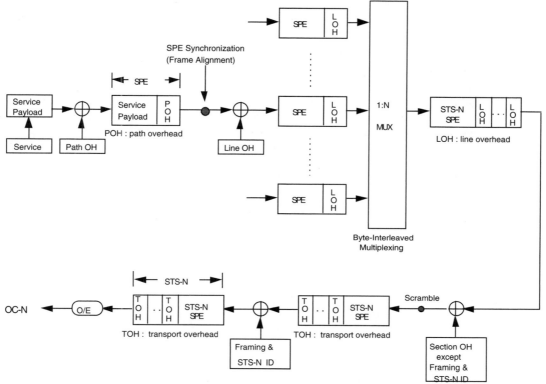

Figure 2-6: Forming an OC-48 signal.

2.3 SONET versus SDH

SONET and SDH are specified by the North American T1 Committee and ITU-T, respectively. SONET is designed to reflect the problems and priorities of North American operators; however, SDH discussions are broader in scope. Thus, it should be no surprise that proposals optimized for SONET may not be agreed upon by ITU-T. For example, STS-1 used as a building block for SONET is not agreed upon by ITU-T, which uses the much larger Synchronous Transport Module Level 1 (STM-1) as the building block.

SONET and SDH are similar, but not identical, digital hierarchies. They have similar sets of overheads and functions, but with differences in their use of overhead structures. A preliminary comparison of SONET and SDH has been reported in ANSI T1X1.2/93-24R2 [6], where T1X1.2 is responsible for Digital Transmission Network Architecture standards, including SONET and SDH interworking schemes. Reference [5] discusses the differences between SONET and SDH with respect to overhead bytes and performance monitoring information discussed in T1X1.2/93-24R2 [6].

In the past, SONET standards have been written as self-contained documents. The trend is to reference ITU standards for selected specifications. SONET/SDH defines a progressive hierarchy of optical signal and line rates. The basic SONET building block is the STS-1 signal operating at 51.84 Mbps. Higher rate signals (STS-*N*) are multiples (*N*) of this basic rate. As shown in Table 2-1, there are six line rates (OC-1, OC-3, OC-12, OC-24, OC-48, and OC-192) defined in SONET, whereas there are only three high-speed line rates (STM-1, STM-4, and STM-16) defined in SDH. Figure 2-7 depicts a frame structure of SDH building block, STM-1.

Table 2-3 depicts a relative comparison between SONET and SDH path payloads [6], where VCs means Virtual Containers, which are equivalent to VTs in SONET.

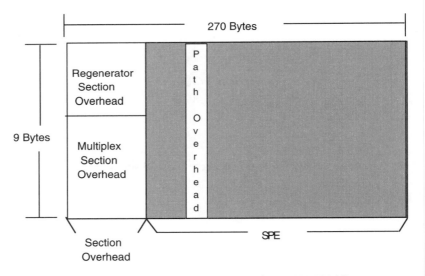

9 x 270 bytes x 8 bits/byte x 8000 frames/sec = 155.520 Mbps

Figure 2-7: ITU-T STM-1 frame structure.

Table 2-3: SONET and SDH payloads.

SONET Paylloads			
Payload	Container	Actual Payload Capacity	Payload & POH
DS1 (1.544 Mbps)	VT1.5 SPE	1.648 Mbps	1.728 Mbps
E1 (2.048 Mbps)	VT2 SPE	2.224 Mbps	2.304 Mbps
DS1C (3.152 Mbps)	VT3 SPE	3.376 Mbps	3.392 Mbps
DS2 (6.312 Mbps)	VT6 SPE	6.832 Mbps	6.848 Mbps
DS3 (44.736 Mbps)	STS-1 SPE	49.536 Mbps	50.112 Mbps
E4 (139.264 Mbps)	STS-3c SPE	149.760 Mbps	150.336 Mbps
ATM (149.760 Mbps)	STS-3c SPE	149.760 Mbps	150.336 Mbps
ATM (599.040 Mbps)	STS-12c SPE	599.040 Mbps	601.344 Mbps
FDDI (125.000 Mbps)	STS-3c SPE	149.760 Mbps	150.336 Mbps
DQDB (149.760 Mbps)	STS-3c SPE	149.760 Mbps	150.336 Mbps

SDH Paylloads			
Payload	Container	Actual Payload Capacity	Payload & POH
DS1 (1.544 Mbps)	VC11	1.648 Mbps	1.728 Mbps
DS1 (1.544 Mbps)	VC12	2.224 Mbps	2.304 Mbps
E1 (2.048 Mbps)	VC12	2.224 Mbps	2.304 Mbps
DS2 (6.312 Mbps)	VC2	6.832 Mbps	6.912 Mbps
E3 (34.368 Mbps)	VC3	48.384 Mbps	48.960 Mbps
DS3 (44.736 Mbps)	VC3	48.384 Mbps	48.960 Mbps
E4 (139.264 Mbps)	VC4	149.760 Mbps	150.336 Mbps
ATM (149.760 Mbps)	VC4	149.760 Mbps	150.336 Mbps
ATM (599.040 Mbps)	VC4-4c	599.040 Mbps	601.344 Mbps

In addition to the line rate and payloads, other major differences between SONET and SDH include use of overhead, payload mapping, and multiplexing. Table 2-4 summarizes relative differences between SONET overheads and SDH overheads, as described in T1X1.2/93-24R2 [6]. Table 2-5 describes a payload mapping between SONET and SDH systems that has also been discussed in T1X1.2/93-24R2 [6].

Table 2-4: SONET overhead vs. SDH overhead.

Layer	Overhead Bytes	Functions	Differences
Section	C1	STS-1/STM-1 identifier	SONET uses STS-1 interleaving, while SDH uses STM-1 interleaving
	E1	Orderwire	Signaling on E1 in SONET is for further study
Line	H1, H2, H3	Payload pointer	Byte assignment associated with multiplexing is different
	K1, K2	Line automatic protection switching	The same K1/K2 format for bidirectional line switched rings is used, but SONET rings have more functions than SDH rings
	Z1	Synchronization status messages	Number of messages is different
	E2	Orderwire	Signaling on E2 in SONET is for further study
Path	J1	Path trace	Formats are different
	B3	Path error monitoring	Calculation is different
	G1 (bit 5)	Performance status	SONET defines more error conditions than SDH

Table 2-5: SONET-SDH payload mapping [6].

Payload	STS-1	STS-3c	AU3/STM-1	AU4/STM-1
DS1 (1.544 Mbps)	(VT1.5)	none	(VC11) or VC12*	VC11 or VC12*
E1 (2.048 Mbps)	(VT2)	none	(VC12)	VC12
DS1C (3.152 Mbps)	VT3	none	none	none
DS2 (6.312 Mbps)	(VT6)	none	(VC2)	VC2
E3 (34.368 Mbps)	none	none	VC3	VC3
DS3 (44.736 Mbps)	(STS-1)	none	(VC3)	VC3
E4 (139.264 Mbps)	none	(STS-3c)	none	(VC4)
ATM (149.760 Mbps)	none	(STS-3c)	none	(VC4)
ATM (599.040 Mbps)	none	STS-12c	none	(VC4-4c)
FDDI (125.000 Mbps)	none	(STS-3c)	none	(VC4)
DQDB (149.760 Mbps)	none	(STS-3c)	none	(VC4)

( ) Compatible SONET/SDH mappings
* In SDH, a Ds1 may be carried in a VC12 (2.224 Mbps)

Another major difference between SONET and SDH systems is in their multiplexing structures, the signal rates of which are lower than either STS-3 or STM-1. In SONET, the subrate multiplexing method is unique, but there are several possible ways to multiplex subrate signals into an STM-1 in SDH. Figure 2-8 depicts such a difference in subrate multiplexing between SONET and SDH.

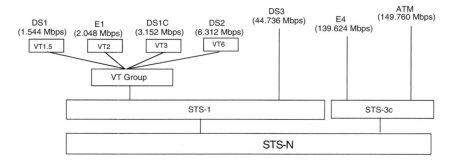

(a) SONET Multiplexing Structure

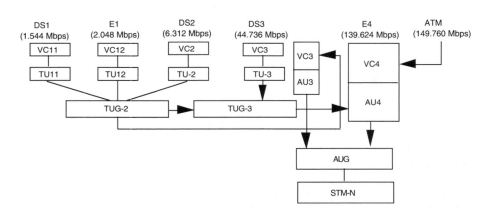

(b) SDH Multiplexing Structure

Figure 2-8: SONET multiplexing vs. SDH multiplexing.

The level of interworking between SONET and SDH depends generally on the payloads and the payload mappings chosen. Although SDH does provide SONET-compatible mappings in most cases, these mappings may not be widely used. Please refer to ITU-T Rec. G.708 for these mappings.

2.4 SONET Network Architecture and Systems

2.4.1 SONET Target Network Architecture and End-to-End Model

Figure 2-9 depicts a target SONET network architecture that has been described in a Bellcore Special Report, SR-TSV-002387. This SONET target network architecture is composed of a SONET service transport network, an Operations Support System (OSS), a network synchronizer, and an Operations Communications Network. A SONET service transport network may be a set of point-to-point systems (using Terminal Multiplexers), rings [using Add–Drop Multiplexers (ADMs)], mesh [using Digital Cross-connect Systems (DCSs)], or a combination of these. The Operations Systems (OSs) may be a centralized system with or without OSs for subnetworks. Communication among network elements and the OSs proceed via an operations communications network, which could be a wide-area Data Communications Network (DCN), a SONET embedded Data Communications Channel (DCC), or a Local-Area Network (LAN).

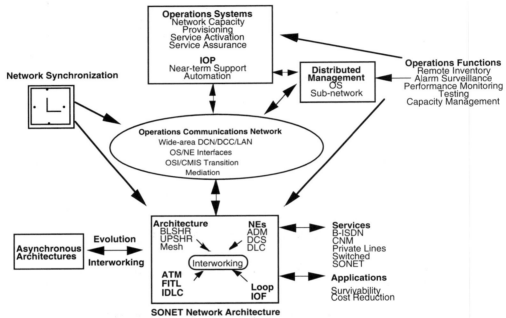

Figure 2-9: A target SONET network architecture.

Figure 2-10 depicts an end-to-end network operation model that may be used to implement the target SONET network architecture depicted in Figure 2-9. This end-to-end model consists of three layers. The lowest layer is the service transport layer, which is used to support services such as POTS, data, PCS, video, and other broadband services. The middle layer is the control transport layer, which processes and transports control and signaling messages to ensure that a network functions properly even when the network is under stress (i.e., congestion and/or failures). The control messages carried in the control transport layer include OAM&P messages for SONET networks. The control transport network here is equivalent to the operations communications network shown in Figure 2-9.

The upper level of the model is the service control and intelligent layer, which consists of service control systems (e.g., SCP), OAM&P supporting systems, and database systems needed to support advanced intelligent services (e.g., PCS mobility management and AIN services). The communication path from the control transport network to the intelligent node systems is sometimes referred to as network management, and communications between intelligent nodes is called service management.

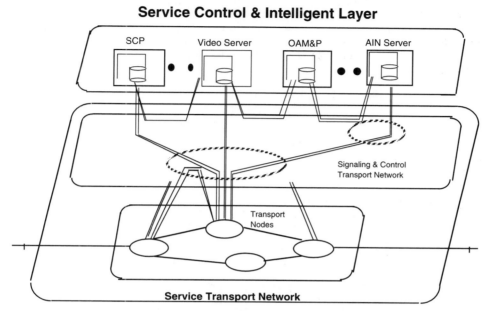

Service Control & Intelligent Layer

SCP: Service Control Point
OAM&P: Operations, Administration, Maintenance, and Provisioning
AIN: Advanced Intelligent Network

Figure 2-10: An end-to-end network operations model.

The service transport network, as well as the control transport network, can be conceptually implemented using a layered model described in ITU-T Rec. G.803, as depicted in Figure 2-11. In Figure 2-11, the bottom layer, the physical facility network, provides bit transport and transmission functions for delivering signals over the physical media between network nodes. This facility network may use fiber, radio, and/or other media. The top layer, the service switching layer, provides an end-to-end connection established in real time for any single call. Between these top and bottom layers, the path network provides resource management, efficient utilization of network resources, and decoupling of the physical (fixed) and switching (rapidly changing) layers. The path layer is further divided into two functions: multiplexing and cross-connection. Multiplexing combines several low-speed signals into a single high-speed signal to obtain more economical transport in the facility network. Examples of multiplexing equipment include Terminal Multiplexers (TMs) and ADMs. Cross-connection provides routing, on a semipermanent basis, for bundles of circuits, or aggregated traffic streams. One example of cross-connect equipment is a DCS.

Traditionally, the switching layer in LEC public networks, which predominantly provides voice traffic switching, has been realized with circuit-switched, or equivalently STM techniques. However, with the introduction of the ATM technique, and while corresponding standards are being developed, public carriers may deploy ATM technology for service switching applications if the cost can be justified.

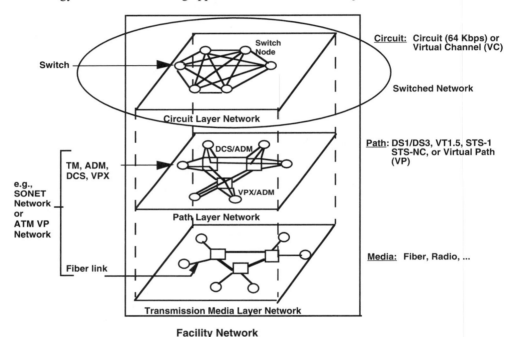

Facility Network
Figure 2-11: A layered transport network model.

There are two possible transport design concepts for supporting switched services, as depicted in Figure 2-12. The first design is usually referred to as a "Virtual CO" or "Large Switch" design that brings the remote customer traffic to larger switches in some strategic locations for switching and processing through high-speed transport systems such as ADMs and/or DCSs. This transport network design may allow larger switches to be used to process aggregated traffic so that the number of switches needed to support switched services can be reduced. The second design concept is sometimes referred to as "Traditional" or "Small Switch" network design. This "Small Switch" transport network design brings the switch to the customers by deploying smaller switches closer to the customer sites. In this design, switches are physically interconnected directly; thus, there are no high-speed ADMs and/or DCSs between switching nodes. The use of "Small Switch" or "Large Switch" design depends on network size, traffic patterns and volumes, costs, and reliability and performance requirements.

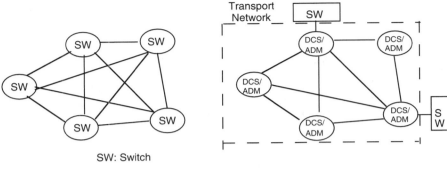

SW: Switch

Traditional (Small Switch) Network Design

- A large set of smaller switches with
  distributed control
- Bring the switch to customers

Virtual CO (Large Switch) Network Design

- A small set of larger switches
- Use less expensive transport
  to bring customers to the switch

Figure 2-12: Two possible transport network designs supporting switched services.

The SONET transport network is considered to be best implemented using the "Virtual CO" (or "Big Switch") transport network design because the cost of SONET transport network equipment is much less expensive than the switching equipment. This "Large Switch" transport network design under the SONET transport platform has been implemented by many LECs that use less expensive SONET transport equipment to support switched voice services.

2.4.2 SONET Transport Network Architectures

There are three network configurations that have been implemented in SONET transport networks. These are point-to-point systems, rings, and mesh networks, with different equipment used for each transport network architecture as depicted in Figure 2-13. There are three major types of equipment used in the broadband transport network, depending on the network configuration being considered. This equipment includes TMs, ADMs, and DCSs. A mapping of transport network configurations and equipment used is also depicted in Figure 2-13. In general, TMs are primarily used in point-to-point systems. ADMs, having two possible configuration modes (terminal mode and add–drop mode), can be used in several configurations, including the point-to-point, chain, tree, and ring configurations. Technically, the TM can be viewed as an ADM with the terminal mode. DCSs are typically used in a mesh core network in which large traffic volume and high connectivity are involved.

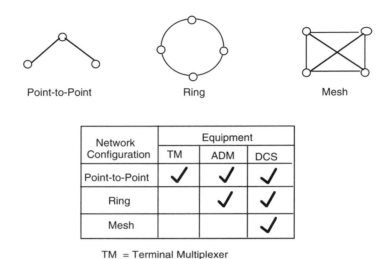

<table>
</table>

Network Configuration	Equipment		
	TM	ADM	DCS
Point-to-Point	✓	✓	✓
Ring		✓	✓
Mesh			✓

TM = Terminal Multiplexer
ADM = Add-Drop Multiplexer
DCS = Digital Cross-Connect System

Figure 2-13: Transport network configurations and equipment type.

2.4.2.1 SONET Point-to-Point System

A point-to-point system is a system in which the signal is added to one end of the system and is dropped at the other end without any processing and multiplexing in the middle of the system. The point-to-point systems can be extended to linear and tree configurations, as depicted in Figure 2-14. The linear system is typically used in the feeder, where the fiber connectivity is very sparse. The tree configuration is an extension of the linear configuration where several loop branches are served by a single higher-speed ADM. This configuration is typically applied to the distribution network [e.g., Digital Loop Carrier (DLC)] [7]. In this configuration, each branch interfaces with a high-speed optical line (to CO) at the optical domain, where the sum of the branches' optical rates is up to the high-speed optical rate.

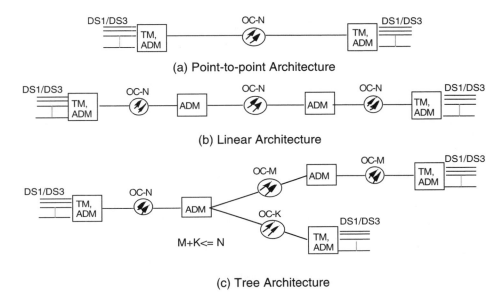

(a) Point-to-point Architecture

(b) Linear Architecture

(c) Tree Architecture

Figure 2-14: Possible network configurations for point-to-point systems

There are two possible signal transfer modes for ADMs: terminal and add-drop modes, where the terminal mode is a special configuration of the add–drop mode. The terminal mode is usually used in a point-to-point configuration described in Figure 2-14(a), and the add–drop mode is used for network configurations such as linear and tree systems, as described in Figs. 2-14(b) and 2-14(c), and rings (see Section 2.4.2.2). Figure 2-15 depicts a generic SONET ADM functional architecture in the add-drop mode. In the add–drop mode, the SONET ADM terminates two full-duplex OC-N signals and provides multiplex functions between the OC-N level and the DSn (n = 1, 2, 3) or OC-M ($M < N$) level. The incoming (OC-N) information payloads that are not received locally are passed through the SONET ADM and transmitted by the OC-N interface on the other side. Each DSn or OC-M interface reads data from an incoming OC-N and/or inserts data into an outgoing OC-N stream when appropriate. Figure 2-15 also shows a synchronization interface for a CO application with external timing and an operation interface that provides local craftsperson access, local alarm indications, and an interface to remote OSs. Detailed use of SONET overhead in the ADM can be found in Bellcore Technical Reference TR-TSY-000496 [8].

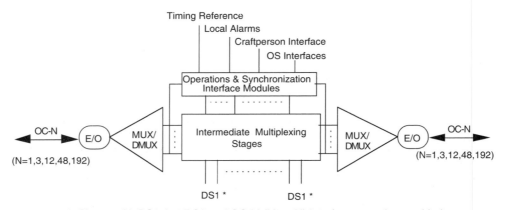

* Shown with DS1, but DS3 and OC-M (M < N) interfaces may be provided.

Figure 2-15: A generic ADM functional architecture (add–drop mode).

Figure 2-16 depicts an integrated loop network configuration that integrates SONET, Integrated Digital Loop Carrier (IDLC), Fiber-In-The-Loop (FITL), and B-ISDN. This integration configuration represents four major applications for ADMs in loop networks. The first application is a simple add–drop signal to other remote node or customer premises. The second application is as an interface with Remote Digital Terminal (RDT) used for IDLC systems. The third application is used to interconnect the ADM with Optical Network Units (ONUs) in the FITL system. The last application provides high-speed optical pipelines (e.g., OC-3 or OC-12) to ATM switches or multiplexers for B-ISDN applications.

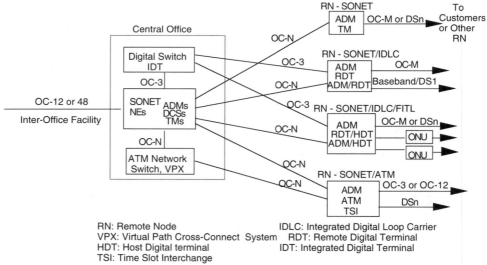

RN: Remote Node
VPX: Virtual Path Cross-Connect System
HDT: Host Digital terminal
TSI: Time Slot Interchange

IDLC: Integrated Digital Loop Carrier
RDT: Remote Digital Terminal
IDT: Integrated Digital Terminal

Figure 2-16: An integrated loop network configuration.

2.4.2.2 SONET Rings

A ring network is a collection of nodes forming a closed loop, where each node is connected via a duplex communications facility. The multiplexing devices used in the ring architectures are ADMs that add and drop local channels and pass through transit channels.

SONET rings can be classified as unidirectional and bidirectional rings based on the routing principle during normal network conditions [4,9]. A ring is called a unidirectional ring if its duplex channel travels through opposite physical routes around a ring. Figure 2-17(a) depicts an example of a unidirectional ring. In Figure 2-17(a), a channel from Node 1 to Node 3 travels through path 1 Æ 4 Æ 3, while the returning channel (from Node 3 to Node 1) travels through path 3 Æ 2 Æ 1. Thus, a unidirectional ring requires only one fiber to support its duplex communications.

In contrast, a ring is called a bidirectional ring if its duplex channel travels through the same physical routing path, as shown in Figure 2-17(b). In Figure 2-17(b), the duplex channel between Nodes 1 and 3 travels through the same physical path: 1 ´ 4 ´ 3. Thus, a bidirectional ring requires two fibers to support its duplex communications. Based on these channel routing characteristics, a bidirectional ring in general may carry more channels than a unidirectional ring, at the expense of a more complex control system that could be needed when the self-healing feature is introduced (see Section 2.7).

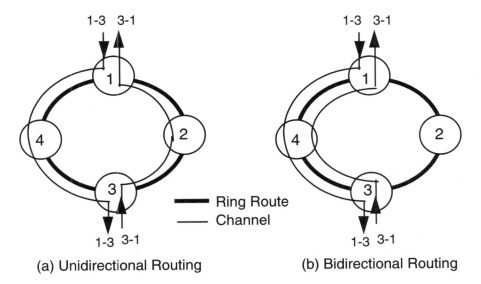

(a) Unidirectional Routing   (b) Bidirectional Routing

Figure 2-17: Unidirectional ring vs. bidirectional ring.

A description of SONET ring operations follows. In each ring node, an incoming optical signal, say OC-48, is demultiplexed into a maximum of 16 STS-3 (155.52 Mbps) channels. Some of the STS-3 channels may pass directly through the SONET ADM. These "through" STS-3 channels, together with other STS-3 channels added from this ring node, are then multiplexed and converted to an OC-48 optical signal and sent to the next ring node. If the payload of an STS-3 channel is terminated at this ring node and the service being considered is a private line service, an STS-3 demultiplexer may be used to demultiplex the STS-3 signal to a maximum of three STS-1 (51.84 Mbps) channels and then to convert the STS-1s to DS3s (44.736 Mbps) via the STS-1 interface cards.

The self-healing architectures and protocols for SONET rings will be discussed in Section 2.7.

2.4.2.3 SONET Mesh Networks

A mesh network is a network the path layer (see Figure 2-11) configuration of which can be dynamically adjusted based on applications or network design criteria. Such a path reconfigurable network is made possible by DCSs and network controllers. The major function of a DCS is to perform SONET path (e.g., STS or VT) rearrangement so that the path can be dynamically routed to its appropriate destination. Compared with the asynchronous DCS (e.g., DCS 3/1 or DCS 3/3), the SONET DCS integrates multiple functionalities of asynchronous networks (e.g., optical-to-electrical conversion, add-drop capabilities, and the cross-connect function) into a single system. Thus, it eliminates back-to-back multiplexing and reduces internal overhead processing. It also reduces the need for intermediate electrical distribution frames and labor-intensive jumpers between frames.

Figure 2-18 depicts a SONET DCS functional architecture. As shown in the figure, a SONET DCS system consists of three major modules: Input/Output interface module (I/O), Control and Operations Module (COM), and Cross-Connect Module (CCM).

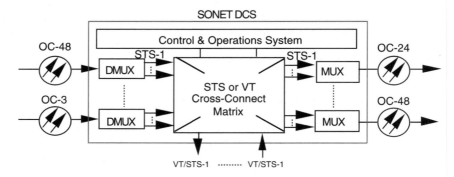

Figure 2-18: A SONET DCS functional architecture.

The I/O module includes (1) multiplexers, which terminate at fiber at one end and lower-speed STS paths at the other end; (2) interface ports, which terminate STS signals and pass them to the CCM module; and (3) a microprocessor-based controller, which monitors the I/O connections. The COM module controls the entire cross-connect system and interfaces with external operations systems or local operators. The COM module includes main memory and nonvolatile memory (e.g., hard disk), a CPU-based system controller with some secondary memory, alarm interface, and an OS interface. The system controller within the COM stores all application programs (in ROM) and controls and coordinates I/O and CCM modules via a LAN-based bus or star medium. These COM modules communicate with each other via a serial or parallel bus depending on system design. The main memory within the COM stores all configuration maps, which may be preplanned or calculated on-line, and other key information.

The CCM module includes a Switching Matrix Controller (SMC) and a Cross-Connect Switching Matrix (CCSM), which is typically protected on a 1:1 basis. The SMC stores the present cross-connect matrix configuration and monitors the accuracy of the cross-connections. The CCSM is used for data (e.g., STS-1) transport and is typically composed of several (three or five) switching stages.

Basic operations for the DCS system during normal mode are as follows. In the normal mode, the STS-1 channel enters the input module (i.e., I/O) and then is forwarded to the CCSM of the CCM module for transport based on the residing cross-connection map. The configuration map may be recomputed and downloaded from a central operation system or can be recomputed on-line when needed (e.g., when the network fails).

SONET DCSs are categorized into Broadband DCS (B-DCS) and Wideband DCS (W-DCS) based on the signal level being cross-connected. A W-DCS interfaces at the STS rate and cross-connects tributary channels at the VT1.5 or VT group rate. A B-DCS interfaces at STS-1, and/or STS-Nc rates and cross-connects channels at the STS-1 and/or STS-Nc rates. SONET B-DCSs are primarily deployed at major hubs that perform facility grooming at higher rates (at least the STS-1 rate). The W-DCS is primarily deployed in a location where service grooming is needed.

The example of using a B-DCS for facility grooming based on TSI switching principle has been discussed in Chapter 1 (see Figs. 1-4 and 1-5). Major applications for SONET DCS include efficient network resource utilization, fast provisioning, automatic network restoration and, customer control and management. We will discuss network restoration in Section 2.8. Details of other applications can be found in [9].

2.4.3 SONET Operations Communications Architecture

The SONET operations communications network is a support network that provides operations communication paths among SONET Operations System (OS), network elements (NEs), mediation devices, and operator consoles (workstations). Figure 2-19 depicts application areas where the SONET operations communications network would be used [3].

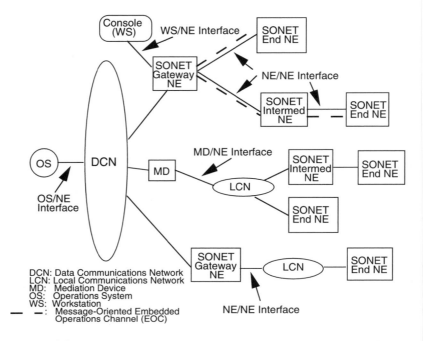

Figure 2-19: SONET operations communications: Example network element and interface types.

The design criteria for the SONET operations communications network are consistent with the Telecommunications Management Network (TMN) concept, which is specified in ANSI T1.210 [5]. The SONET operations communications network can be implemented using SONET Embedded Operations Channel (EOC), Data Communications Channel (DCC), LAN, and/or WAN such as X.25 networks. The DCC supports operations communications, including alarm, performance monitoring, maintenance, control, and configuration management messages, as well as other functions including software download. The Section DCC is a 192 Kbps (3 bytes), message-based, EOC; the Line DCC is a 576 Kbps (9 bytes), message-based EOC. Figure 2-20 depicts a generalized view of a SONET operations communications architecture that includes an X.25 wide area Data Communications Network (DCN), an intrasite LAN, and DCC tree and ring configurations. This example can be viewed as three operations communication subnetworks. Each operations communication subnetwork is connected to OSs via the X.25 DCN. Communication between the subnetworks (such as NE–NE communications) can be done using DCC links.

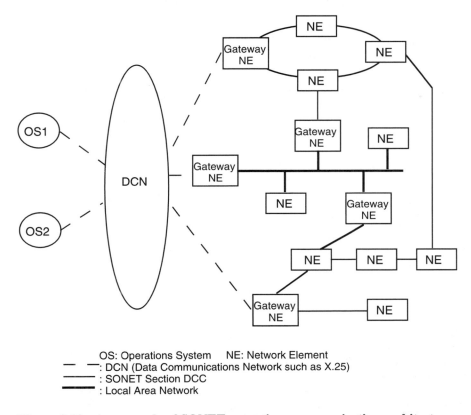

OS: Operations System   NE: Network Element
— ⎤: DCN (Data Communications Network such as X.25)
——— : SONET Section DCC
▬▬▬ : Local Area Network

Figure 2-20: An example of SONET operations communication architecture.

Figure 2-21 depicts the protocol stacks for the X.25 DCN, the DCC, and the intrasite LAN. These three cases use the Connectionless Network layer Protocol (CLNP ISO8473) for routing. Thus, interworking the three protocol stacks is done by standard routing and relaying functions. Figure 2-22 depicts routing and communications using a Common Management Information Service Element (CMISE) for the DCC.

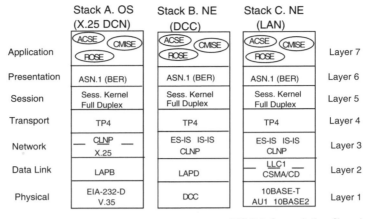

Figure 2-21: Interactive protocol stacks for SONET operations communications.

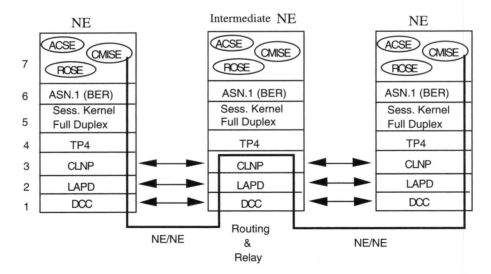

Figure 2-22: DCC routing and communications using CMISE.

2.5 SONET Network Integrity

Network integrity is a network capability to ensure that the network provides needed QoS not only in normal network conditions, but also under network stress. The conditions that cause network stress include network congestion and network failures. Since SONET network is a provisioned network with preplanned constant bandwidth allocated to services, the traffic exceeding the network capacity will be dropped, not queued as the packet-switched network is. Thus, the network stress for SONET networks is primarily due to network failures. In general, the network capability that would still need QoS for services even when the network failure occurs is referred to as *network survivability*.

Network survivability is an issue of great concern to a telecommunications industry that is eager to deploy high-capacity fiber networks, because loss of services in high-capacity fiber systems due to disasters and catastrophic failures could be devastating and result in significant revenue loss as well. However, providing protection against fiber network failures could be very expensive due to the high costs of fiber transmission equipment. Thus, how to reduce network protection costs and maintain an acceptable level of survivability has become an important challenge for network planners and engineers. Technology advancement is crucial to meeting this challenge. In the remainder of this section and in subsequent sections, we will discuss the class of network protection schemes available for providing different levels of network survivability and how to use SONET technology to implement these survivable network architectures.

2.5.1 Definitions of ITU-T Network Protection Schemes

ITU-T has defined three network protection schemes based on criteria such as the method of protection route planning (i.e., preplanned or dynamic) and the involvement of a network management system [10]. These network protection schemes can be implemented in SONET/SDH networks as well as in ATM networks. Three network protection schemes are defined as follows [10].

1.   *Protection switching*: Protection switching is the establishment of a pre-assigned replacement connection with equipment but without a network management control function. The equipment may reside in either the connecting or terminating points of the related path level. An example of protection switching is Automatic Protection Switching (APS) systems.

2.   *Rerouting*: Rerouting is the establishment of a replacement connection by the network management control connection. When a connection failure occurs, the replacement connection is routed depending on the network resources available at that time. An example of rerouting is the centralized control DCS network restoration.

3.   *Self-healing:* Self-healing is the establishment of a replacement connection by a network without a network management control function. When a connection failure occurs, the replacement connection is found by the network elements and rerouted according to the network resources available at that time. Examples of self-healing include the distributed control DCS network restoration and SONET

self-healing rings. Note that some proposed DCS restoration schemes use both rerouting and self-healing capabilities during network restoration.

Each network protection scheme offers different restoration speeds and costs. In SONET networks, the protection switching scheme may restore services within tens of ms; the self-healing scheme may restore services in few seconds; and the rerouting scheme may take a few minutes to tens of minutes to restore services. On the other hand, the spare capacities needed for the rerouting scheme and protection switching are the least and the most among the three schemes, respectively.

In SONET networks, the network protection would be provided through the line layer on a per-link basis or the path layer on an end-to-end basis. For example, protection switching is typically implemented at the line layer, rerouting is implemented at the path layer, and self-healing could be implemented at either line layer (e.g., bidirectional self-healing rings) or path layer (e.g., DCS self-healing mesh network). These examples are also valid for SDH networks. Several survivable SDH network architectures specified by ITU-T can be found in ITU-T COM 15-101-E [11].

2.5.2 Customer Impacts Due to Network Failures

Which network protection schemes are appropriate depends on the services being considered. Figure 2-23 depicts the customer impact analysis due to network failures, which has been reported in T1A1.2/93-001R3 [12]. As shown in the figure, different services have different outage duration tolerances. For example, voiceband call tolerances (i.e., the interval in which the calls in progress are dropped) can vary anywhere from 150 ms to 2 seconds, whereas data (packet) session time-out can vary from 2 to 300 seconds. Today, session-dependent applications, such as file transfer, commonly use System Network Architecture (SNA) and TCP/IP protocol architectures. With SNA, in particular, a session-dependent application has a software programmable session time-out of from 1.1 to 255 seconds and can be specified by users based on the file size, for example. Lost data from shorter interruptions is retransmitted, a process that in some applications is triggered by receiver time-outs. Receivers may begin to time-out based on twice the round-trip delay. X.25 packet networks often incorporate idle channel state condition timers that are settable from 1 to 30 seconds, in 1 second increments (with a suggested time of 5 seconds). When links are lost between switches, these timers may expire. If they do, they will trigger disconnection of all virtual calls that were up on those links. The customers must then restart their sessions. Based on the figure, most data applications fall into Range 4 and Range 5 (R4 and R5). This is the range that both network layer and the network management system can support.

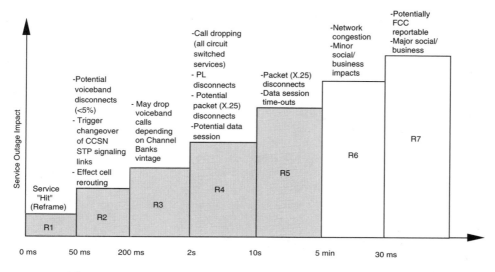

Figure 2-23:   Customer impact analysis due to network failures.

2.5.3 Network Failure Detection and Propagation

A network restoration process that provides network survivability capability has two phases. The first phase is the network failure detection and protection switching message propagation. This phase can be implemented by using the network maintenance feature, which is part of network management system. The second phase is the restoration protocol activation and execution after triggering by the maintenance signals (this indicates a network failure). We discuss an example of the phase one process here; the process of the second phase will be discussed in Sections 2.6 through 2.8.

Figure 2-24 depicts an example of maintenance signal operation following the detection of an LOS (Loss of Signal). In this example, three maintenance signals are used: Alarm Indication Signal (AIS), Far End Receive Failure (FERF), and Yellow Alarm. The AIS is used to alert downstream equipment that an upstream failure has been detected. Line FERF alerts the upstream LTE that LOS, Loss of Frame (LOF), or Line AIS has been detected along the downstream line. STS Path Yellow alerts the upstream STS PTE that a downstream failure indication has been declared along the STS path.

In Figure 2-24, STE-1 detects an LOS, generates a line AIS, and sends it to the downstream LTE-1. After LTE-1 receives a line AIS from STE-1, it initiates protection switching and tries to restore the failed link between PTE-1 and LTE-1. In this example, LTE-1 fails to complete protection switching. Thus, LTE-1 generates an STS path AIS, sends a line FERF to PTE-1, and reports the protection switching failure and the AIS received to the OS. After receiving an STS path AIS from LTE-1, PTE-2 converts that STS path AIS to a DS3 AIS for termination reporting and generates an STS Yellow signal to PTE-1.

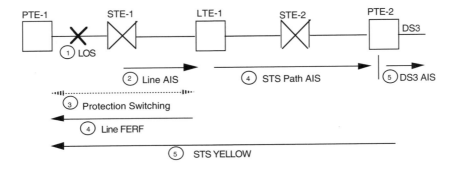

① STS-1 detects LOS caused by laser failure.

② STE-1 generates line AIS to LTE-1 and reports LOS alarm to OS via section DCC.

③ LTE-1 attemps but fails protection switching.

④ LTE-1 generates STS path AIS, sends line FERF to PTE-1, and reports switch failure and AIS received to the operations system.

⑤ PTE-2 converts STS path AIS to DS3 AIS for termination reporting and generates STS YELLOW signal to PTE-1.

Figure 2-24: An example of SONET fault detection and propagation.

2.5.4 Survivable SONET Network Architectures

Fiber-hubbed architecture is an economical transport architecture for LEC networks. This architecture can best utilize the economical scale of high-capacity fiber systems by reducing the amount of expensive fiber terminating equipment needed. In this architecture, each Central Office (CO) is connected to a hub through a fiber-optic system. At the hub, a DCS partitions incoming traffic by destination and routes channels, over fiber, to the appropriate end office. Aggregating interoffice traffic onto a single fiber pair greatly reduces the costs for small COs that require links to other offices. Fiber-hubbed architecture is economically attractive, but at the expense of service vulnerability, because a single fiber cut or a hub office failure would isolate a large area served by the failed facility or prevent the CO from communicating with other communities.

To alleviate survivability concerns caused by the fiber-hubbed network architecture, several modifications to the hub architecture and alternative survivable architectures have been explored in the past few years. These survivable architectures include Automatic Protection Switching (APS) with diverse protection (APS/DP), self-healing rings (SHRs), and dynamically path rearrangeable mesh architecture. Among them, SONET APS/DP systems and SONET SHRs have been widely deployed by the telecommunications industry around the world.

These survivable network architectures are generally divided into two categories: dedicated facility restoration and dynamic facility restoration. Dedicated facility restoration uses the dedicated protection facility for service restoration; dynamic facility restoration uses the spare capacity within working facilities for service restoration. The former restoration category includes APS and rings; the latter includes self-healing mesh architecture and dual homing. There are trade-offs between the flexibility (thus, system complexity) and the additional spare capacity required for each restoration category. In general, the more sophisticated techniques require less spare capacity but slow down the restoration procedure. Figure 2-25 depicts four survivable fiber network architectures, which are described in the text that follows.

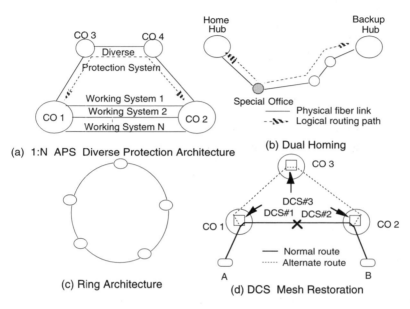

(a) 1:N APS Diverse Protection Architecture

(b) Dual Homing

(c) Ring Architecture

(d) DCS Mesh Restoration

Figure 2-25: Class of survivable fiber network architectures.

2.5.4.1 APS Diverse Protection (APS/DP)

The APS approach shown in Figure 2-25(a) is totally automatic and commonly used to facilitate maintenance and protect working services. The 1:N diverse protection structure is an alternative to the commonly used 1:N protection structure, where N working fiber systems share one common protection fiber system. The only difference between these two structures is the location of the fiber protection system. The 1:N protection structure places the protection fiber in the same route as that of working systems; the 1:N diverse protection structure places the protection fiber in a physically diverse route. In a 1:N system, a cable cut may damage the protection fiber as well as the working fibers. If a fiber cable cut occurs and a 1:N diverse protection scheme is used, some service can survive because one of the N working systems can be restored through the diverse, protected route. This diverse protection scheme is attractive in LEC networks because electronics costs dominate total costs and remain unchanged when attempting to achieve higher survivability. A 1:1 diverse protection arrangement, which provides 100 percent survivability for fiber cable cuts, requires more facilities and equipment than the 1:N diverse protection arrangement. We will discuss how SONET is used to implement the APS/DP system in Section 2.6.

2.5.4.2 Dual Homing

In contrast to the single-homing approach, which aggregates demands from a CO to destination COs through an associated home hub, "dual homing" is an office backup concept that assigns two hubs to each office and requires dual access to other offices. In the dual-homing approach, demand originating from a special CO is split between two hubs: a home hub and a designated foreign hub. In the case of a home hub failure, an office that uses dual homing can still access other offices through the backup hub. Figure 2-25(b) shows such a dual-homing architecture. Dual homing does not automatically accomplish restoration by itself, but it may be used in conjunction with DCSs that restore services at the path layer.

2.5.4.3 Self-Healing Rings (SHRs)

The SHR [see Figure 2-25(c)], like the 1:1 diverse protection structure, is totally automatic and provides 100 percent restoration capability for a single fiber cable cut and equipment failure through its ring topology and simple, but fast, protection switching scheme. It can also provide some survivability for hub DCS failures or major hub failures (e.g., flooding or fires). We will discuss in Section 2.7 three types of SONET SHRs being deployed.

2.5.4.4 Self-Healing Mesh Network

The self-healing mesh network uses DCSs to reroute demands around a failure point at the SONET path layer. Unlike APS/DP and rings, DCS restoration does not require standby protection facilities dedicated to working systems for restoration. Instead, it uses spare capacities within working systems to restore affected demands. Figure 2-25(d) shows an example of DCS restoration. In Figure 2-25(d), demands between locations A and B are normally routed over a link between DCS#1 and DCS#2, but they are rerouted through DCS#3 if a cable cut occurs on that link. A centralized or distributed control system may optimize the use of the available spare capacities by referring to a database that contains the status of the network (both working and spare capacities). The penalties for this efficient use of spare capacities, compared with other architectures, are the time and complexity needed for the controller(s) to communicate with the network DCSs, as well as maintenance of the database. We discuss how to use SONET technology to implement the self-healing network in Section 2.8.

2.6 SONET Automatic Protection Switching (APS) System

The SONET APS protocol was initially proposed in 1986 and was standardized in ANSI T1X1.5 in 1992 [13]. Two types of APS architectures (1:N APS and 1+1 APS) are defined in the SONET standard. The 1:N APS architecture allows one of the N (permissible values for N are from 1 to 14) working channels to be bridged to a single protection channel. The SONET APS protocol uses in-band signaling for protection switching through K1 and K2 bytes within the SONET line overhead. The K1 byte requests a channel for the switch action; the K2 byte confirms that the channel is bridged onto the protection line.

The SONET 1:1 APS protocol uses a three-phase protocol for bidirectional protection switching operations, which can be summarized as follows. When a failure is detected or a switch command is received at the tail end (i.e., the receiving end), the protection logic compares the priority of this new condition with the request priority of the working channel (if any) that requests the use of the protection channel. This comparison includes the priority of any bridge order (i.e., of a request on the received K1 byte). If the new request is of higher priority, the K1 byte is loaded with the request and the ID number of the channel requesting the use of the protection line. The tail end then sends out the K1 byte on the protection line. Figure 2-26 depicts the APS protocol specified by T1X1.

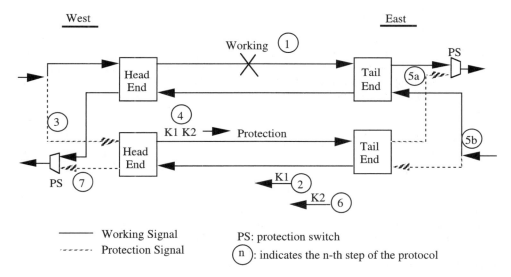

Figure 2-26: SONET APS protocol.

When this new K1 byte has been verified (i.e., received identically for three successive frames) and evaluated (by the priority order) at the head end (i.e., the transmitting end), the K1 byte is sent back to the tail end with a reverse request (to confirm the channel requesting the use of the protection channel). A bridge is also ordered at the tail end for that channel. This action initiates a bidirectional switch. At the head end, the indicated channel is bridged to protection. When the channel is bridged, the K2 byte is set to indicate the number of the channel being protected.

At the tail end, when the channel number on the received K2 byte matches the number of the channel requesting the switch, that channel is selected for protection. This completes the switch to the protection channel for one direction. The tail end also performs the bridges, as ordered by the K1 byte and indicates the bridged channel on the K2 byte. The head end completes the bidirectional switch by selecting the channel from protection when it receives a matching K2 byte. Note that K1 and K2 bytes always travel over the protection line in present SONET standards. Protection switching, including K1/K2 operations and switch reconfiguration, must be completed within 50 ms [3].

Another type of APS also defined in the SONET standard is 1+1 protection switching, which is a form of 1:1 APS with the head end permanently bridged. Thus, a decision to switch is made solely by the tail end. For bidirectional switching, the K1 byte is used to convey the signal condition to the other side, and the actual switching is decided by the tail end. Detailed APS protocols for SONET 1:N systems can be found in References [9,13].

2.7 SONET Self-Healing Rings (SHRs)

A ring is called a self-healing ring (SHR) if it provides redundant bandwidth and/or network equipment so that disrupted services can be restored automatically following network failures [14]. As discussed in Section 2.4.2.2, a SONET ring may be either a unidirectional or bidirectional ring, depending on how the duplex channel is routed under normal network conditions. A ring is called a unidirectional ring if its duplex channel travels through opposite physical routes around a ring; a ring is called a bidirectional ring if this duplex channel travels through the same physical routing path (see Figure 2-17). Note that a unidirectional ring and a bidirectional ring requires one fiber and two fibers to support its duplex communications, respectively, in normal network conditions.

The self-healing protocol of a SONET ring can be implemented using overheads at the SONET line or path layer [4,9]. An SHR is a path switched self-healing ring if its protection switching is triggered by the SONET path layer signal (e.g., Path AIS). In contrast, an SHR is called a line-switched SHR if its protection switching is triggered by the SONET line layer signal (e.g., K1 and K2 bytes). Each ring type (i.e., unidirectional or bidirectional) can be protected through path protection switching operating at the SONET path layer, or line protection switching operating at the SONET line layer. Today, only unidirectional rings with path protection switching (called UPSHR) and bidirectional rings with line protection switching (called BLSHR) are specified by ANSI T1X1.5 and Bellcore requirements [15,16]. They are commercially available in the marketplace.

2.7.1 Unidirectional Path-Switched SHR (UPSHR)

In a UPSHR, two fibers are needed between adjacent nodes: one for working and the other for protection. Each ring node is equipped with one ADM, which adds/drops local channels and passes through transit channels. Figure 2-27 shows an example of UPSHR operation under the normal and failure scenarios. A path (e.g., STS-1) in the transmitting side (e.g., Node 1) is duplicated and sent to both fibers in opposite directions. Thus, the receiving end always receives two identical path signals with different delays. The receiving end (e.g., Node 3) always chooses a path from the primary ring (e.g., the outer ring) but switches to the secondary ring if the ADM detects an alarm signal (e.g., path AIS) or a poor Bit Error Rate (BER). The unidirectional, path switched ring may perform protection switching at the VT1.5, STS-1, or STS-Nc level depending on applications. Generic requirements for the UPSHR can be found in Bellcore GR-1400-CORE [15].

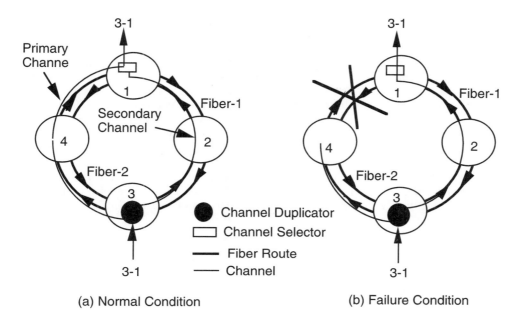

(a) Normal Condition   (b) Failure Condition

Figure 2-27: An example of UPSHR operation under normal and failure scenarios.

2.7.2 Bidirectional Line-Switched USR (BLSHR)

Bidirectional line switched self-healing rings (BLSHRs) are further divided into two classes: 2-fiber BLSHR (BLSHR/2) and 4-fiber BLSHR (BLSHR/4), depending on spare capacity provisioning. For a 4-fiber BLSHR, two fibers are used for working traffic, while another two fibers serve as standby fibers that provide 1:1 protection against equipment or fiber facility failures. The 2-fiber BLSHR uses only half the capacity of the fiber system for working traffic and reserves the other half as protection capacity. To perform protection switching against network failures for a 2-fiber BLSHR, a form of the TSI capability is needed to move signals from time slots of the affected fiber to the reserved (spare) time slots of the other fiber. Note that standby fibers or capacities in BLSHRs (as well as in other 1:1 systems) are allowed to carry lower-priority traffic that will be dropped during the network restoration process.

In BLSHRs, the ring node needs to determine whether the incoming K1 or K2 byte is a local or transit byte in order to trigger line protection switching at the right place. Since there are only 4 bits in K1 and K2 used for addressing, the maximum number of nodes that can be supported by a SONET ring is 16 (i.e., 4 bits), if SONET paths supported are STS-1 paths.

In the following, we discuss generic BLSHR network architectures and their self-healing operations during the network failure condition. Details on BLSHRs can be found in References [4,16,17].

2.7.2.1 Four-Fiber BLSHR (BLSHR/4)

For a 4-fiber BLSHR (BLSHR/4), two fibers are used to carry normal services and the other two fibers are used for protection. Figure 2-28 depicts an example of the BLSHR/4 operations under the normal and failure scenarios. In the normal scenario, demand routing occurs as it does in today's point-to-point systems, but the service channels are looped back from working fibers to protection fibers via facility protection switches in the case of network component failures. The loopback scheme is used as the protection switching scheme to simplify the ring reconfiguration complexity; only two nodes adjacent to the failed network component are involved in protection switching. In addition to loopback, the self-healing function also includes APS span protection. The line loopback function protects against cable cuts, and the span protection provides protection against fiber failures and equipment failures. The BLSHR/4 with both loopback and span protection has the highest system reliability among SONET SHR alternatives.

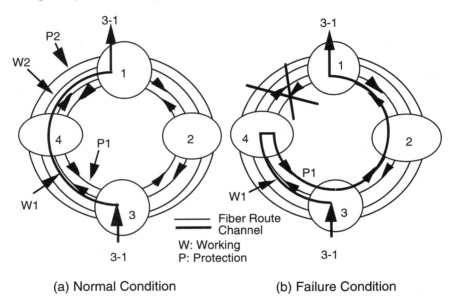

(a) Normal Condition   (b) Failure Condition

Figure 2-28: An example for BLSHR/4 operations under normal and failure scenarios.

2.7.2.2 Two-Fiber BLSHR (BLSHR/2)

Unlike the BLSHR/4, working and protection channels for BLSHR/2s are routed on the same fiber with a portion of bandwidth reserved for protection. To provide a self-healing capability (1:1 protection) and simplify ring control system complexity, half of the fiber system bandwidth is reserved for protection (e.g., for an OC-12 BLSHR/2, STS-1 channels 1 to 6 may be assigned to working STS-1s and channels 7 to 12 may be dedicated to protection). Figure 2-29 depicts an example of BLSHR/2 operations under normal and failure conditions.

In any normal situation, traffic is evenly split into the outer ring and the inner ring by filling the first half of the STS-1 time slots (or even and odd numbers of time slots, respectively) [see Figure 2-29(a)]. When a fiber breaks or equipment fails, traffic is automatically switched from this fiber (e.g., time slots 1–6 of fiber W1) into corresponding reserved (protection) time slots of the other fiber (e.g., time slots 7–12 of fiber W2) in the opposite direction using TSI technology to avoid the fault [see Figure 2-29(b)]. Note that these protection channels can then be reused by other links that carry low-priority channels when the protection channels become available again.

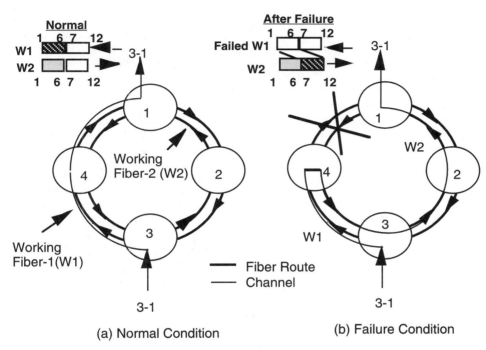

Figure 2-29: An example of BLSHR/2 operation under normal and failure scenarios.

2.7.3 Application Areas for SHRs

Determining which ring architecture is appropriate for network providers depends on several factors and their relative weight in strategic planning. One major factor is the distribution of pairwise demands carried on the ring, because this directly affects ring capacity engineering. This application area analysis is needed primarily because of the difference of ring capacity engineering. The ring capacity requirement is defined to be the largest STS-1 (or VT1.5, STS-Nc) cross section in the ring, providing the path transported on the ring is STS-1 (or VT1.5, STS-Nc, . . .). The line rate of the ring is selected based on its capacity requirement. The capacity requirement of a UPSHR is the sum of demands (e.g., STS-1s) for all demand pairs carried on that ring. The ring capacity requirement for a BLSHR is determined by the ring demand assignment algorithm used and the type of BLSHRs (e.g., BLSHR/2 or BLSHR/4). In the following, we use one example to demonstrate how ring capacity is determined. We briefly explains how this demand assignment algorithm works.

Figure 2-30 shows how the ring capacity requirement for both the UPSHR and BLSHR is calculated. In this example, we assume that only OC-12 and OC-48 rates are used, because they are only two available SONET line rates commercially available for interoffice networks. In addition, we assume demand is not split for the case of BLSHRs. In other words, all demand from one ring node to another ring node uses the same routing path from BLSHRs. For example, in Figure 2-30(b), four STS-1s between Nodes 2 and 4 are all routed through a single path: Path 2–3–4. For the considered demand requirement, the ring needs an OC-48 UPSHR because the ring capacity requirement for a UPSHR is the sum of all demands carried on the ring, that is, 13 STS-1s [see Figure 2-30(a)]. To calculate the ring capacity requirement for BLSHRs, we first sort the demand pair so that demand is in decreasing order. Then we distribute demand in as balanced a way as possible to minimize the number of STS-1s passing through each link, as depicted in Figure 2-30(b). The STS-1s carried on each link in this example are (3, 7, 7, 6, 3, 6), and the ring capacity requirement is the largest STS-1 cross section, namely, seven STS-1s. Thus, the BLSHR/4 requires an OC-12 line rate, and the BLSHR/2 needs an OC-48 rate because it uses only half the capacity of the ring. Note that the algorithm used in this example is just one of the algorithms that would be used to compute the ring capacity of BLSHRs. Some other demand assignment algorithms for BLSHRs can be found in Reference [4].

Application roles between UPSHRs and BLSHRs have been extensively studied. It is suggested that a UPSHR may be more economically attractive in areas where demands are homed to one central location, but a BLSHR may be attractive in areas where demands are more uniformly distributed [4,9]. Thus, the UPSHR may be more appropriate in feeder networks and peripheral networks (e.g., from the CO to the hub), and the BLSHR may be more economically attractive in interoffice networks with a nonhubbing structure. In a realistic network environment, a SONET network may use different types of survivable network architectures due to economics and demand distribution. Thus, how these different survivable network architectures interwork together becomes crucial to ensuring end-to-end service integrity during network failure conditions. The architecture interworking issues have been studied in ANSI T1X1.2 and some results have been reported in [18]. The purpose of this T1X1.2 report [18] is to help network providers better understand the potential impact of ring interworking on various architectures (e.g., ring-to-ring, ring-to-DCS mesh), network interfaces, and applications and to provide recommendations on ring interconnection practices. Technical issues covered in [18] include (1) ring interconnection architecture alternatives, including path-to-path, line-to-line, line-to-path, (2) OAM&P interworking impact and considerations, and (3) network architecture recommendations. Similar activities have also been conducted in ITU-T (see ITU-T Rec. G.842), and outputs for SDH rings standards and SDH ring interconnections are expected to be available in late 1997.

Demand Pair	STS-1s
(1,4)	3
(2,4)	4
(3,5)	3
(3,6)	3
Total	13

Assumptions:

(1) Only OC-12 and OC-48 are considered.

(2) Demand is not split for the BLSHR.

◀─── UPSHR ring capacity requirement
(Needs an OC-48 ring)

(a) UPSHR Ring Capacity Requirement Computation

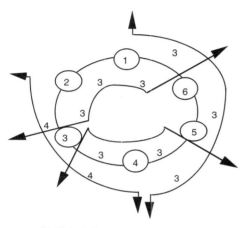

(i) Sort demand requirement

Demand Pair	STS-1s
(2,4)	4
(1,4)	3
(3,5)	3
(3,6)	3
Total	13

(ii) Assign demand in the clockwise and the counterclockwise directions as balanced as possible.

--> Link capacity requirement vector
= (3,7,7,6,3,6).

(iii) Capacity requirement for BLSHR = 7 STS-1s.
For BLSHR/4 --> OC-12 ring required
For BLSHR/2 --> OC-48 ring required

(b) BLSHR Capacity Requirement Computation

Figure 2-30: Calculating ring capacity requirements.

2.7.4 Ring Interworking

In some applications, geographical limitation or the need for extra bandwidth dictates a network solution to the survivability problem that requires multiple, interconnected rings. Figure 2-31 shows some possible ring interconnection configurations for single-access and dual-access applications [4]. For the single-access application, two rings are interconnected via an access node, which may be a hub in the fiber-hubbed network. The simplest nodal configuration for this interconnection access node is to use two ADMs, one for each ring. Connection is made between two SONET ADMs on the low-speed side [see Figure 2-31(a-i)]. The second approach is to use a Broadband DCS (B-DCS) to interconnect inter-ring demands that are dropped from ADMs, as depicted in Figure 2-31(a-ii). The third alternative is to use a DCS (W-DCS or B-DCS) to interconnect multiple rings [see Figure 2-31(a-iii)]. For the dual-access application, the ring interconnection needs two common access points as depicted in Figure 2-31(b). The possible nodal configurations for each access node are similar to those described for the single-access application.

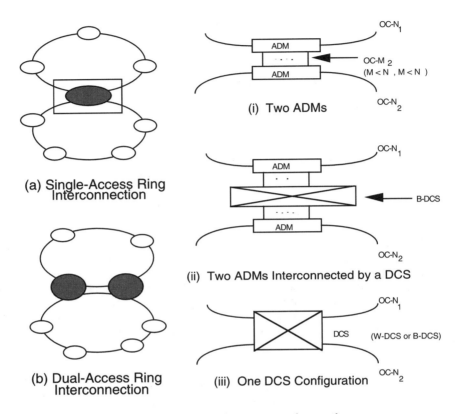

Figure 2-31: Ring interconnection options.

Figure 2-32 shows a design that uses a DCS as a node of one or more SONET SHRs [see Figure 2-31(a-iii)]. This DCS ring design minimizes the cost and complexity for ring interconnections by using TSI capability within the DCS. The DCS in this architecture provides both transport and interconnection functions for the rings. A Bellcore view of the proposed generic functional criteria for the use of W-DCSs and B-DCSs to provide self-healing ring functionality for ring interconnection can be found in Bellcore GR-1375-CORE [19].

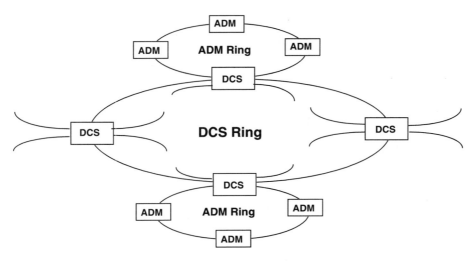

Figure 2-32: DCS ring.

Figure 2-33 shows three ring interconnection configurations specified by T1X1.2 [18]. Figure 2-33(a) shows an interworking configuration between two unidirectional path-switched rings. In this configuration, the signal associated with inter-ring traffic is dropped at one serving node and continued onto the second serving node using the drop-and-continue function. At both serving nodes, path selector functions independently choose the best of two incoming signals to drop into the next ring. Note that the path selectors can be set up to select the same path signal or set up opposite of each other as a default.

Figure 2-33(b) shows a configuration and a proposed method for interconnecting two bidirectional line-switched rings. Similar to interconnected path-switched rings, each ring uses a primary serving node and a secondary serving node for interconnection. The node at which the service would normally exit the ring is designated the primary node and has an associated secondary node. This designation is made on a per-service basis (e.g., STS-1). The service exiting the ring is dropped at the primary node and is also passed through the node (drop-and-continue) toward the secondary node. In the current proposed method, only primary nodes have service selectors that are required for inter-ring traffic entering or being added to the bidirectional line switched ring. In the bidirectional ring, the service selector selects between two copies of the service signal entering the ring at different points to determine which signal should be permitted to continue around the ring. If the service selector detects a signal failure on the primary signal, the service selector will "gang switch" to the appropriate signals coming from the secondary node. The secondary add signal is cross-connected toward the drop-and-continue (primary) node. Note that duplicate copies of the signal are dropped from the primary and secondary nodes of the ring. This allows interconnection between a unidirectional path-switched ring and a bidirectional line-switched ring, as shown in Figure 2-33(c), without any changes to path-switched ring requirements.

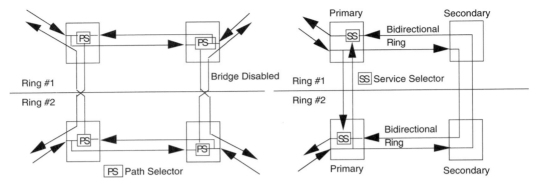

(a) Path-Switched Ring to Path-Switched Ring   (b) Line-Switched Ring to Line-Switched Ring

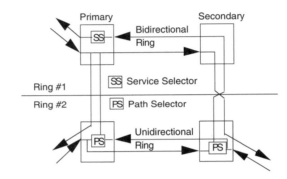

(c) Path-Switched Ring to Line-Switched Ring
Figure 2-33: Ring interworking configurations [18].

2.8 SONET DCS Reconfigurable Mesh Networks

A network is called a mesh network if each end-to-end connection can be supported through at least three link-diverse routing paths between two end nodes of this end-to-end connection. This definition is different from that of SONET self-healing rings, which have two link-diverse routing paths for each end-to-end connection. The dynamic path rearrangeable mesh architecture uses DCSs to reroute demands around a failure point. Unlike point-to-point automatic protection switching systems and rings, DCS restoration does not require standby protection facilities dedicated to working systems for restoration. Instead, it uses spare capacities within working systems to restore affected demands. The operations of DCS mesh network reconfiguration can be performed in a central or distributed manner. The centralized DCS control scheme has a central controller performing the network reconfiguration process, whereas the distributed DCS control scheme executes the network reconfiguration procedure at each DCS node. A centralized or distributed control system may optimize the use of the available spare capacities by referring to a database that contains the status of the network (both working and spare capacities). The penalties for this efficient use of spare capacities, compared with other architectures, are the time needed for the controller(s) to communicate with the network DCSs, complexity, and maintenance of the database. Note that, corresponding to ITU-T's network protection definitions as described in Section 2.5, the centralized DCS network restoration schemes basically use the rerouting techniques, and the distributed DCS network restoration scheme uses the self-healing technique, as defined in ITU-T Rec. I.311.

2.8.1 Path Restoration vs. Line Restoration

For both the centralized and distributed DCS network restoration schemes, there are two restoration techniques possible, line restoration and path restoration, depending on which layer is being used as the protection layer. Line restoration uses the line layer information to trigger the restoration process and restores all affected paths in the affected facility regardless of the sources and destinations of these affected paths. In contrast, path restoration restores affected STS paths on an end-to-end basis. Figure 2-34 depicts examples of line and path restoration methods. In these examples, the link between Nodes 2 and 3 carries two STS-1s [STS-1 (1,6) and STS-1 (4,6)]. If that link fails and the line restoration method is used [see Figure 2-34(a)], route 2–3 is replaced by route 2–5–3, and all channels use the new route when they pass from Node 2 to Node 3. On the other hand, if the path restoration method is used [see Figure 2-34(b)], each channel [i.e., STS-1 (1,6) and STS-1 (4,6)] affected by the link failure selects a new route for restoration. For example, STS-1 channel (4,6) may select new route 4–5–6 and STS-1 (1,6) may select route 1–4–5–6.

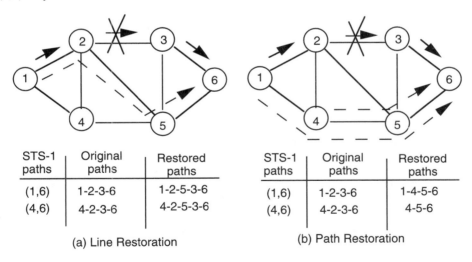

STS-1 paths	Original paths	Restored paths
(1,6)	1-2-3-6	1-2-5-3-6
(4,6)	4-2-3-6	4-2-5-3-6

(a) Line Restoration

STS-1 paths	Original paths	Restored paths
(1,6)	1-2-3-6	1-4-5-6
(4,6)	4-2-3-6	4-5-6

(b) Path Restoration

Figure 2-34: Line restoration vs. path restoration.

2.8.2 Centralized DCS Network Restoration

Figure 2-35 depicts an example of a centralized DCS reconfigurable mesh network configuration and its restoration operations. In this example, demands between Nodes A and B are normally routed over a link between DCS#1 and DCS#2, but they are rerouted through DCS#3 if a cable cut occurs on that link. Once a failure is detected at DCS#2, the DCS#2 sends alarm messages to the central controller as well as its adjacent node (i.e., DCS#1) to inform them of the failure of the link between DCS#1 and DCS#2. Once the central controller receives the alarm message, it will recompute a new routing path that would allow services to be rerouted around the failed link based on the updated network topology and the network resources available at that time. Control messages will then be sent from the central controller to DCSs, which lie along this new routing path (i.e., DCS#1, DCS#2, and DCS#3), to change their DCS configurations and reserve necessary bandwidth. Once this process is completed, the service originally routed through the direct link between DCS#1 and DCS#2 is now rerouted through DCS#3.

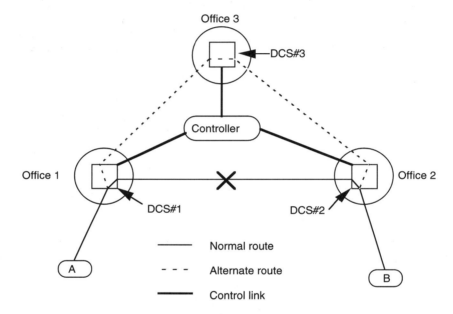

Figure 2-35: An example of centralized DCS network restoration.

The centralized DCS mesh network restoration systems have been implemented by some local and long-distance exchange carriers. The restoration speed is on the order of minutes for 100 STS-1 (or DS3) paths restored. This restoration time can be significantly reduced by using the distributed control approach, which is discussed in the next section.

2.8.3 Distributed Self-Healing Mesh Network

A mesh network is a self-healing mesh network if its end-to-end path can be rerouted around the failed area automatically and dynamically based on the network resources available at that time; this rerouting process is done by DCSs without the involvement of the network management system [10]. The self-healing network can be controlled in a centralized (see Section 2.8.2) or distributed manner.

The concept of distributed DCS network restoration was first proposed by W. D. Grover in 1987; it was later modified and reported in Reference [20]. This proposal uses the physical layer signaling. Another scheme was later proposed in Reference [21] that uses the Section DCC for signaling and a restoration protocol similar to that of the X.25 protocol. Later, several other schemes were proposed based on these two proposals [22–25]. A summary of these self-healing network proposals can be found in Reference [26].

This section describes a generic line restoration algorithm for distributed control self-healing mesh networks. The restoration algorithm for path restoration is similar to that for line restoration. For the distributed self-healing control architecture, each DCS stores local information that includes working and spare capacities associated with each link terminating at that DCS. The actual implementation could vary depending on the specific algorithm considered. Due to interdependence between the routing assignment and the capacity allocation in SONET DCS mesh networks, the distributed restoration system generally executes a three-phase protocol during a complete restoration cycle. Figure 2-36 depicts a three-phase distributed self-healing protocol.

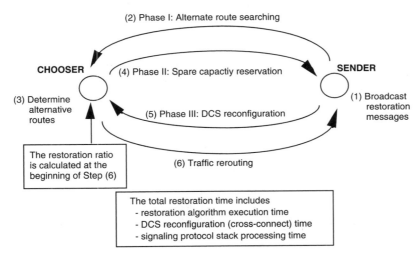

Figure 2-36: Distributed SONET DCS self-healing protocol.

For the distributed self-healing control architecture using path restoration, each DCS stores local information that includes working and spare capacity associated with each link terminating at that DCS. When a failure is detected, one of two ends of the affected STS path on the failed facility is designated as the Sender and the other is designated as the Chooser. All other nodes that participate in the restoration process are called tandem nodes.

In the restoration process, the Sender first broadcasts (floods) restoration messages to all adjacent nodes. To restrict the number of restoration messages and limit the algorithm execution time, selective message flooding is implemented, for example, by the hop count limit. The tandem node updates received restoration messages and rebroadcasts them to other adjacent nodes based on the particular flooding algorithm used. When the message reaches the Chooser, it implies that one or more rerouting paths for restoration is identified, and acknowledgment (ACK) messages are conveyed back to the Sender to reserve the spare capacity for the selected restoration route. Note that reservation of DCS ports for the alternate routes is made either in the first or second phases in each of the DCS nodes involved, whereas cross-connections are made in the last phase.

When the Sender node receives the ACK and verifies the restoration path, it sends a confirmation message back to the Chooser node via the selected restoration path. When the tandem node receives the confirmation message, it reconfigures its DCS switching matrix according to instructions stored in the confirmation message. After the Chooser receives the confirmation message, it changes its DCS switching matrix and cross-connects the affected STS path from the failed facility to newly identified alternate routes. The process described here applies separately for each affected STS path. The restoration process is complete when all affected paths are restored (providing that enough spare capacity exists). Note that reservation of DCS ports for the alternate routes can be made either in the first or second phases in each of the DCS nodes involved, whereas cross-connections are made in the last phase.

2.8.4. Spare Capacity Planning

In the DCS self-healing network, the restoration performance of the self-healing protocol depends on the spare capacity engineered into the network. Thus, in most cases, spare capacity assignments have been part of the DCS self-healing network design process [22,25]. The spare capacity assignment problem is to place the minimum amount of spare capacity needed to restore a predetermined percentage of demands (i.e., the restoration ratio) from single or multiple network component failures, providing the network topology and the working channel assignment meet all of the end-to-end demand requirements that are given. The restoration ratio, which is the ratio of restored demands to affected demands, depends on the self-healing protocol and types of failures (e.g., single-link or multilink) being considered. Note that the spare capacity assignment designed for line restoration is also sufficient for path restoration, because restoration paths for line restoration may also be used for path restoration if other paths are not available. Thus, this section discusses only the spare capacity assignment for the DCS self-healing mesh network using line restoration.

As discussed earlier, the line restoration method identifies K alternate paths to restore all or some of the affected demands (depending on the preset restoration ratio) passing through the failed link as soon as possible. This design concept leads to some design heuristics for assigning the near-minimum spare capacity needed. A simple design concept is to partition the problem into an initial assignment problem, for obtaining an initial feasible solution, and an optimized assignment, for minimizing the spare capacity obtained from the first feasible solution.

In the initial assignment phase, the algorithm may assign the alternate path for each failed link based on some simple criteria (e.g., shortest alternative path) [22] or a more complex linear programming problem with a dual-simplex method for solving the problem [25].

The spare capacity assigned to one link for restoration may also be shared by other links for restoration according to the initial spare capacity assignment algorithm. Thus, the second phase (optimized phase) is to identify such spare capacity redundancies and delete as many as possible. In this phase, the link's spare capacity in the alternate path is updated to the required restored demand if its present spare capacity is less than that required for the restored channel (i.e., affected working channels on the affected link). It remains unchanged if the present spare capacity is sufficient to restore the required demands. For example, one simple procedure is to first choose one spare channel from the highest-spare-capacity assignment and delete it. This identifies corresponding working channels that need the deleted spare channel. This process can be designed in a serial or parallel way, which results in very different computing complexities with a different level of required spare capacity. The same process is repeated until some stopping criterion (e.g., the restoration ratio) is achieved.

Detailed algorithms for and examples of designing spare capacities into DCS self-healing mesh networks can be found in References [22,25].

2.8.5 Impact of DCS Technology on SONET Self-Healing Networks

Use of the distributed control DCS self-healing mechanisms in LEC networks was thoroughly studied by Bellcore in 1992. Economics, operations, and reliability were considered; study results were summarized in Bellcore SR-NWT-002514 [27]. This report concludes that the DCS self-healing network is economically attractive in areas where high demand and high connectivity are involved. In addition, References [27,28] indicated that currently proposed SONET distributed DCS restoration systems may not completely restore services within 2 seconds[1] in large metropolitan LEC networks, as long as the present DCS system architecture (i.e., serial processing and serial cross-connection) and its switching hardware technology remain unchanged. The network restoration time here includes the distributed control restoration algorithm (protocol) execution time and the DCS reconfiguration time, where the DCS reconfiguration time may be a dominant aspect of the total restoration time in metropolitan LEC networks [27,28]. The slow cross-connection time of present DCS systems is due primarily to the inherent serial processing/cross-connect architecture and a slow cross-connection hardware system (e.g., approximately 100–300 ms for a complete process of each STS-1 path reconfiguration).

To meet the 2-second restoration objective, two basic requirements have been identified in [27,28]. The first requirement is to design a DCS self-healing algorithm with a minimum set of restoration messages. The other requirement is to enhance the DCS performance by using a parallel CPU-based processing architecture with a parallel path cross-connection capability. Alternately, the 2-second service restoration objective may be met by implementing priority service restoration. This restoration objective may also be relaxed based on hybrid network restoration architectures. These hybrid restoration

---

[1] The 2-second restoration time is taken as the service restoration objective for DCS distributed restoration [29], because all public switched voice and data services will be dropped when the outage lasts two or more seconds, and a Carrier Group Alarm (CGA) will be generated. Once the CGA is generated, the switch will take approximately 10–20 seconds to restart the transmission system so that it can accept new calls.

architectures deploy fast restoration mechanisms (APS or rings) to meet specific customer needs, as well as DCS mesh networks with distributed control that can provide high survivability, enhanced protection against node failures, and adequate restoration time for most other customers. Which approach should be used depends on the cost for system enhancement and the revenue expected from services supported by the distributed control DCS network restoration system.

Furthermore, Bellcore SR-NWT-002514 [27] suggests that using the SONET Section DCC channel for signaling will contribute several seconds to the restoration execution time just for its OSI protocol stack handling, where traveling seven-layer OSI protocol stack only without processing may take approximately 25 ms, based on current commercial technology. Several alternatives for efficient signaling suggested in SR-NWT-002514 [27] include the use of the unassigned SONET overhead byte or out-of-band signaling such as DS0 or DS1.

2.9 SONET Network Architectures Interworking

2.9.1 Comparisons of Survivable SONET Architectures

Table 2-6 shows comparisons among the SONET APS/DP, SHR, and DCS self-healing networks [26]. The APS/DP and rings have a similar capability to restore services very quickly (i.e., within 50 ms). Their major difference is in growth impact and costs. Rings have a greater concern for system exhaustion, because all ADMs on the ring need to be upgraded if exhaustion occurs. Thus, APS/DP systems are more appropriate in areas where point-to-point demand is extremely high and rings are appropriate for areas where the growth rate is stable and relatively slow. Compared with the SHR, the DCS self-healing network needs less protection capacity at the expense of longer restoration time, and more complex planning and operations systems needed. The restoration time for the DCS self-healing network may range from seconds to minutes, whereas the restoration time for the SHR is within 50 ms. The spare capacity savings for the DCS self-healing network are due primarily to sharing of spare capacities across the entire network. This sophisticated system provides tremendous advantages when it functions properly, but may cause problems in a much wider area (compared with APS and rings) when a software failure occurs. To avoid bringing the entire DCS network down from software failures, sectionalization may need to be incorporated to improve the DCS network's reliability.

Table 2-6: Comparison of SONET survivable network architectures.

Attributes	APS/DP	SHR/ADM	DCS Mesh
Network size	2 nodes	up to few tens of nodes	global
Spare capacity needed	most	moderate	least
Per-node cost	moderate (OLTM*/APS)	lowest (ADM)	highest (DCS)
Fiber counts	highest	moderate	moderate
Connectivity needed	lowest	moderate	most (mesh)
Restoration time	< 50 ms	< 50 ms	seconds–minutes
Mixed line rates	possible for 1:N/APS	no	yes
Software complexity	least	moderate	most
Areas affected by software failure	2 nodes	nodes on the ring	wider areas
Protection against major failures	worst	medium	best
Scalability	easiest	most difficult	moderate
Planning/operations complexity	least	moderate/least+	most

* Optical Line Terminating Multiplexer.
+ Assume that the UPSHR with path protection switching is used. Note that this comparison does not include the ring interconnection feature.

2.9.2 Two-Tier SONET Network Interworking Architecture

The differences among survivable architectures, summarized in Table 2-6, have resulted in a well-accepted two-tier transport network model. Figure 2-37 shows a two-tier network architecture using different survivable network architectures based on the demand distribution, network connectivity, and demand growth conditions [4]. This two-tier network comprises a core network and several peripheral networks. The core network usually has a high degree of connectivity and high demands aggregated from smaller Central Offices (COs) and a substantial growth rate. Examples of the core and peripheral subnetworks in today's fiber facility network are the hub-to-hub subnetwork (having a DCS mesh structure) and the CO-to-hub subnetwork (having either a star/hubbing structure with diversely protected routes or a ring structure), respectively.

In this core network, the DCS self-healing mesh network architecture and/or point-to-point systems with diverse-route APS protection are among the candidate architectures. For the peripheral network, which usually involves lower demand with a lower growth rate, and low connectivity, SHRs and/or point-to-point APS systems have been considered to be cost-effective candidate systems. Note that bidirectional self-healing rings have been considered for both the core network and the peripheral network.

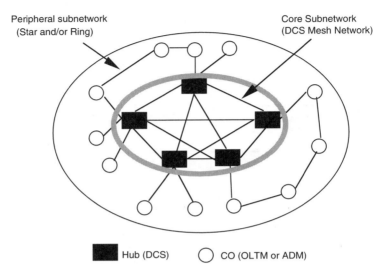

Figure 2-37: A two-tier fiber transport network model.

The two-tier network model requires network architecture interworking capability in order to provide end-to-end service integrity during network failure conditions. The architecture interworking issues were studied in ANSI T1X1.2 [18]. The purpose of this T1X1.2 report [18] is to better understand the potential impact of ring interworking on various architectures (e.g., ring-to-ring, ring-to-DCS mesh), network interfaces, and applications, and to provide recommendations on ring interconnection practices.

2.10 SONET Network Deployment and Standards Status

2.10.1 SONET Network Deployment Status

SONET point-to-point systems are the first SONET systems to be deployed in broadband telecommunications networks, and the SONET line rate of OC-192 has been available since 1996. SONET rings have been widely deployed in LEC networks since 1991. This ring deployment is primarily on a SONET unidirectional, path-switched ring due to its simplicity and the availability of the Bellcore generic requirements completed in 1992–93. In December 1993, T1X1 standards for bidirectional line-switched rings were approved. OC-48 unidirectional, path-switched rings were made commercially available in 1993, and OC-48 bidirectional line-switched rings became commercially available in 1994. Figure 2-38 summarizes the status of survivable SONET network deployment status and associated network protection objectives. As shown in the figure, all SONET network architectures described in this chapter have been deployed in both LEC and Interexchange Carrier (IXC) networks, except for the distributed self-healing mesh network architecture. The distributed SONET self-healing mesh network architecture remains in the research stage and may not be practically attractive because of its technology limitations and the expected availability of ATM VP self-healing network technology.

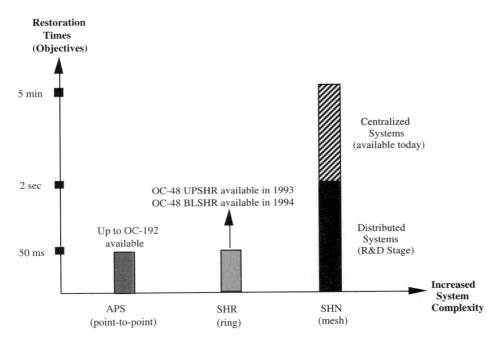

UPSHR = Undirectional, path-switched self-healing ring
BLSHR = Bidirectional, line-switched self-healing ring

Figure 2-38: Survivable SONET network protection deployment status.

SONET self-healing rings were originally designed to support both voice services and private lines. However, the same ring structure has been implemented to support reliable transport for Cable TV signals. Figure 2-39 shows an example of the ring-to-ring broadcast architecture. If this broadcast ring architecture uses SONET technology, the SONET ADM at each node requires a drop-continue feature that drops signals for local processing and passes through the identical signals to the downstream node.

This "ring-of-rings" architecture used for the Cable TV network was planned by Time Warner Cable Group, the second largest U.S. Cable TV operator, to provide reliable transport for CATV services. This architecture features signal redundancy at three levels: head end to head end, head end to hub, and hub to feeder.

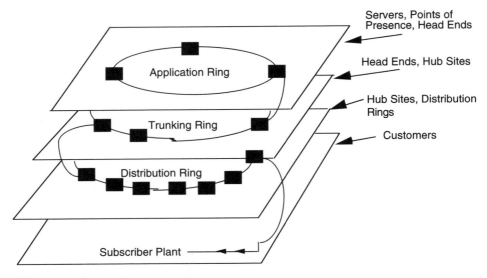

Figure 2-39: A ring-to-ring architecture for Cable TV transport.

2.10.2 SONET Network Standards and Requirements

Table 2-7 summarizes available standards and Bellcore requirements for survivable network architectures at the SONET and ATM layers. As shown in Table 2-7, ANSI T1 standards for SONET APS and bidirectional rings have been completed. Requirements for the SONET unidirectional path-switched ring architecture were defined by Bellcore [15], because it does not require standards work. The SONET DCS distributed restoration scheme is still in an early R&D stage, and it is not being considered in current ANSI standards. SONET APS systems and self-healing rings of up to OC-48 (both unidirectional and bidirectional rings) have been deployed in LEC networks.

Table 2-7: Standards and Bellcore requirements for survivable SONET network architectures.

Protection Layer	Architectures		Standards				Bellcore Requirements		Deployment Status (D, T, or R&D)
			ANSI		ITU-T				
			Doc. No.	D/C	Doc. No.	D/C	Doc. No.	Date	
SONET	S H R	APS	T1X1.5/93-057	C	G.783	C	GR-253-CORE	12/95	D
		UPSHR	–	–	G.841	C	GR-1400-CORE	10/95	D
		BLSHR	T1.105.1	C	G.841	C	GR-1230-CORE	12/96	D
	Mesh		No Standards Available				FA-NWT-1353	12/92	R&D

D/C: D = Draft, C = Complete.
(D,T,R&D): D = Deployed , T = Trial, R&D = Research & Development.

2.11 Summary and Remarks

We have reviewed SONET technology and its network architecture, particularly for its survivable network architectures. Most of these SONET network architectures have been deployed and are operational. Due to the merging services of computer and communications and the new telecommunications laws, SONET network capability may not be sufficient to transport emerging broadband services efficiently, especially for bursty high-speed data and real-time video, which require a long holding time and a significant bandwidth. ATM technology provides an integrated, on-demand switching and multiplexing platform to implement broadband network architectures that may support required QoS of emerging broadband services cost-effectively. Thus, how the existing SONET network infrastructure evolves as a new SONET/ATM network infrastructure cost-effectively presents a great challenge for network planners who are seeking an integrated network platform to support all emerging broadband services.

References

[1]   ITU-T Recommendations G.803, "Architectures of Transport Networks Based on Synchronous Digital Hierarchy," 1992.

[2]   C. A. Siller, Jr. and M. Shafi, *SONET/SDH: A Sourcebook of Synchronous Networking*, IEEE Press, New York, 1996.

[3]   Bellcore GR-253-CORE, "*SONET Transport Systems: Common Generic Criteria*," Issue 2, December 1995.

[4]   T.-H. Wu, *Fiber Network Service Survivability*. Artech House, May 1992.

[5]   ANSI T.210, "Operations, Administration, Maintenance and Provisioning (OAM&P)—Principles of Functions, Architectures, and Protocols for Telecommunications Magazine Network (TMN) Interfaces," 1993.

[6]   "T1 Technical Report on a Comparison of SONET and SDH," T1X1.2/93-024R2, May 1994.

[7]   Bellcore GR-303-CORE, "Integrated Digital Loop Carrier System Generic Requirements, Objectives, and Interfaces," December 1996.

[8]   Bellcore TR-TSY-000496, "SONET Add–Drop Multiplex Equipment (SONET ADM) Generic Requirements and Objectives," Issue 2, September 1991.

[9]   T.-H. Wu and R. C. Lau, "A Class of Self-Healing Ring Architectures for SONET Network Applications," *Proc. IEEE GLOBECOM'90,* pp. 403.2.1–403.2.8, San Diego, December 1990, .

[10] ITU-TRecommendation I.311, "B-ISDN General Network Aspects," Temporary Document 5G (XVIII), January 1993.

[11] ITU-T COM 15-101-E, Draft Recommendation G.SHR-1, "Types and Characteristics of SDH Network Protection Architectures," July 1994.

[12] "Draft Proposed Technical Report on Network Survivability Performance," T1A1.2/93-015R1, March 1, 1993.

[13] Draft proposed American National Standard SONET Automatic Protection Switching contained in ANSI T1X1.5/93-057, 1993.

[14] T.-H. Wu, and M. Burrowes, "Feasibility Study of a High-Speed SONET Self-Healing Ring Architecture in Future Interoffice Fiber Networks," *IEEE Commun. Mag.,* Vol. 28, No. 11, pp. 33–42, November 1990.

[15] Bellcore GR-1400-CORE, "SONET Dual-Fed Unidirectional, Path Switched Ring (UPSR) Equipment Generic Criteria," Bellcore, Issue 1, October 1995.

[16] "SONET Bidirectional Line Switched Rings Standard Working Document," T1X1.5/92-004, February 3, 1992.

[17] Bellcore GR-1230-CORE, "SONET Bi-Directional Line-Switched Ring Equipment Generic Criteria," Issue 3, December 1996.

[18] "T1 Technical Report on SONET Ring Interworking — Baseline Document," T1X1.2/93-006R1, August 2, 1993.

[19] Bellcore GR-1375-CORE, "Self-Healing Ring-Functionality in Digital Cross-Connect Systems Generic Criteria," Issue 1, August 1995.

[20] W. D. Grover, B. D. Venables, M. H. MacGregor, and J. H. Sandham, "Development and Performance Assessment of a Distributed Asynchronous Protocol for Real-Time Network Restoration," *IEEE J. Selected Areas in Commun.*, Vol. 9, No. 1, pp. 112–125, January 1991.

[21] C. H. Yang, and S. Hasegawa, "FITNESS: A Failure Immunization Technology for Network Service Survivability,'" *Proc. IEEE GLOBECOM'88*, pp. 1549–1554, Ft. Lauderdale, Florida, December 1988.

[22] H. Komine, T. Chujo, T. Ogura, K. Miyazaki, and T. Soejima, "A Disributed Restoration Algorithm for Multiple-Link and Node Failures of Transport Networks," *Proc. GLOBECOM'90*, pp. 0459-0463, 1990.

[23] H. Fujii, T. Hara, and N. Yoshikai, "Characteristics of Double Search Self-Healing Algorithm for SDH Networks," *IEICE CS91-48*, 1991.

[24] C. E. Chow, J. Bicknell, S. McMaughey, and S. Syed, "A Fast Distributed Network Restoration Algorithm," *Proc. 12th Int'l Phoenix Conf. on Computers and Communications*, pp. 261–267, Phoenix, March 1993.

[25] H. Sakauchi, Y. Nishimura, and S. Hasegawa, "A Self-Healing Network with an Economical Spare-Channel Assignment," *Proc. IEEE GLOBECOM'90*, pp. 403.1.1-403.1.6, San Diego, December 1990.

[26] T.-H. Wu, "Emerging Technologies for Fiber Network Survivability," *IEEE Commun. Mag.*, pp. 58–74, February 1995.

[27] Bellcore SR-NWT-002514, "Digital Cross-Connect Systems in Transport Network Survivability," Issue 1, January 1993.

[28] T.-H. Wu, H. Kobrinski, D. Ghosal, and T. V. Lakshman, "The Impact of SONET DCS System Architecture on Distributed Restoration," *IEEE J. Selected Areas in Commun.* pp. 79–87, January 1994.

[29] J. Sosnosky, "Service Applications for SONET DCS Distributed Restoration," *IEEE J. Selected Areas in Commun.*, pp. 59–68, January 1994.

Chapter 3

ATM Transport Networks

3.1 Introduction

The increasing demand for Frame Relay (FR) services, Personal Communications Services (PCSs), emerging networked multimedia applications, and high-speed data services has made the traditional transport network infrastructure based on Synchronous Transfer Mode (STM) technology (e.g., SONET/STM networks) insufficient to support these emerging telecommunications services cost-effectively. In addition, with the explosive growth rate of Internet and World Wide Web sites, the existing Internet backbone network may not be able to support such tremendous traffic and bandwidth growth. The Internet is not reliable, secure, or fast, but it is global, standardized, and less expensive. Efforts are underway in the IETF and IP community to improve the performance of the Internet and, as with ATM, integrate multiple services over the network. High-speed, scalable ATM has become increasingly attractive as a broadband transport platform supporting emerging telecommunications and IP services.

ATM is a standards-based transport system that combines the bandwidth efficiency of packet technology with the predictable performance of circuit technology. This technology is scalable from a megabit to gigabits, and it is expected to have a very long life cycle in the emerging telecommunications industry. ATM can be used to support LANs as well as to support worldwide networks. In this way, it contributes to the enterprise goal of developing a single, seamless network infrastructure. It enables the merging of separate networks onto a single platform, resulting typically in improved price/performance. In particular, service integration capability enables operations costs to be reduced by implementing a single operations system and a single type of operations interface, instead of several heterogeneous network operations systems with many types of interfaces. Note that network operations costs are a significant share of total network costs for providers.

ATM is being developed under the ITU-T with the support of the ATM Forum, a consortium of industry manufacturers and service providers. The continuing work of the ATM Forum is an important part of these international development efforts. Unlike ITU-T, which develops ATM standards only for public networks (i.e., B-ISDN), the ATM Forum develops standards for private enterprise networks as well. The major distinction between enterprise ATM switches and public ATM switches is that Central Office (CO) switches in public ATM networks must meet stringent reliability requirements for redundancy and automatic operation, which can significantly increase the cost.

This chapter reviews ATM technologies and shows how this technology is used to build a scalable, high-speed ATM transport network capable of supporting emerging broadband services as well as serving as a next-generation Internet backbone network. Interworking of the ATM-based B-ISDN with existing transport networks will also be discussed.

3.2 ATM Technology Overview

ATM is a high-speed, integrated multiplexing and switching technology that transmits information through uniform (or fixed) cells in a connection-oriented manner. ATM has been adopted by ITU-T as the core technology for B-ISDN. The physical interfaces of 155 and 622 Mbps for the User–Network Interface (UNI) provide integrated support for high-speed information transfer and various communication modes, such as circuit and packet modes, and constant/variable/bursty bit-rate communications.

3.2.1 B-ISDN Reference Model

Figure 3-1 depicts the B-ISDN protocol reference model specified in ITU-T Rec. I.321 [1]. This model reflects the principles of layered communications defined in ITU-T Rec. X.200 [the reference model of Open Systems Interconnection (OSI) for ITU-T applications]. It consists of three planes, user, control, and management. The User Plane, with a layered structure, transfers user application information. The Control Plane, also with a layered structure, handles the call and connection control functions, including signaling. The Management Plane provides management application functions and a mechanism for information exchange between Control Plane and User Plane processes. The Management Plane has two sections: Layer Management and System Management (Plane Management). The Layer Management performs the layer-specific Operation and Maintenance (OAM) functions, and it interacts with User and Control Plane entities and System Management. System Management performs the management functions related to the system as a whole, the coordination between all the planes and the various Layer Management entities.

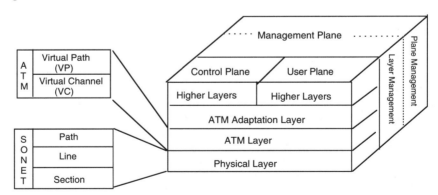

Figure 3-1: Broadband ISDN reference model.

The reference model is further divided into three layers for signal transport. These layers are the Physical, ATM, and ATM Adaptation (AAL) Layers, where the Physical and ATM layers form the ATM transport platform. In North America, the Physical Layer uses SONET standards; in ITU-T, it uses the SDH standards or the cell-based transmission form. Figure 3-2 depicts functions performed at each B-ISDN transport layer. The Physical Layer is responsible for bit timing and physical medium transmission, cell rate decoupling, Header Error Control (HEC) sequence generation and extraction, cell delineation, transmission frame adaptation, and transmission frame generation and recovery. The ATM Layer functions include generic flow control for UNI, cell header generation and extraction, cell routing, and multiplexing.

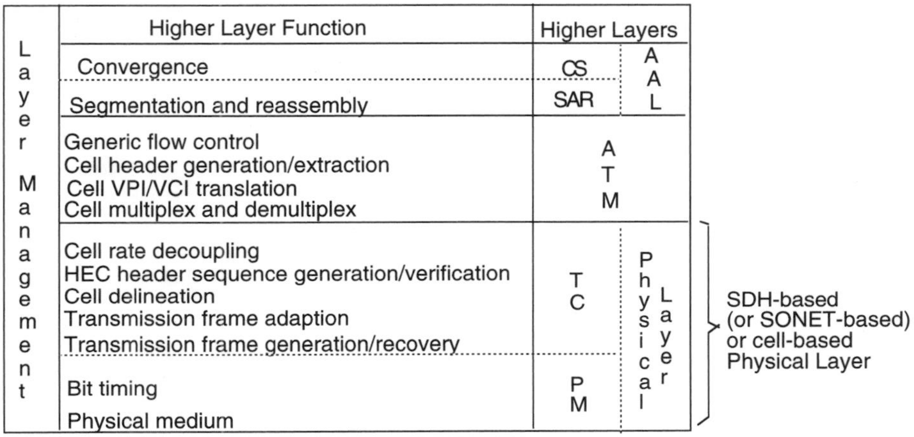

Layer Management	Higher Layer Function	Higher Layers	
	Convergence	CS	AAL
	Segmentation and reassembly	SAR	
	Generic flow control Cell header generation/extraction Cell VPI/VCI translation Cell multiplex and demultiplex		ATM
	Cell rate decoupling HEC header sequence generation/verification Cell delineation Transmission frame adaption Transmission frame generation/recovery	TC	Physical Layer
	Bit timing Physical medium	PM	

SDH-based (or SONET-based) or cell-based Physical Layer

Figure 3-2: B-ISDN layered functions.

The Physical Layer and the ATM Layer are common to both the User Plane and the Control Plane. The Physical Layer provides for the transport of Physical Layer Service Data Units between two ATM entities. The ATM Layer provides for the transparent and sequential transfer of fixed-sized data units between a source and the associated destination(s) with an agreed Quality of Service (QoS). These two layers, the Physical Layer and the ATM Layer, are common to all services.

The AAL functions can adapt the services (including QoS) provided by the ATM Layer to the services (including QoS) required by the different service users. Therefore, the AAL functions are service-specific. The functions of the AAL include service-related functions and the segmentation and reassembly of the data units so that they can be mapped into the fixed-length payloads of the ATM cells.

3.2.2 ATM Cell Format and Encoding

Figure 3-3 depicts the ATM cell formats and encoding. The ATM cell has two parts: the header and the payload. The header has 5 bytes, and the payload has 48 bytes. Within the header, there are fields to perform routing, flow control, and other functions. The transmission sequence for the ATM cell is to (1) process bits within an octet in decreasing order, starting with bit 8, and then to (2) process octets in increasing order, staring with octet 1. For all fields, the first bit sent or processed is called the Most Significant Bit (MSB). The header's primary role is to identify cells belonging to the same VC on an asynchronous time-division multiplexed stream. As depicted in the figure, two different header coding structures are defined: the User–Network Interface (UNI) and the Network–Node Interface (NNI). In the UNI header format, 4 bits are assigned to the Generic Flow Control (GFC) field. The GFC protocol is applied within the customer premise network, which is intended to assist in controlling the flow of traffic of ATM peer–peer connections with various QoS at each UNI by scheduling the information transfer among multiple sources and destinations and then coordinating medium access among multiple broadband terminals at the customer premises and across the UNI.

Twenty-four bits are assigned to routing fields, which are identified by a VCI (16 bits) and a VPI (8 bits). The VCI and VPI are used for routing that is established by negotiation between the user and the network. Three bits are assigned to the Payload Type (PT) field to identify whether information is user information or network information, and 1 bit is assigned to the Cell Loss Priority (CLP) field to determine if the cell should be discarded based on network conditions. Finally, 8 bits are assigned to the HEC field to monitor header correctness and perform single-bit-error correction. One bit is reserved for future use. The NNI header structure is almost the same as the UNI's except that it has (1) no GFC field, and (2) 28 bits for routing (12 bits for VPI and 16 bits for VCI).

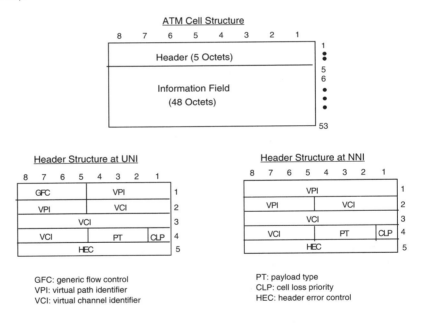

Figure 3-3: ATM cell and header structure.

3.2.3 ATM Layered Structure

The major function of the ATM layer is to provide fast multiplexing and routing for data transfer based on information included in the 5-byte header. This 5-byte header includes information not only for routing (i.e., VPI and VCI fields), but also for fields used to (1) indicate the type of information contained in the cell payload (e.g., user data or network operations information), (2) assist in controlling the flow of traffic at the UNI, (3) establish priority for the cell, and (4) facilitate header error control and cell delineation functions. One key feature of this ATM layer is that ATM cells can be independently labeled and transmitted on demand. This allows facility bandwidth to be allocated as needed, without fixed hierarchical channel rates. Since ATM cells may be sent either periodically or randomly (i.e., in bursts), both constant- and variable-bit-rate services are supported at a broad range of bit rates.

The ATM layer in the B-ISDN model is divided into VP and VC sublayers, as depicted in Figure 3-4. The connections supported at the VP sublayer (i.e., VPCs) are either permanent or semipermanent connections that do not require call control, bandwidth management, or processing capabilities. The connections at the VC layer (i.e., VCCs) may be permanent, semipermanent, or switched connections (i.e., SVCs). SVCs require the signaling system to support its establishment, tear-down, and capacity management.

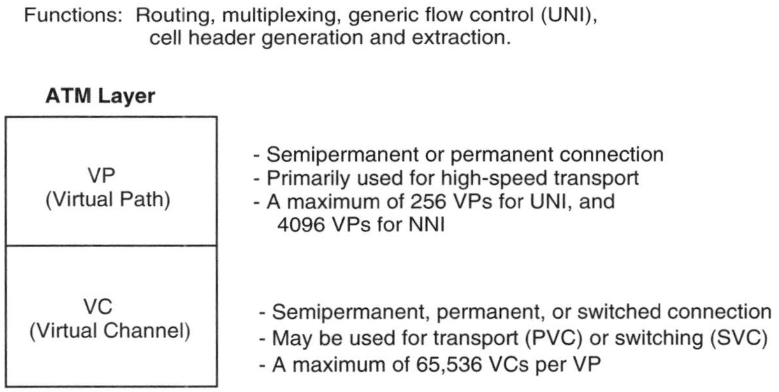

Functions: Routing, multiplexing, generic flow control (UNI),
cell header generation and extraction.

ATM Layer

VP (Virtual Path)	- Semipermanent or permanent connection - Primarily used for high-speed transport - A maximum of 256 VPs for UNI, and 4096 VPs for NNI
VC (Virtual Channel)	- Semipermanent, permanent, or switched connection - May be used for transport (PVC) or switching (SVC) - A maximum of 65,536 VCs per VP

Figure 3-4: ATM layered structure.

3.2.4 ATM Cell Mapping and Multiplexing

In ANSI/T1S1 standards, ATM cells are carried through a SONET transport platform; in ITU-T, ATM cells may be carried through an SDH- or cell-based transport platform. SONET is the North American standard synchronous transmission network technology for optical signal transmission that terminates from OC-1 (51.4 Mbps) to OC-256 (13.2 Gbps), with up to OC-192 available today. The functions of the SONET layer in the B-ISDN reference model is to carry ATM cells at high speed and provide very fast protection switching capability to ATM cells whenever needed.

Figure 3-5 shows the relationship among the VC, VP, SONET path, and SONET link. The SONET link provides a bit stream to carry ATM cells that are logically carried on VCs within VPs. The VPs are then physically accommodated by SONET STS paths. A VC is a generic term used to describe a unidirectional communication capability for the transport of ATM cells. A VP is a group of VCs. Each ATM cell header contains an

ATM label that uniquely identifies the VP and the VC to which the cell belongs. The ATM cell label is the concatenation of the VPI and VCI. VCs on different VPs may have the same VCI value (i.e., local significance), but they have different VPI values. Cells associated with a particular VP are identified by a VPI value in the ATM cell label and by the physical SONET path and link over which the cell is carried.

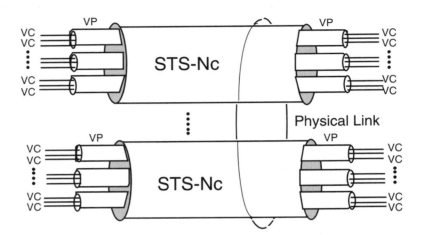

Figure 3-5: Mapping among VC, VP, STS, and the physical link.

The transport technologies used for SONET and ATM layers are completely different. SONET transport uses STM based on Time Slot Interchange (TSI) technology to cross-connect SONET paths; ATM transport uses cell switching technology to switch and/or cross-connect cells. The implications of these transport technologies on the emerging broadband transport infrastructure will be discussed in Section 3.3.

As specified by ANSI/T1S1, ATM cells are transported through SONET STS paths. SONET path formats that have been standardized to carry ATM cells are STS-1, STS-3c, STS-12c, and STS-48c, with STS-192c likely in the near future. In addition to these service mappings, the ATM Forum has specified the ATM cell mapping formats for DS1, DS3, and 25.6 Mbps (over twisted pair) interfaces, as shown in Table 3-1.

Table 3-1: Standardized physical transport systems.

Standards Organization	Interface Speeds
ITU-T	STM-1, STM-4, STM-16
ANSI	DS1, DS3, STS-1, STS-3c, STS-12c
The ATM Forum	DS1, DS3, STS-3c, STS-12c, 25.6 Mbps (over twisted pair)

Figure 3-6 depicts an example of ATM cell mapping and cascaded SONET multiplexing. In this figure, ATM cells are placed into the payload of an STS-3c [see Figure 3-6(a)], and this STS-3c is then multiplexed with other STS-3c's and STS-3s to form an STS-12 stream. Similarly, this STS-12 path can be multiplexed with SONET paths to form an STS-48 multiplex stream for OC-48 transmission, providing the underlying SONET system is an OC-48 system.

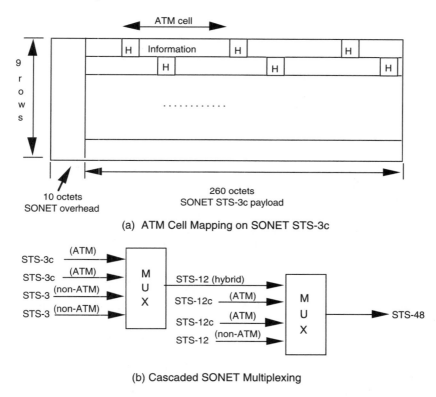

(a) ATM Cell Mapping on SONET STS-3c

(b) Cascaded SONET Multiplexing

Figure 3-6: ATM cell mapping and multiplexing.

Figure 3-7 shows a hierarchical path structure of a SONET-based ATM connection. The highest layer is an end-to-end VC connection, which is terminated at two VCCEs (Virtual Channel Connection Endpoints) and carried over a set of VC links terminated at ATM VC switches. One example of a VCCE is Customer Premise Equipment (CPE) that terminates VCs. One VC link is then accommodated on a VP connection that is realized by a set of VP links terminated at ATM VP terminating equipment, such as ATM VP Cross-Connect Systems (VPXs). Each VP link is accommodated by one or more SONET STS paths that are terminated at SONET PTE (Path Terminating Equipment), such as SONET DCSs or ADMs, when there is one or more SONET PTEs along this VP link. Each SONET STS path is then accommodated by one or more SONET lines that is terminated at SONET LTE (Line Terminating Equipment), such as SONET ADM, and each SONET line is realized by a set of SONET sections if there is one or more SONET regenerators along this SONET line.

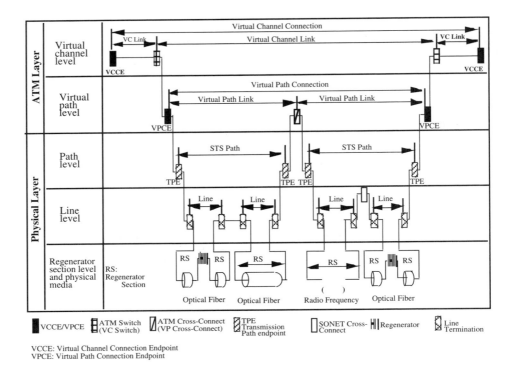

VCCE: Virtual Channel Connection Endpoint
VPCE: Virtual Path Connection Endpoint

Figure 3-7: Hierarchical path structure of a SONET-based ATM connection.

3.2.5 ATM Adaptation Layer (AAL)

The ATM service class is defined through the AAL. The AAL is a protocol layer that performs the function of adapting services onto the ATM protocol. It represents the link between particular functional requirements of a service and the generic, service-independent nature of ATM transport. Depending on the type of service, the AAL can be used by end customers only (i.e., CPEs having ATM capability), or it can be terminated in the network (e.g., SMDS server at the ATM node for connectionless routing).

Figure 3-8 depicts the AAL layer structure. As shown in the figure, the AAL layer is divided into two sublayers: Segmentation and Reassembly (SAR) and Convergence Sublayer (CS). The CS handles higher layer messages with a header and a trailer for every message. CS is further divided into two sublayers: SSCS (Service-Specific Convergence Sublayer) and CPCS (Common Part Convergence Sublayer). The coding of CPCS is shared by both connection-oriented (COS) and connectionless services (CLS), whereas SSCS may be needed only for connection-oriented services for connection management, flow control, error recovery, and so on. The function of SAR is to segment the Protocol Data Unit (PDU) from the CS by adding a 5-byte header to each 48-byte data segment and reassembling 53-byte ATM cells back to the SSCS PDU. Details of the AAL functional description and specifications can be found in References [2,3].

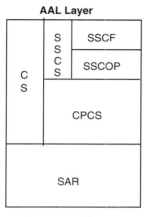

AAL Layer

CS: Handles higher layer messages with a header
    and a trailer for every message (AAL PDU)

SSCS:

- Needed for Connection-Oriented services (COS) for
  connection management, flow control, error
  recovery, and so on
- Not needed for Connectionless services (CLS)

CPCS

- Coding shared by both COS and CLS

SAR

- Related to cell-based operations using headers and/or
  trailers for every cell payload (ATM PDU)
- Coding shared by both COS and CLS

CS: Convergence Sublayer
SSCS: Service-Specific Convergence Sublayer
CPCS: Common Part Convergence Sublayer
SSCF: Service-Specific Coordination Functions
SSCOP: Service-Specific Connection-Oriented Protocol
SAR: Segmentation and Reassembly

Figure 3-8: AAL structure.

Figure 3-9 shows the service class for AAL defined by the ATM Forum, and service examples associated with this AAL service class. This ATM service class is consistent with ITU-T standards.

Three parameters forming the AAL service classes are delay, bit rates, and connection modes. Class A deals with connection-oriented services with constant bit rates; their timing at the source and receiver are related. These services involve an uninterrupted flow of digital information (e.g., DS1/DS3) at a constant bit rate. They are time-sensitive, need a constant and predetermined amount of capacity, and will be supported primarily through circuit emulation at the ATM layer. Circuit emulation is a method to emulate the circuit switching capability on ATM networks. A timing recovery scheme is needed to recover timing at the receiver side in order not to lose received information.

Class B represents connection-oriented services with variable bit rates, and with related source and receiver timing. These services are real-time. A Class B service type is the VBR video. The required capacity of a VC with Class B services necessarily varies between an absolute maximum and an absolute minimum. As with Class A services, this class typically has stringent cell delay, cell delay variation, and cell loss requirements.

	CBR	VBR-RT	VBR-NRT-COD	VBR-NRT-CLD	UBR	ABR
Service Class	Class A	Class B	Class C	Class D	Class X	Class Y
Timing Relationship between Source & Destination	Required (Real-Time)		Not required (Non-Real-Time)			
Bit rate	Constant	Variable				
Connection Mode	Connection-oriented (PVC and SVC)			Connection-less	Connection-oriented	
Service Examples	Voice & Private lines	Variable-bit-rate packet video	X.25/ Frame relay	SMDS/IP/LAN Interconnection/ Multimedia	LAN	LAN/ Internet
AAL Types	AAL-1	AAL-2 (ITU-T) AAL-1/5 (ATM Forum)	AAL-3/4 AAL-5	AAL-3/4	Any	AAL-3/4 & AAL-5

CBR: Constant Bit Rate; VBR: Variable Bit Rate; RT: Real Time;
COD: Connection-Oriented Data; CLD: Connectionless Data
UBR: Unspecified Bit Rate; ABR: Available Bit Rate

Figure 3-9: ATM service class.

Class C deals with bursty connection-oriented services with variable bit rates that do not require a timing relationship between the source and the receiver. The main performance metrics of this class of services are mean cell delay and/or a percentile of the cell delay (e.g., 95th percentile), throughput and the service data unit loss rate. Examples of Class C services are connection-oriented data services such as X.25 and file transfer. Class D is essentially the same as Class C, but the services are connectionless. When services are transported through the network, traffic belonging to Class A and Class B services is assigned a higher cell dropping priority during congestion than that belonging to Class C and Class D services.

In Figure 3-9, CBR (Constant Bit Rate) offers the consistent delay predictability of leased line services with guaranteed bandwidth allocation and uses AAL Type 1 with an available cell payload of 47 bytes for data. VBR (Variable Bit Rate) offers the variable bit rate for services needing more efficient use of network resources and uses either AAL Type 3/4 (available cell payload for data is 44 bytes) or AAL Type 5 (available cell payload is 48 bytes). AAL1 is used primarily for circuit emulation in order to make the ATM connections behave like guaranteed circuits. On the other hand, AAL5 tends to be more useful for data because it allows for more efficient transfer than AAL1. For example, AAL5 allows for cyclic redundancy checking of data integrity because the loss of data is normally unacceptable. On the other hand, one can easily tolerate losses of a few bits of voice and video because human users can adjust. AAL1 does not provide for cyclic redundancy checking but it does provide for sequence integrity, which is crucial for voice and video applications. Note that for video services, ITU-T chose AAL Type 1, whereas the ATM Forum chose AAL Type 5. Interoperability between systems using AAL1 and AAL5 is under study in the ATM Forum. In addition, a null AAL can be used to provide the basic capabilities of ATM switching and transport directly, as in Cell Relay Service (CRS). End customers may use proprietary AALs for special applications as well.

ABR (Available Bit Rate) is similar to CBR, except that it provides variable data rates based on whatever is available through the use of the end-to-end flow control system. ABR is primarily used in LAN and TCP/IP environments. The use of UBR (Unspecified Bit Rate) does not specify the required bit rate, and cells are transported by the network whenever the network bandwidth is available. No flow control is used for the UBR service class. Some specifications for CBR, VBR, ABR, and UBR can be found in the ATM Forum [4].

3.2.6 ATM Multiplexing and Switching

In this section, we will use an example to show how ATM connections are multiplexed and switched through an ATM statistical multiplexer and an ATM switching platform, respectively. Figure 3-10 depicts a high-level architectural diagram for ATM connection multiplexing and switching.

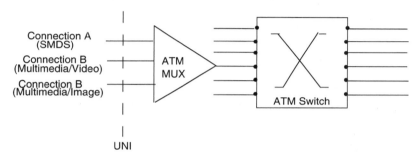

Figure 3-10: ATM multiplexing and switching.

Figure 3-11 depicts a generic protocol model of the UNI as described in Bellcore GR-1110-CORE [5]. The model partition shown is not meant to indicate implementation; rather, it serves as a convenient way of organizing various types of information that flow across the physical UNI. This diagram should be interpreted as showing the complete set of protocols that must be terminated by the Broadband Switching System (BSS) to provide service to a single Customer Premise Network (CPN). The protocol model shown in the figure divides functions into core functions and service specific functions. The core functions are common to all services supported on the interfaces and are organized into the ATM Layer and the Physical Layer. As also shown in the figure, access signaling uses AAL Type 5, CRS users may use a null AAL, circuit emulation users adopt AAL Type 1, SMDS users use AAL Type 3/4 and FR service users use AAL Type 5 for the message exchange protocol.

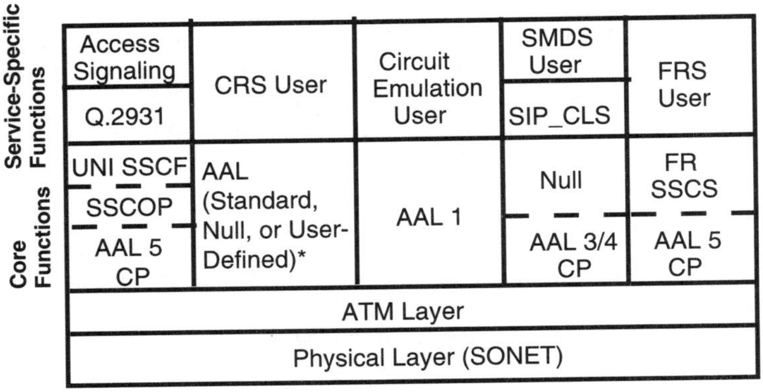

	Access Signaling Q.2931	CRS User	Circuit Emulation User	SMDS User SIP_CLS	FRS User
Service-Specific Functions	Access Signaling Q.2931	CRS User	Circuit Emulation User	SMDS User / SIP_CLS	FRS User
Core Functions	UNI SSCF / SSCOP / AAL 5 CP	AAL (Standard, Null, or User-Defined)*	AAL 1	Null / AAL 3/4 CP	FR SSCS / AAL 5 CP
	ATM Layer				
	Physical Layer (SONET)				

* Transparent to the network

CP: Common Part
CRS: Cell Relay Service
FRS: Frame Relay Service
SSCF: Service-Specific Coordination Functions
SSCOP: Service-Specific Connection-Oriented Protocol
SSCS: Service-Specific Convergence Sublayer

Figure 3-11: An example of protocol model of UNI.

Figure 3-12 shows an example of the SMDS's PDU formatting process at UNI using SMDS Interface Protocol (SIP) based on the UNI protocol model discussed in Figure 3-11. The PDU formatting process starts from the higher layer user data that is added to the overhead for the SIP access protocol. Then the CPCS header and trailer are added to the resultant SIP PDU at CPCS of the AAL Layer. The resultant PDU is then divided into small equal-size cells. The AAL's SAR sublayer starts to add SAR's header and trailer to each cell to form a 48-byte ATM cell (i.e., AAL Type 3/4). The resultant 48-byte ATM cells are then passed to the ATM layer, where the 5-byte header is added to each cell to form a standard 53-byte ATM cell. This 5-byte header is used for ATM network multiplexing and routing.

Figure 3-13 depicts a multiplexing example of ATM connections with different rates. This multiplexing is typically performed using an ATM statistical multiplexer. For example, in the figure, there are two connections, A and B, where Connection B has two substreams with different rates (e.g., audio and video on the same multimedia connection). When performing connection multiplexing, the AAL data unit for each connection or substream is segmented to form fixed-size ATM cells with headers and each ATM cell carries the timing indication obtained from its AAL data unit. The relative timing indication associated with ATM cells forms an order of ATM cells in the multiplexed cell stream within a physical SONET STS path, as shown in the figure. This example illustrates that ATM connection multiplexing may cause Cell Delay Variation (CDV).

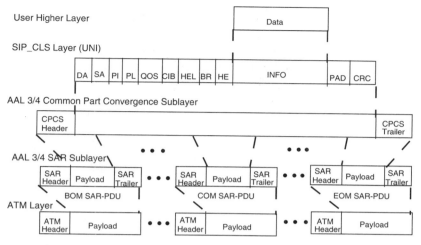

BOM = Begin of Message, COM = Continued of Message, EOM = End of Message

Figure 3-12: An example of SMDS PDU format at UNI.

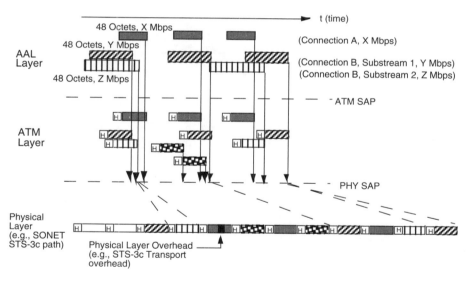

Figure 3-13: An example of ATM multiplexing.

Figure 3-14 illustrates a simplified ATM cell switching system configuration and a switching principle. The ATM cell entering input port #3 of the switch arrives with VPI = 9 and VCI = 4 (e.g., the first ATM cell in the figure). The call processor has been alerted through the routing table that the cell must leave the ATM switch with VPI = 4 and VCI = 5 on output port # 4. The call processor directs the virtual channel identifier converter to remove the "9" VPI and "4" VCI and replaces them with "4" and "5," respectively. If the switch is a large multistage switch, the call processor further directs the VCI converter to create a tag that travels with the cell, identifying the internal multistage routing within the ATM switch matrix. Note that ATM cells belonging to the same VP must have the same VPIs in the input port (e.g., VPI = 9 in input port #3) or the output port (e.g., VPI = 4 in output port #4).

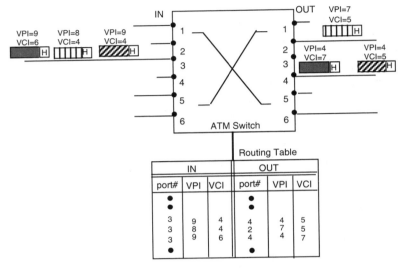

Note: The first and the third cells are in the same VP.

Figure 3-14: An example of ATM cell switching principle.

A large ATM-based telecommunications network may require ATM switches that can scale up to large switches easily and still meet the stringent delay requirements of emerging telecommunications services. The self-routing ATM switch is one of the systems that may meet these requirements [6]. An ATM cell is self-routing if the header of the cell contains all the information needed to route through a switching network. In the following, we use a simple example to explain the self-routing concept, as depicted in Figure 3.15. Note that ATM switches designed for enterprise networks may not use the self-routing switching system.

The example switching network shown in Figure 3-15 is an 8 x 8 shuffle exchange network. An N x N shuffle exchange network has $N = 2^n$ inputs and 2^n outputs, interconnected by n stages of 2^{n-1} binary switch nodes. We assume that the ATM cell is to be routed from input port #4 (i.e., code 011) to output port #2 (i.e., code 001). When the cell arrives at input port #4 (011), the call processor checks with VPI/VCI routing table, identifies the destination output #2 (001), and then generates a routing label that combines the source and destination addresses: 011001. With this code, the system identifies edges in each stage, where each edge is represented by three consecutive digits from the left to the right side of the routing label. In this case, four edges are identified: Edge #1: 011; Edge #2: 110; Edge #3: 100; Edge #4: 001. Each edge (*ijk* connects switch node (ij) in the $(M - 1)$ stage and switch node (jk) in stage M, if this edge is Edge #M. For example, Edge#2 connects switch node (11) in stage 1 to switch node (10) in stage 2. Thus, a self-routing path is established as follows: Input port (011) – switch node (11) of stage 1–switch node (10) of stage 2–switch node (00) of stage 3–output port (001), as shown in the figure.

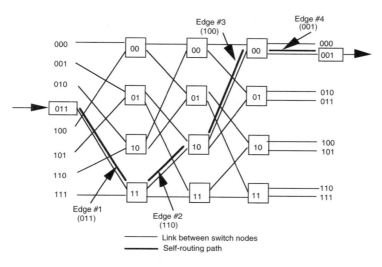

Figure 3-15: An example of the self-routing concept.

3.2.7 ATM Protocol Processes

The purpose of the ATM Layer is to provide a common set of services to the upper layer that can be used to support the large variety of B-ISDN services. The fundamental network services of the ATM Layer transfer information over "virtual connections" between ATM Layer endpoints. These virtual connections are established over the available Physical Layer connections that interconnect the ATM Layer equipment. The characteristics of each virtual connection are determined at the time of its establishment and are referred to as the negotiated connection parameters. These include the throughput of the virtual connection, the service-data-unit loss ratio, and the delay characteristics. The negotiated connection parameters are constrained by the capabilities of the Physical Layer connections and influenced by the ATM Layer equipment characteristics and the previously established virtual connections in the given route. Any originating ATM system may support multiple virtual connections and, therefore, must provide the multiplexing of these connections onto the Physical Layer connections. At intermediate ATM systems, they relay cells received from one physical connection to the appropriate physical connection(s) for transmission. The terminating ATM system demultiplexes the arriving virtual connections and delivers the service data related to each connection to the appropriate user(s). No retransmission of lost or corrupted information is performed by the ATM layer.

To allow the end systems using ATM to coordinate their transmissions and avoid temporary network congestion, the ATM Layer provides an indication of the level of congestion along the path of the ATM peer-to-peer connection within the network to each terminating upper layer. We will discuss the ATM flow control schemes in Chapter 5.

Before discussing the ATM protocol processes, we need to understand which parameters to use for these processes. The following summarizes parameters that are used to characterize the expected traffic on an ATM link and its expected service performance:

- Cell transfer delay acceptable for the ATM link
- The cell loss ratio for cells marked with a Cell Loss Priority (CLP) field of 0 for the ATM link
- Cell Delay Variation (CDV) for the ATM link
- Shape descriptor of the cells marked with a CLP field of 0
- Transfer capacity of all cells of the ATM link
- Copy indication, which controls whether cells from the ATM peer-to-peer connection are copied and delivered to the ATM management system
- Status indication, which indicates whether the ATM link is holding cells or processing them in the normal fashion
- Link identifier(s) associated with the ATM link
- Connection endpoint(s) associated with the ATM link
- Connection endpoint(s) of the Physical Layer associated with the ATM link

These parameters are derived from the negotiated ATM user-to-user connection parameters. Note that cell transfer delay is the time interval in which an ATM cell successfully passed between two designated boundaries. Cell Delay Variation (CDV) is a quantification of cell clumping for a connection, in which the cell clumping is defined as the difference between a cell's expected reference arrival time and its actual arrival time. The shaping descriptor is an ordered triple of numbers (i, m, Rs) used to specify the maximum number of requests (i) to send an ATM protocol data unit at the physical service access point within any time interval of length m cell times, where a cell time (r) is the time required to send 53 octets at the data rate of the ATM Layer. The realized shaping rate Rs is equal to $(i)/(m \times r)$. The ATM Layer protocol is to ensure that these parameters are performed within limited network resources to meet traffic performance requirements. We will discuss these traffic parameters in more detail in Chapter 5.

According to functions performed and parameters specified at the ATM Layer, ATM protocol processes consist of the following major processes: connection establishment process; connection removal process; link parameter update process; ATM cell transfer sending process; and ATM cell transfer receiving process. These processes are briefly explained in what follows. However, for in-depth discussion, see Chapter 4 (ATM Signaling Networks) and Chapter 5 (ATM Network Traffic Management).

3.2.7.1 Connection Establishment Process

This connection establishment process is initiated by a request from the ATM management system and establishes local associations between the ATM system and the specified connection endpoints. It also sets the following parameters: cell transit delay, cell loss ratio for cells with CLP of zero and aggregated traffic, cell delay variation for cells with CLP of zero and aggregated traffic, and the aggregated shape descriptor.

3.2.7.2 Connection Removal Process

This connection removal process releases the local association between the ATM system and the connection endpoint of the existing ATM peer-to-peer connection upon receiving a request from the ATM management system.

3.2.7.3 Link Parameter Update Process

This process changes parameters associated with a specific ATM link upon receiving the request from the ATM management system. These parameters are copy indication, status indication, and capacity. The update of copy indication is the result of performance monitoring management or fault management activation. The update of the status indication is due to certain types of fault management function activation. The capacity update is the result of capacity renegotiation for an existing ATM peer-to-peer connection.

3.2.7.4 ATM Cell Transfer Sending Process

The ATM Layer may receive transit ATM cells and local ATM cells from multiple sources. The ATM entity needs to schedule the resulting collection of ATM cells for service by the Physical Layer (this process is referred to as the multiplexing process). The scheduling of these ATM cells must be performed within the limited resources of the equipment, such as the data rate at which the cells can be served and the buffer available to hold cells before they are served by the Physical Layer. The goal is to avoid cell loss and delay. To meet user service requirements, each ATM user-to-user connection negotiates the acceptable cell delay and cell loss rate when the connection is established, and these performance characteristics remain unchanged during the duration of the ATM peer-to-peer connection. The user may also be able to control the cell loss level (this may be coupled with the subscription rate) by explicitly indicating the level of cell loss priority in the Cell Loss Indication (CLI) within the ATM cell header. The ATM Layer multiplexing process must be performed to meet these performance levels.

In order not to affect the receiving capability of the next downstream ATM system within the same ATM peer-to-peer connection, some transmitters are expected to ensure that the multiplexed stream output from the ATM Layer does not exceed the values of a negotiated traffic shaping descriptor, as already discussed, of each ATM peer-to-peer connection. The multiplexing function performing traffic shaping must ensure that the resulting cell stream for each ATM peer-to-peer connection does not violate the agreed shaping rate.

While queuing ATM cells for service, the ATM system needs to monitor its local resources. If the resources at an intermediate ATM system are close to exhaustion, the sending ATM system may set an Explicit Forward Congestion Notification (EFCN) indicator in the cell header so that this indicator may be examined at the destination. When an ATM system is not in a congestion state, it will not modify the EFCN indicator.

When the ATM Layer has no cell to transmit, it inserts unassigned cells into the flow of assigned cells to be transmitted. This transforms a noncontinuous stream of assigned cells into a continuous stream of assigned and unassigned cells.

3.2.7.5 ATM Cell Transfer Receiving Process

The receiving system determines if the received cell header has an invalid pattern (see [5] for invalid cell header patterns). If the cell header is invalid, this error is reported to the management entity and the cell is discarded. Otherwise, the receiving ATM entity discards the unassigned cells from the flow of cells received, where unassigned cells are identified by a standardized header field. For cells with assigned header values, the ATM system determines whether the VPI or VPI/VCI of the received cell is active. If the received cell is not associated with an active connection, the cell is discarded and the event is reported to ATM Layer management. If the cell is associated with an active connection, the ATM system determines if the cell should be relayed or delivered to either the upper level or the management system. If it should be relayed, it is passed to the Cell Generation and Multiplexing functions; if it is not relayed, the cell is passed to the upper layer along with indications of Service Data Unit (SDU) Type and whether congestion was experienced (or to the management system, along with the Received Loss Priority indication, and an indication of whether congestion was experienced).

It is likely that the source of traffic may not comply with negotiated traffic characteristics; a receiving ATM system at an intermediate node may perform Usage Parameter Control (UPC) or Network Parameter Control (NPC) by monitoring each ATM peer-to-peer connection to identify connections that are not in compliance with the negotiated traffic parameters. The UPC and NPC parameters used by these functions are derived from the negotiated connection parameters agreed on when an ATM peer-to-peer connection is established. These UPC and NPC parameters need to account for cell delay variation that may be introduced by previous multiplexing stages. Within an administrative guideline, the loss priority indication of an ATM cell may be altered by the UPC or NPC to assist the network in properly managing the offered QoS of an ATM peer-to-peer connection. This function of changing the explicit loss priority indication of a cell from high-priority level to low-priority level is referred to as tagging.

3.3 Impact of ATM Technology on Broadband Transport Infrastructure

3.3.1 ATM Technology Characteristics

SONET transport networks using STM nodal transfer technology have been widely deployed in both LEC networks and IXC networks. These SONET transport networks are primarily used to support Plain Old Telephone Services (POTS) and private lines. However, emerging broadband services, such as high-speed data and real-time video, require a much more flexible network infrastructure that can carry both bursty and constant-bit-rate traffic with the quality of a service guaranteed. ATM technology may be a good candidate to fulfill this role due to its unique characteristics described here:

- Nonhierarchical multiplexing structure
- Flexible and scalable switching
- Simpler nodal complexity
- Separation between path and capacity assignments
- On-demand OAM features

3.3.2 Nonhierarchical ATM Multiplexing

A major difference between SONET transport and ATM transport is in the path and multiplexing structure: SONET transport uses a physical path structure; ATM transport uses a logical path structure, as depicted in Figure 3-16. For example, VT1.5 and STS-1 in SONET transport have fixed capacities of 1.5 and 52 Mbps, respectively. In contrast, VPs and VCs used in ATM transport have no fixed capacity physically associated with them. The capabilities of VPs and VCs may be varied in a range from zero (e.g., for protection switching) to the physical link rate (e.g., OC-48), depending on applications.

In SONET transport, transmitting the optical signals over fiber (e.g., OC-48 rate) requires several stages of multiplexing lower-rate signals (e.g., VT1.5s or STS-1s) to the line rate. For example, it may take four stages to multiplex VT1.5s signals to an OC-48 signal: STS-1, STS-3, STS-12, and STS-48. If both VT1.5s and STS-3s (for example) need to be rearranged in an intermediate node along the SONET transport path (e.g., facility grooming), that node needs to equip a B-DCS for STS path cross-connects and a W-DCS for VT path cross-connects, as depicted in Figure 3-16. In contrast, because the VP/VC capacity is not fixed, multiplexing can be performed in a nonhierarchical way. Thus, VP/VC service or facility grooming can be performed by a single type of ATM equipment (e.g., ATM Cross-Connect System). The nonhierarchical multiplexing structure makes ATM transport more efficient in transmission than its SONET counterpart.

Thus, the hierarchical STM path structure and multistage multiplexing/demultiplexing may result in poor link utilization. For example, even if the multiplexing efficiency of the lower stage path into next stage is 0.8 for each multiplexing stage, three multiplexing stages result in a final link utilization of only 0.51 (= 0.8^3). ATM transport, on the other hand, allows direct multiplexing of paths into transmission links and, therefore, better fill of these links. Lower link capacity requirements are also possible. The cost savings are most significant when this capacity requirement difference falls into the two-line-rate breaking point (e.g., one with 11 STS-1s, which needs an OC-12 system and another with 14 STS1-s, which then needs two OC-12 systems or one OC-48 system). Note that for interoffice fiber systems, only OC-12 and OC-48 systems are commercially available and deployed in LEC networks.

Note that the nonhierarchical multiplexing characteristics for ATM transport are applied only to the boundary of the SONET path that carries ATM VPs and/or VCs. For example, if VPs/VCs are carried on an STS-12c, VPs/VCs can be multiplexed directly onto this STS-12c. However, if the transmission line rate is OC-192, it still requires two stages of multiplexing (i.e., STS-48 and STS-192) for optical transmission. This observation leads to the argument that ATM may be much more efficient in multiplexing and switching if the ATM cells can be directly mapped into the payload of a SONET path having an optical line rate [7].

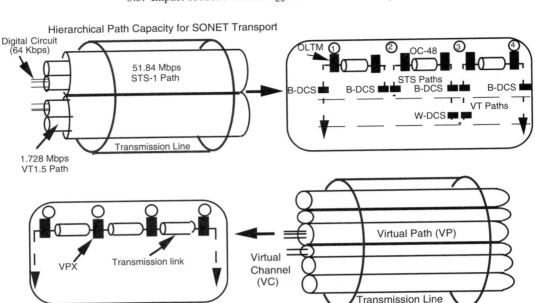

Figure 3-16: Hierarchical SONET multiplexing vs. nonhierarchical ATM multiplexing.

3.3.3 Flexible and Scalable ATM Switching

Table 3-2 shows differences between ATM switching and SONET cross-connects based on the path relationship among the virtual channel (VC), virtual path (VP), SONET's STS path, and SONET link. The switching (cross-connect) method used for SONET's STS paths is STM using a hierarchical TSI concept; the switching method for VPs/VCs uses a nonhierarchical ATM switching concept. The number of path capacities that can be cross-connected through SONET transport equipment (e.g., SONET DCS) is limited (e.g., it can only cross-connect VT1.5 of 1.728 Mbps, STS-1 of 51.84 Mbps, STS-3 of 155.52 Mbps, and so forth), but the current ATM switching technology allows the ATM switch cross-connects VPs/VCs at any bit rate up to 2.4 Gbps. For network restoration, STM performs network rerouting through physical network reconfiguration; ATM performs network rerouting using logical network reconfiguration through the update of the routing table.

Equipment used for SONET transport includes SONET Add–Drop Multiplexers (ADMs) and SONET cross-connect Systems (i.e., DCSs). Similarly, equipment used for ATM transport includes ATM Add–Drop Multiplexers and ATM cross-connect systems. Here, we compare only SONET DCSs and ATM cross-connect systems. Most concepts and discussions for DCSs/VPXs are applicable to the case of ADMs.

Table 3-2: ATM switching vs. SONET cross-connect.

Switching technology	SONET/STM	ATM
Units	VT/STS-1/STS-Nc	ATM cells in VCs/VPs
Method	Time Slot Interchange (TSI)	Self-routing with routing table control
Signal routing path	Hierarchical	Nonhierarchical
Network rerouting characteristic	Physical network reconfiguration	Logical network reconfiguration

Like SONET DCSs, ATM cross-connect systems can be classified as VP/VC cross-connect systems or VP cross-connect systems (VPXs), as specified in ITU-T Rec. I.311 [8]. As with SONET Broadband DCSs (B-DCSs), the VPXs shown in Figure 3-17(a) cross-connect VPs without changing the sequence of VCs accommodated within those VPs. The VPX has routing and multiplexing functions, but it lacks call control and call management capabilities. In a similar fashion to SONET Wideband DCSs (W-DCSs), the VP/VC cross-connect system terminates VPs and crossconnects VPs and VCs, as shown in Figure 3-17(b). As with the ATM VPX, the VP/VC cross-connect system does not have call control and bandwidth management capabilities.

Figure 3-18 depicts an example of a stand-alone switch configuration with an $n \times n$ ATM switch module as the core fabric and different SDH line interfaces terminating different Synchronous Transfer Mode (STM) signals that carry ATM cells in their payloads [9]. The ATM cells from different line rates are internally multiplexed up to the core fabric speed (i.e., 2.4 Gbps) for switching. This 8 x 8 core switch fabric uses a shared-memory design. With this 2.5 Gbps 8 x 8 core switch fabric, the switch can be configured in many ways depending on services and the line cards used. Similar VPX configurations can also be found in Sato *et al.* [7] and Yamaguchi *et al.* [10].

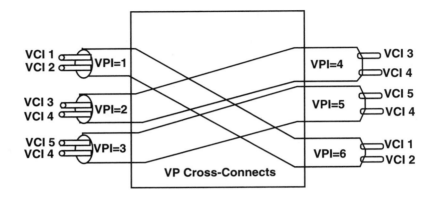

(a) VPX Functional Diagram

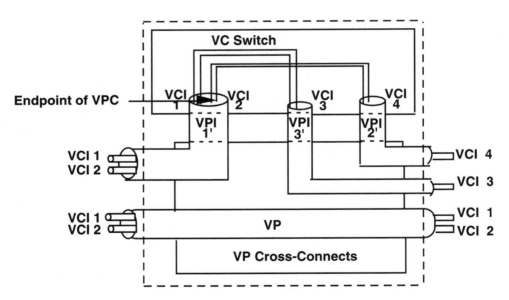

(b) VP/VC Cross-Connect System Functional Diagram

Figure 3-17: ATM cross-connect systems.

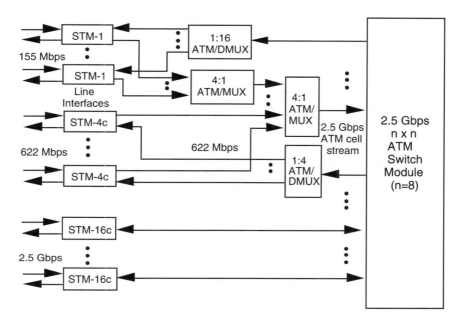

Figure 3-18: An example of an ATM VPX configuration [9] (©1992 IEEE).

Table 3-3 presents a comparison between the VPX and the SONET DCS. The SONET DCS cross-connects physical paths with fixed capacities (e.g., STS-1, VT1.5), whereas the VPX cross-connects VPs with virtually any possible capacity up to the transmission line rate. To cross-connect low-speed signals from a high-speed signal stream (e.g., cross-connect VT1.5s from an STS-3 signal), the SONET DCS must demultiplex the high-speed signal to low-speed signals on a hierarchical basis. This hierarchical multiplexing/demultiplexing structure is not needed for the VPX to cross-connect VPs, because each VP may accommodate any capacity. For the SONET DCS, any change in path capacity (i.e., from STS-1 to STS-3s) requires a change in the switching matrix configuration because the path capacity is physically associated with I/O ports of the switching matrix. In contrast, a change in the VP path capacity will not affect the switching matrix configuration in the VPX, because the VP path capacity can be dynamically adjusted as long as the total capacity of all VPs being cross-connected does not exceed the VPX capacity. Due to its nonhierarchical path structure, the VPX network is better utilized than the SONET DCS network. It also has faster provisioning and restoration capabilities, because its path structure is nonhierarchical and the OAM cells needed for provisioning and restoration can be conveyed within the STS-Nc payloads in any time cycle. The SONET DCS network has limited available bandwidth of DCCs in a fixed cycle of 125 ms.

Table 3-3: ATM VPX vs. SONET DCS.

Attributes and Applications	SONET DCS	ATM VPX
Path structure	physical path	logical path
Path cross-connected capacity	limited	more
Path multiplexing hierarchy	hierarchical	nonhierarchical
Physical switching matrix reconfiguration when capacity or route assignment changes	yes	no
Switching port speed	lower	higher
Facility utilization	lower	higher
Service provisioning (Path)	slower	faster
Alternate routing (Restoration)	slower	faster
Operations and control	simpler	complex
Transmit delay	shorter	longer

3.3.4 Simpler Nodal Complexity

This section compares the nodal complexity of the SONET and ATM transport technologies. This nodal complexity difference is due primarily to the path structures (e.g., physical path vs. logical path, and hierarchical vs. nonhierarchical) and switching technologies (e.g., TSI vs. ATM cell switching). One premise of the ATM VPX is simplification of the cross-connection node and path design. On the other hand, the internal cell fabric requires additional interfaces to convert between STM TDM and ATM cell formats. Figure 3-19 is used here to illustrate these points qualitatively, and also in a simple example summarized in the table. Figure 3-19(a) shows a cascade of STM B-DCS and W-DCS that supports SONET OC line interfaces and STS-1, DS3, and DS1 add-drop interfaces. Figure 3-19(b) shows an ATM VPX with the same interfaces.

As indicated in Figure 3-19(b), the internal VPX fabric assumed here is based on 2.4 Gbps cell streams. This implementation is one reason that the ATM VPX interfaces may cost less than equivalent STM DCS interfaces. For example, an OC-48 incoming signal has to be demultiplexed down to the STS-1 level before its constituent signals can be cross-connected by a B-DCS or a W-DCS. These additional STM demultiplexing (and multiplexing at the output) stages roughly double the cost of an OC-48 STM interface relative to an OC-48 ATM interface.

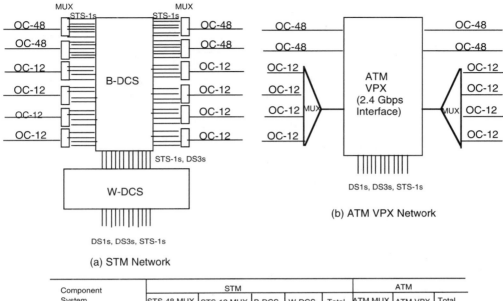

(a) STM Network

(b) ATM VPX Network

Component System	STM					ATM		
	STS-48 MUX	STS-12 MUX	B-DCS	W-DCS	Total	ATM MUX	ATM VPX	Total
Number of Systems	4	8	1	1	14	2	1	3
Number of STS1 INF	192	96	384*	192*	576	-	96	96
Number of STS12 INF	-	8	-	-	8	8	-	8
Number of STS48 INF	4	-	-	-	4	2	4	6

* Assume that 33% of Trunking STS1s needs VT1.5 grooming via the W-DCS.

(c) Composition of Cross-Connect Node for STM and ATM Networks

Figure 3-19: Nodal complexity for SONET DCSs and ATM VPX.

Figure 3-19(c) depicts a nodal complexity comparison between an STM node and an ATM node in terms of numbers of systems and interfaces needed. This example assumes the same number of line (four OC-48s and eight OC-12s) and add–drop (96 STS-1s) interfaces. It also assumes 96 STS-1 interfaces between the B-DCS and W-DCS for DS1 grooming and signal add-drop. Two cell multiplexers/demultiplexers from 600 Mbps to 2.4 Gbps are needed to match the port speed of the ATM VPX. The portion of the trunk interface corresponding to Optical-to-Electrical (O/E) conversion and SONET overhead processing is not accounted for in the table, since it is assumed to be equal in both implementations. As indicated in the table, 576 STS-1 interfaces are required for the STM cross-connect node compared with 96 for the ATM cross-connect node. Similarly, the STM implementation requires 14 system units compared with 3 ATM units. The number of system units is directly related to cost, size, required power supplies and cabling. These factors all contribute to the simplification of the ATM cross-connecting nodes.

3.3.5 Separation between Path and Capacity Assignments

In SONET transport networks, an end-to-end connection, or a path, is established by assigning it a particular time slot in a Time-Division Multiplex (TDM) frame. For example, a particular VT1.5 path going through OC-3 physical links is assigned 27 bytes in the Synchronous Payload Envelope (SPE) of an STS-3 path. The allocation of these specifically positioned 27 bytes in a synchronous frame allows access to this VT1.5 signal without having to demultiplex the entire STS-3 signal to its constituent VT1.5s. However, it also imposes well-defined bandwidth capacity and physical interface requirements (1.728 Mbps and a VT1.5 interface, in the foregoing example) on this end-to-end connection. In other words, in a TDM frame there is an inherent coupling between the connection (determined by time slot positions in the frame) and its bandwidth (determined by the number of time slots in a frame). Thus, with STM DCSs, any change in path capacity (e.g., from STS-1 to STS-3) requires a change in the switching matrix configuration, or a change of the path establishment, because the path capacity is physically associated with I/O ports of the switching matrix. This results in a coupling between physical route assignment and capacity allocation. The three-phase protocol used for SONET self-healing mesh networks (see Section 2.8) is one of the results of these characteristics.

In contrast, a change in the VP path capacity will not affect the switching matrix configuration in the ATM VPX because the VP path capacity can be dynamically adjusted as long as the total capacity of all VPs being cross-connected does not exceed the output port capacity or the ATM VPX capacity. Thus, the capacity allocation and the physical route assignment are decoupled in ATM transport networks. For example, the ATM network may preassign an alternative physical route for the ATM connection without allocating any capacity to those protection VPs or VCs. These protection VPs or VCs will share a pool of spare capacity when needed. Thus, the separation of capacity allocation from physical route assignment for VPs/VCs would reduce required spare capacity, compared with its SONET transport counterpart.

3.3.6 On-Demand OAM Feature

Bandwidth on demand is a basic capability that is needed for some advanced network control functions, including rerouting for network congestion and network restoration and customer bandwidth management for private line customers. The bandwidth-on-demand feature is a network capability that would allocate needed bandwidth to meet the QoS requirement whenever and wherever needed during the call setup and call delivery processes.

Path management functions in SONET transport are performed via its SONET transport overheads (i.e., section, line, and path); ATM VP transport uses OAM cells for path management. For example, an Alarm Indication Signal (VP-AIS) OAM cell, which has been defined in ITU-T Rec. I.610 as one of the maintenance signals, can be used as a triggering signal to stop billing, to set blocking at an ATM switch, or to operate VP level automatic protection switching. One major difference between SONET transport overhead usage and VP OAM cell usage is the frequency of signal generations and transmission during a time interval. In STM transport, transport overhead is generated and transmitted every 125 μs, but ATM transport allows for many OAM cells to be generated and transmitted within a 125 μs time interval as long as the capacity is available.

Unlike the STM transport's OAM fixed capacity, the ATM layer OAM capacity can be assigned dynamically based on the need of the particular maintenance activity or procedure invoked. Thus, ATM transport can convey necessary network management messages faster than its STM counterpart. Also, it has been suggested that ATM VP transport may provide a significant improvement over STM transport for the speed of failure detection for BERs in the range between 10^{-3} to 10^{-5} [11] (e.g., it takes approximately 2 seconds and 1 ms to detect BERs of 10^{-4} or better for STM transport and ATM VP transport, respectively). This potentially allows for faster network response to "soft" failures in ATM transport. Thus, performance degradation due to line protection switching may be reduced by using line protection switches only for "hard" failures (i.e., BER > 10^{-3} or Loss of Signal, Loss of Frame), and using ATM VP OAM mechanisms for "soft" failure protection.

3.4 ATM Transport Network Architectures

3.4.1 Broadband Transport Network Architecture Alternatives

Figure 3-20 depicts a B-ISDN transport reference model corresponding to the B-ISDN layered model depicted in Figure 3-1. This reference network model, specified in ITU-T Rec. I.311 [8], includes an ATM transport network layer and a service control and intelligent layer. The ATM transport layer network is further divided into two sublayer networks: a service transport network and a control transport network that correspond to the user plan and the control plan, respectively, in the B-ISDN reference model. The service transport network carries user traffic such as video, data, and voice; the control transport network carries only traffic associated with network control and signaling, such as call control, resource management, and protection switching. The service control and intelligence system is the source of service generation and control that drives network intelligent requirements. This system is a database-processing-oriented system and includes service modules (such as 1-800, PCS, AIN, and VoD), service control modules (such as SCPs, Video Level 1 gateway, and so on), and OAM modules (such as network management nodes, traffic management nodes, and the like).

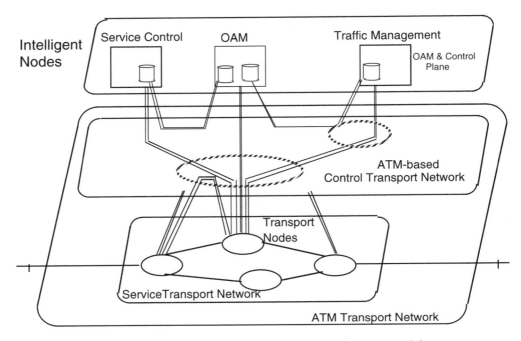

Figure 3-20: B-ISDN transport network reference model.

To use ATM transport capability cost-effectively in the B-ISDN environment, services and control/signaling messages are carried by the same physical network, but they are logically separated from each other by using different Virtual Channels (VCs) or Virtual Paths (VPs). We will discuss the service transport network in the remainder of this chapter and leave discussions of the ATM control and signaling network to Chapter 4.

As with SONET transport networks, ATM transport also may use either the "Large Switch" or "Small Switch" transport network design concept. The "Large Switch" network design concept is to bring the remote customer traffic to larger ATM VC switches in some strategic locations through high-speed transport systems such as SONET ADMs/DCSs and/or ATM VP-based ADM/VPX. This network design is used to reduce the number of switches needed to support switched services. The "Small Switch" network design concept brings the switch to the customers by deploying smaller ATM VC switches closer to the customer sites. In this design, ATM VC switches are physically interconnected directly, and SONET and ATM VP paths are treated as simply transmission pipelines. The use of the "Small Switch" or "Large Switch" network design depends on network size, traffic patterns and volumes, costs, and reliability and performance requirements. Figure 3-21 depicts these two ATM transport network design concepts [12].

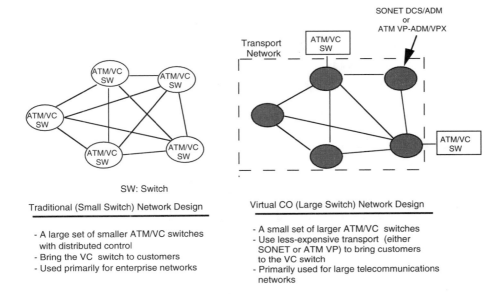

SW: Switch

Traditional (Small Switch) Network Design

- A large set of smaller ATM/VC switches
  with distributed control
- Bring the VC switch to customers
- Used primarily for enterprise networks

Virtual CO (Large Switch) Network Design

- A small set of larger ATM/VC switches
- Use less-expensive transport (either
  SONET or ATM VP) to bring customers
  to the VC switch
- Primarily used for large telecommunications
  networks

Figure 3-21: Two possible ATM transport network designs.

The foregoing design concepts lead to three possible transport network architectures to support both ATM switched services and private line services. Depicted in Figure 3-22, these network architectures are SONET transport, ATM VP transport, and ATM VC transport. In these three network architectures, end-to-end ATM connections are controlled and managed by two end ATM VC switches; the transport for transit connections are supported through a SONET transport network [Figure 3-22(a)], an ATM VP transport network [Figure 3-22(b)], or an ATM VC transport network [Figure 3-22(c)].

3.4.1.1 SONET Transport Networks

The first transport network architecture takes advantage of existing SONET transport infrastructure. For example, SONET self-healing rings have been commonly used by public carriers to support reliable ATM services, as depicted in Figure 3-23. This network configuration works well if the ATM network only supports PVC (Permanent Virtual Connection) services, because routing across the ATM network and the SONET ring can be provisioned to optimize the capacity usage of the SONET ring. However, there could be a problem with capacity allocation when the ATM network supports the SVC service that might route the connections across the provisioned-based SONET ring dynamically. For example, the ATM may use PNNI source routing [13] (see Chapter 4) to route channels across the ATM network and the SONET ring dynamically. In this case, the SONET ring may face an unpredictable traffic pattern that eventually may cause poor ring capacity utilization. Thus, if the SONET transport network is used to support switched ATM connections using dynamic routing, an interworking protocol that would best utilize the dynamic routing feature of the ATM network and the provisioned SONET

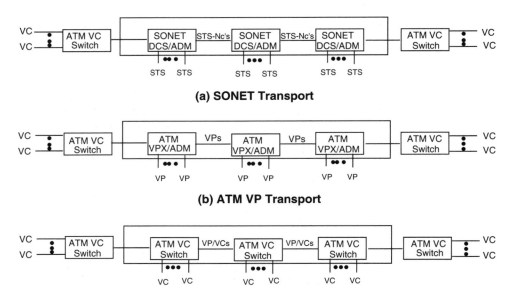

(a) SONET Transport

(b) ATM VP Transport

(c) ATM VC Transport

Figure 3-22: Three possible broadband transport architectures for switched services.

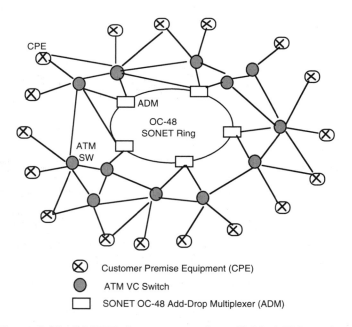

Figure 3-23: SONET ring transport for reliable ATM services.

network capacity should be designed. Issues and designs of such "smart" interworking protocols remain to be studied.

3.4.1.2 ATM VP Transport Networks

Figure 3-24 depicts an ATM VP-based network architecture [14]. In such a network architecture, VCs are processed and assigned to a VP connecting the VC switches at either end (analogous to assigning calls to STM paths) and are transported transparently via an ATM VP-based transport network. An ATM VP-based transport network is made up of ATM VPXs and/or ATM ADMs, depending on the transport network configuration being considered. The end node of a VP connection executes more processing than the transit node within the ATM VP-based transport network. Figure 3-24 also depicts nodal functions of a VP-based network architecture. The ATM VC switch at either end of these ATM connections handles VP assignment setup processing, VP capacity allocation and VP route selection. The ATM VPX in the transit node is much simpler (and thus less expensive) than the ATM VC switch, because it does not perform VP assignment setup, VP capacity allocation, and VP route selection. This simpler equipment design is also applied to ATM VP-based ADM, which is used primarily in a ring topology to add–drop or pass through ATM cells within the ring node.

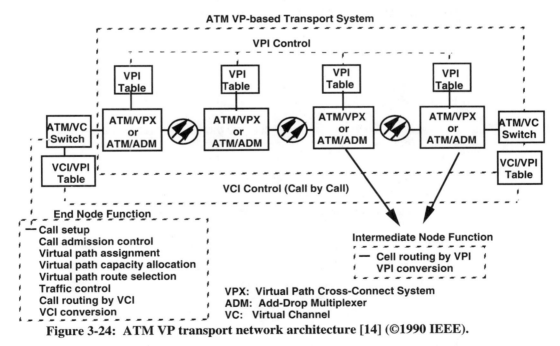

Figure 3-24: ATM VP transport network architecture [14] (©1990 IEEE).

Figure 3-25 depicts the functional configuration of the envisioned NTT's target ATM network that has been reported in by Aoyama *et al.* [15,16]. The network consists of an ATM VP transport network and two types of VC networks: one for private networks and the other for public networks. The VP transport network is shared by the private and public switched VC networks. The private VC network is developed on the VP leased line network by customers and will be used for private business communications. On the other hand, the public VC network is developed by using the ATM switching systems installed in COs that provide B-ISDN public switched services.

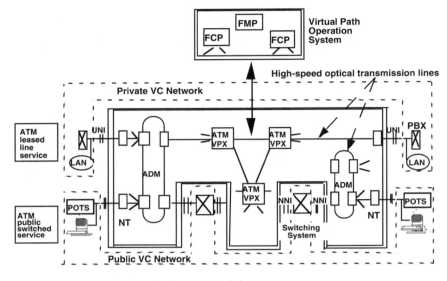

FMP: Facility Management Point
FCP: Facility Control Point

Figure 3-25: An ATM transport network architecture [15] (©1992 IEEE)

The NTT's network evolution plan is divided into three phases [15]. In Phase I, an ATM VP-based transport network is introduced to provide high-speed transport for both N-ISDN and B-ISDN infrastructure. ATM leased line services, including multimedia, for large business customers will be supported. An ATM-based control transport network was introduced during 1993–95 for OAM information and signaling messages. In Phase II, the VP-based transport network will be significantly extended to cover both interoffice and loop networks. End-to-end ATM B-ISDN public switching systems for business customers will be supported in limited areas. In Phase III, the ATM-based B-ISDN network will be fully developed, and remaining SDH-based transport networks will be integrated into the B-ISDN network.

3.4.1.3 ATM VC Transport Network Architecture

A VC network consists of ATM VC switching systems and VP links. The VCs in a VP for a VC network may or may not be statistically multiplexed together. In contrast, a VP network consists of ATM VPXs and physical links (e.g., SONET links). The VPs in a SONET link for a VP network generally are not statistically multiplexed, to ensure the cell layer QoS for each VP.

Table 3-4 summarizes the functional differences between a VC switch and a VPX. There is a trade-off between the statistical multiplexing effect and the call processing requirement from a network point of view, reflecting the functional difference between ATM VC switches and ATM VPXs. The statistical multiplexing effect (performed at the ATM VC switch) in ATM networks improves bandwidth efficiency by sharing the cell intervals among calls. On the other hand, a reduction in required processing capacity is made possible by using VP cross-connect routing (performed at the VPX). This, however, results in less efficient use of statistical multiplexing and a possible decrease in link utilization.

Table 3-4: ATM VC switch vs. ATM VPX.

Attribute	ATM VC Switch (BSS)	ATM VPX
Processing unit	VC	VP
SVC supported	yes	no
Call control and management	yes	no
Bandwidth efficiency	higher	lower
Nodal complexity and cost	higher	lower

3.4.2 Architecture Feasibility Studies

A feasibility study is needed to differentiate among the alternative network architectures and help understand those conditions for which the network architecture may be most suited. This section discusses three architecture case studies that have been reported in References [17–19].

3.4.2.1 Case Study I: DS1 Transport

Figure 3-26 depicts examples of how today's STM and future ATM transport networks support switched services. Figure 3-26(a) is the network infrastructure that has been widely used in LEC networks, where this transport network (e.g., DCS-based network in this example) provides an open transport platform for accessing different types of switches from different vendors. The switch interface to the DCS is typically a DS1. Figure 3-26(b) shows an example of using the ATM VP transport technology on the same transport platform as the one used in today's DCS networks. In this situation, the DS1's ATM/STM conversion can be implemented either at the end ATM VPX or at the circuit switch.

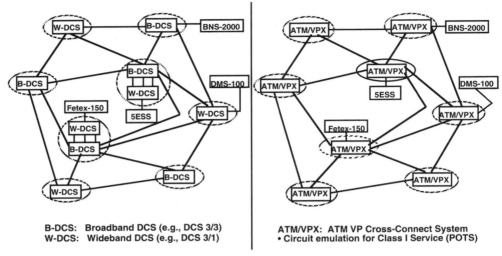

B-DCS: Broadband DCS (e.g., DCS 3/3)   ATM/VPX: ATM VP Cross-Connect System
W-DCS: Wideband DCS (e.g., DCS 3/1)   • Circuit emulation for Class I Service (POTS)

(a) STM Transport Network Architecture   **(b) ATM VP Transport Network Architecture**

Figure 3-26: Network model for a case study.

An economic study and network design methods for comparing ATM VP transport and
SONET DCS transport based on the two network architecture models depicted in Figure
3-26 has been reported in [17]. The network model is a large metropolitan LEC network
that includes 23 nodes and 51 fiber links with a total of 63,226 DS1s for internodal
demands based on projections for POTS and PL deployment. The hubbing location and
strategy for this DCS network were also provided by an LEC. Based on the existing
hubbing strategy, the end-to-end DS1 demand requirement is bundled into the DS3
demand requirement (with a total of 2612 DS3s). Each node is either equipped with a
B-DCS, W-DCS, or both the B-DCS and the W-DCS. If the node has only a B-DCS and
the service unit is accessed at the DS1 rate, an M13-like system is used to multiplex DS1s
to the DS3 (and vice versa), which is then added to/dropped from the B-DCS. For the
ATM VP network, the VPX is placed in each DCS node in this study. Each ATM VPX
has trunking interfaces at the STS-3c rate and DS1 access ports. For POTS, circuit
emulation is needed to convert a DS1 to ATM cells, a function that could be performed at
the CPE or at the VPX access port.

Figure 3-27 depicts cost comparison results for an ATM VP transport network and a
SONET DCS transport network based on the foregoing network model [17]. If the cost of
a fully configured B-DCS is the same as that for the ATM VPX [see Figure 3-27(a)],
ATM VP transport technology has a cost advantage over its SONET/STM counterpart of
approximately 22 percent. The 22 percent cost reduction results from a 13 percent link
cost reduction and a 27 percent node cost reduction. This cost savings is due primarily to
VPX's grooming efficiency.

Figure 3-27(b) compares the relative cost of B-DCS to VPX with the total relative
network cost of VP to SONET/STM transport networks with and without using circuit
emulation. As shown in Figure 3-27(b), if the ATM/STM conversion is performed at

CPEs [i.e., cost(DS1) = 1.0 in Figure 3-27(b)], the ATM VPX network will have a cost advantage over its STM counterpart only if the cost ratio of the VPX to the B-DCS is less than 1.5. For the case of the circuit emulation function performed at VPXs, the break point of economic advantage of VPXs over DCSs decreases from 1.5 to 1.3 when the cost per DS1 line card with the circuit emulation function is 1.4 times this cost without circuit emulation. If the relative cost (per DS1) of the circuit emulation function to no circuit emulation function is increased to 2.0, the VPX may have a cost advantage over its STM counterpart (only if the cost ratio of the VPX to the B-DCS is less than 0.8).

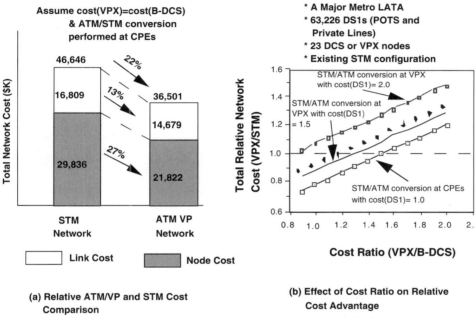

Figure 3-27: A cost study for ATM transport and SONET DCS transport.

Figure 3-28 graphically depicts the results of varying the percentage of DS1 drop in traffic as all other factors are held constant. These results show that DCS has a cost advantage when the dropping level is high. The high relative cost of the ATM VPX DS1 interface causes this effect. The hubbing strategy (the ratio of DS1s hubbed at the DS1 level as opposed to the STS-1 level) causes the crossover point to vary, but the general rule holds true regardless of the hubbing mix. As the amount of STS-1s passing through increases, DCS technology is less expensive over a wider range. However, the break-even point between DCS and VPX for this case is always within the 30–40 percent range in this particular example.

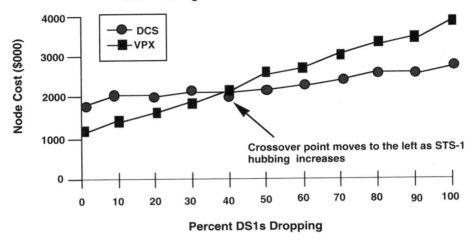

- DCS has dvantage where dropping level is high
- Rule holds true for any hubbing mi x (DS1 or STS-1)
- Crossover range is about 30–40% depending on level of STS-1 hubbing

Figure 3-28: Application areas for SONET DCS and ATM VPX networks.

The study reported in [17] suggested that ATM VP transport may be an economically viable broadband transport system, especially for applications not requiring circuit emulation at the network access points. However, this cost advantage of VP transport over STM transport depends significantly on the cost of DS1 circuit emulation. This study also suggested that the areas that may be attractive for ATM VPX deployment are transit or tandem nodes where the majority of traffic passes through the node. For areas where most traffic is dropped to the local node, the SONET DCS may be an attractive transport system. However, the "boundary" of the ratio of traffic dropping to traffic passing through that determines relative application areas varies depending on the network size and traffic conditions; a detailed network design model is needed. Note that this study was conducted on a highly connected metropolitan area LEC network. It remains to be seen if these results will remain true for nonmetropolitan area LEC networks.

3.4.2.2 Case Study II: DS0 Transport

The second case study reported by Inoue *et al.* [18] explores the feasibility of several ATM-based Public Switched Telephone Network (PSTN) architectures proposed. This low-rate (DS0) ATM network feasibility study was motivated by the assumption (or observation) that the PSTN service is still the dominant telecommunications service today and that ATM deployment will be economically justified only if the PSTN service is integrated into the ATM network.

In a comparison with the DS1 transport case discussed in Section 3.4.2.1, carrying the DS0 channel (i.e., an existing phone line capacity) may require an echo canceller to meet its QoS for real-time voice services. This is because it may take 12 ms to packetize the voice sample (DS0) into ATM cells, and the user will experience echo if the end-to-end delay exceeds 20 ms. Thus, one of the crucial factors that makes the ATM-based PSTN economically feasible is a cost-effective line card that integrates ATM/STM conversion [i.e., cell assembly/disassembly (CLAD)] and echo cancel functions for each subscriber line. Note that the echo canceller may not be needed if the ATM payload carrying voice samples is not fully utilized.

The network model used in this study is an NTT network in the Greater Tokyo area, which comprises Tokyo, Kanagawa, Saitama, and Chiba prefectures, as depicted in Figure 3-29. The overall area is 100 km × 100 km in size and includes 400 access nodes located at the lattice points. Each access node accommodates 30,000 subscribers and originates 2700 erlang traffic; that is, the calling rate is 0.1 and 90 percent of the traffic is assumed to be closed within the region. The gravity model ($x_{ij} = C/d_{ij}$) is used to represent traffic distribution in the region, where x_{ij} is the traffic between nodes i and j, d_{ij} is the distance between nodes i and j, and C is a constant.

Four transit nodes (A, B, C, and D) are implemented to handle traffic in Areas 1, 2, 3 and 4, respectively, and four subtransit nodes (E, F, G, and H) are implemented to handle the four subareas within Area 1. These transit nodes also handle interarea and intersubarea traffic.

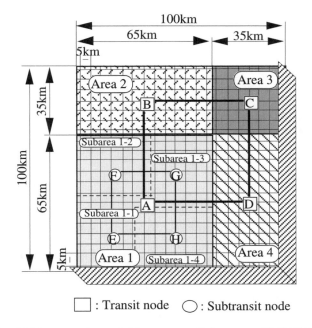

Figure 3-29: NTT network model in greater Tokyo area [18] (©1992 IEEE).

There are two assumptions used in the traffic model. The first assumption is that traffic between neighboring nodes is dense, and traffic is sparse between remote nodes. This assumption is common to all access nodes. The second assumption is that the traffic is heavily concentrated to and from the business center at node A during the daytime.

The three network architectures considered here are depicted in Figure 3-30 along with evaluations from the viewpoints of network resource utilization, controllability, and flexibility.

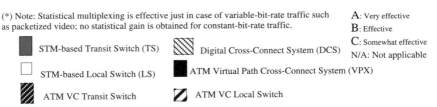

Architecture Option		Option 1	Option 2	Option 3
Transfer mode		STM	ATM	
Network configuration	Transit node			
	Subtransit node			
	Access node	Area 1 Areas 2-4	Area 1 Areas 2-4	Area 1 Areas 2-4
Path-level resource utilization	Effect of fine bit-rate granularity	N/A	B	B
	Statistical multiplexing effect (*)	N/A	C	A
Channel-level resource utilization	Efficiency of large groups	A	A	C
	Statistical multiplexing effect (*)	N/A	A	C
Network controllability & flexibility	# of paths/channels to be controlled	Small	Small	Large
	Call-by-call dynamic routing control	A	A	C
	Path capacity control	N/A	B	B

(*) Note: Statistical multiplexing is effective just in case of variable-bit-rate traffic such as packetized video; no statistical gain is obtained for constant-bit-rate traffic.

A: Very effective
B: Effective
C: Somewhat effective
N/A: Not applicable

◼ STM-based Transit Switch (TS)   ◪ Digital Cross-Connect System (DCS)

☐ STM-based Local Switch (LS)   ◼ ATM Virtual Path Cross-Connect System (VPX)

◪ ATM VC Transit Switch   ◪ ATM VC Local Switch

Figure 3-30: Three network architectures and evaluations [18] (©1992 IEEE).

Network architecture option 1 is an STM-based network with Local Switches (LSs) located at access nodes, Digital Cross-connect Systems (DCSs) at subtransit nodes (i.e., Nodes E, F, G, H), and Transit Switches (TSs) at transit nodes (Nodes A, B, C, D). Since deterministic multiplexing in this architecture results in a multistage transmission hierarchy, resource utilization and flexibility with respect to traffic variation may be limited.

Network architecture options 2 and 3 are ATM-based networks with ATM VC switches located at access nodes with single-channel CLADs on the subscriber side. ATM VPXs and/or ATM VC switches are located at transit nodes and subtransit nodes. Resource utilization is higher than in their STM counterpart due to the fine bit-rate granularity in virtual channel multiplexing into virtual paths, and in virtual path multiplexing into the transmission line. Furthermore, statistical multiplexing in these architectures makes network reconfiguration and adaptive capacity control of virtual channels and virtual paths far easier than for STM channels and paths. Thus, it is expected that architecture options 2 and 3 offer higher flexibility than their STM counterparts in the event of network failures or traffic variation.

For network architecture option 2, traffic concentration at the virtual channel level is achieved on a call-by-call basis at transit ATM VC switches. In contrast, ATM VC switches at access nodes in network architecture option 3 are interconnected by VPs and VPXs at transit nodes concentrate traffic on the virtual path level. There is no call-by-call processing at transit nodes. Network architecture option 2 may achieve higher transmission efficiency than is possible with architecture option 3, but more costly transit node equipment will be needed.

The network costs considered in this study include the ATM VC switch cost, the VPX cost and the link cost. The ATM VC switch cost is the sum of a fixed common cost and the switching cost, which is proportional to the number of 64 Kbps virtual channels handled by the VC switch. Similarly, the VPX cost is calculated as the sum of a fixed common cost and the traffic-dependent cost, which is proportional to the total virtual path capacity handled by the VPX. The transmission link cost is the sum of the fiber cost, which is proportional to the transmission capacity, and the repeater cost, which is inserted every 80 km with the cost proportional to transmission capacity.

The cost values of the ATM VC switch and the VPX were estimated using the values of corresponding STM-based equipment. For example, cost elements of an ATM VC switch were estimated by considering the cost of digital local switches for access nodes and the cost of digital transit switches for transit nodes. Noted that for the access VC switch, only the transit-side cost was taken into account. Cost elements of a VPX were estimated using those of STM DCSs. Cost elements of the transmission link were based on those of SDH transmission equipment.

Figure 3-31 summarizes the cost study reported by Inoue *et al.* [18] based on models and assumptions already described. Network architecture option 2 is simply a replacement of STM-based architecture option 1 with a total cost reduction of 10 percent, due to the elimination of the quantization effect caused by 1.5 Mbps STM path granularity. However, this improvement is not notable, due to the high cost of ATM VC switches in transit nodes. In architecture option 3, the total network cost is drastically reduced, as compared with architecture option 2, by the exclusion of expensive transit VC switches. This cost savings is significant (approximately 50 percent reduction compared with network architecture option 1), even without adaptive control of the virtual path capacity.

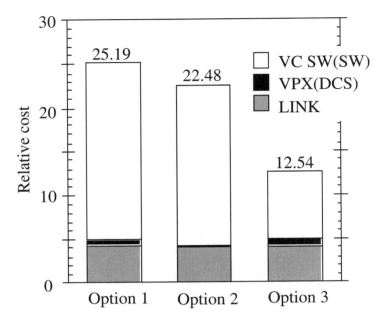

Figure 3-31: Cost study results for ATM-based PSTN architectures [18] (©1992 IEEE).

Further cost savings may be attained by using adaptive control of virtual path capacity based on the noncoincidence of peak traffic between day and night. VP capacity adaptive control applies to the business center during the daytime and to neighboring nodes at night. With adaptive control of virtual path capacity, a network cost reduction of approximately 10 percent has been observed [18]. If the same control based on measured virtual path traffic is used combined with the admission control of VPs, the network will become even more flexible with respect to short-term traffic variation.

Obviously the foregoing architecture feasibility study depends on the cost of the ATM Subscribe Line Card (SLC), which then depends on the progress of Large Scale Integration (LSI) technology. Assuming that one-chip ATM SLC becomes available and less expensive, the ATM-based PSTN will become even more attractive due to the simpler, single operation platform that the network provider needs to maintain. The following discusses the technical feasibility of an ATM SLC with a single-channel CLAD and an echo canceller.

LSI technology, especially submicron Complementary Metal Oxide Semiconductor (CMOS) and Digital Signal Processor (DSP) technology, affects the feasibility of ATM circuit packs. Based on forecasts, it will be technically feasible by the year 2000 to pack a single-channel CLAD and an echo canceller with Battery feed/Over voltage protection/Ringing/Supervision/Coder/Hybrid/Testing (BORSCHT) and N-ISDN line termination function onto one printed circuit board.

Figure 3-32 illustrates the configuration of a line card implemented by LSI technology [18]. In this figure, DSP functions include coding/decoding for POTS, time compression modulation for N-ISDN, and echo canceling. The CLAD assembles/dissembles ATM cells and processes ATM layer 2 protocol, that is, Link Access Procedure for D-Channel (LAPD). The DSP and CLAD will be achieved by one LSI chip. In addition, the integrated circuit pack is estimated to become as compact as the subscriber line card for POTS by the year 2000 if the power feeding function for N-ISDN is omitted.

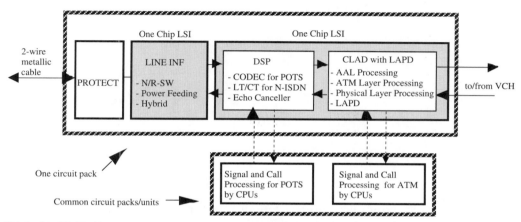

INF = Interface; N/R-SW = Normal/Reverse Switch; CODEC = Coder and Decoder; LT/CT = Line/Central Termination; VCH = Virtual Channel Handler.

LSI	Process	Power Consumption	Chip Size	Note
Line Interface	Bipolar process for high voltage	~ 150 mW	~ 6 mm square	~ 700 Transistors
DSP + CLAD with LAPD	0.5 - 0.6 μm CMOS process	~ 300 mW	~ 10 mm square	~ 30K Gates for CLAD with LAPD

Figure 3-32: Configuration of a line card [18] (©1992 IEEE).

3.4.2.3 Case Study III: Multimedia Transport

The third case study evaluates the capabilities of three alternative transport network architectures supporting multimedia services; this case was discussed by Yoshida and Okazaki [19]. The architecture evaluation in this case study is from the routing perspective. Figure 3-33 depicts three possible routing designs for an ATM network. The first one is VC direct routing, which carries traffic directly between two VC switches. The second one is VC indirect routing, which carries traffic between local switching nodes in VCs using tandem switching. This routing scheme has the effect of reducing link capacity due to statistical multiplexing of concentrated VCs. The VC routing design should ensure acceptable cell and call quality (e.g., cell loss probability and call blocking probability). The third possible design is VP routing, which carries traffic between local switching nodes in a VP using the cross-connect function. When tandem switching is replaced by cross-connect using VP routing, call processing for tandem switching in VC routing will be decreased. Thus, a reduction of the node cost is expected.

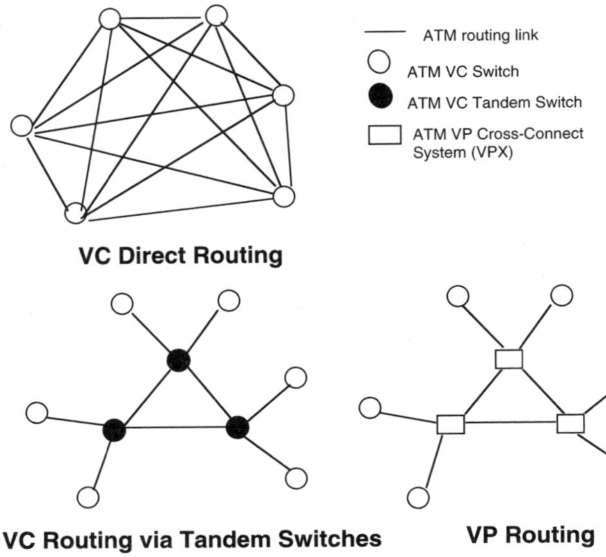

VC Direct Routing

Legend:
— ATM routing link
○ ATM VC Switch
● ATM VC Tandem Switch
□ ATM VP Cross-Connect System (VPX)

VC Routing via Tandem Switches   **VP Routing**

Figure 3-33: ATM routing alternatives.

A study that compares the costs and performance of the foregoing routing schemes has been reported by Yoshida and Okazaki [19]. The study results are summarized in Figure 3-34. The results shown in Figure 3-34(a) suggest that, in general, VC routing has cost advantages over VP routing when there is either a low traffic load or a high ratio of the link cost to the node cost. The node cost here is assumed to be a linear function of VP capacities connected to that node; the link cost is assumed to be a linear function of VP capacities connected to that link. In this study, the ratio of switching cost to cross-connect cost per VP capacity is assumed to be 2:1.

Assume: The ratio of VC switching cost to VP cross-connect cost per VP is 2

Conditions		VC Routing	VP Routing
Traffic load	low	✔	
	high		✔
Ratio of link cost to node cost	low		✔
	high	✔	

Assume: 1. The ratio of VC switching cost to VP cross-connect cost per VP is 2
2. Uniformly offered traffic of 10 (Erlangs).

Conditions		VC Direct Routing	VC Tandem Routing
Traffic load	low		✔
	high	✔	
Ratio of maximum to average bandwidth	low	✔	
	high		✔
Network scale	small	✔	
	large		✔

(a) VC Routing vs. VP Routing.   (b) VC Direct Routing vs. VC Tandem Routing.

Figure 3-34: A study for three ATM routing designs [19].

Figure 3-34(b) depicts an application area for VC direct routing versus VC tandem routing. In general, VC tandem (hierarchical) routing has a cost advantage over VC direct (nonhierarchical) routing under conditions of light traffic load, large-scale networks, or a large ratio of maximum to average bandwidth (which is service-related). Under light traffic conditions, the statistical multiplexing gain obtained by VC direct routing, which gathers only VCs with the same source and destination, is much smaller than that obtained by VC tandem routing. However, under heavy traffic conditions, even VC direct routing provides significant statistical multiplexing gain, when the number of VCs in a VP is large enough. Moreover, when the number of nodes becomes large, more VCs for different source-destination pairs are gathered on a VP by VC tandem routing. Thus, the statistical multiplexing gain is large enough to compensate for the increase in routing hops. In addition, the statistical multiplexing gain in VC tandem routing increases when the ratio of maximum bandwidth to average bandwidth (which is related to services) is large.

The study results reported in Yoshida and Okazaki [19] suggested a vertical routing interworking scheme, as depicted in Figure 3-35. This vertical routing interworking scheme uses VP routing for a core network with high traffic demand and high connectivity, but it uses VC tandem routing to concentrate local ATM traffic to be carried through high-speed VP transport. This study result is consistent with what was found for the two-tier broadband transport model discussed in Section 2.9.2; however, the problem is explored from a different perspective (i.e., from the routing perspective). Using this vertical relationship, network design would be simplified through separation of VP and VC network level designs.

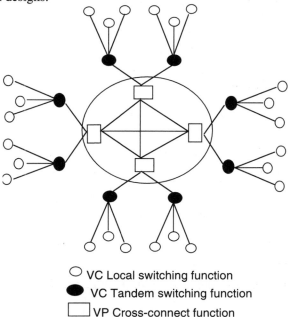

○ VC Local switching function
● VC Tandem switching function
▢ VP Cross-connect function

Figure 3-35: A vertical VP/VC routing hierarchy for ATM networks.

3.4.3 VDT Transport

3.4.3.1 VDT Network Architecture

Video Dial Tone (VDT) is the term used by the FCC in referring to their concept of a common carrier network service that provides the transport and gateway functionalities that would enable end users access to a variety of video information. These VDT network services can be grouped into five classes: (1) Distribution, (2) Interactive Distribution, (3) Retrieval, (4) Interactive, and (5) Conversational; all of them are shown in Table 1-3 (see Section 1.2.2.6).

Figure 3-36 depicts a simplified end-to-end VDT network architecture that includes several major components: a VDT backbone network, the VIP (Video Information Provider), and an access network. VIPs are connected to the VDT backbone network via a VIP–Network Interface (VNI), and Video Information Users (VIUs) are connected to the access network via a Customer–Network Interface (CNI).

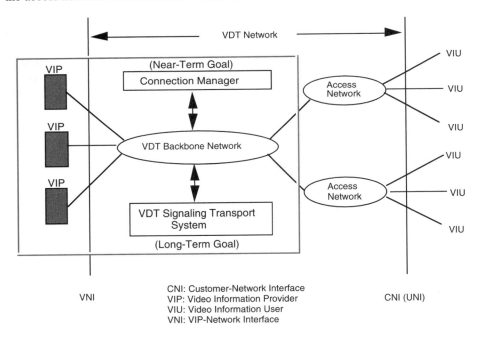

Figure 3-36: A simplified end-to-end VDT network architecture.

The VDT backbone network should support two types of ATM connections between VIPs and VIUs: point-to-point and point-to-multipoint connections. Point-to-point ATM connections are used to carry video-on-demand applications to VIUs as requested. Point-to-multipoint ATM connections are used to carry basic, subscription, and pay-per-view video applications to VIUs.

The VDT signaling and control messages may be carried on X.25, TCP/IP, SS7, or ATM networks, where the X.25, TCP/IP, or SS7 network is considered a near-term, interim transport platform. The signaling/control bandwidth allocated by X.25, TCP/IP, or SS7 may range from 16 to 56 Kbps and it may range from 16 Kbps to 1.5 Mbps in ATM signaling networks. For video transport, SONET systems (e.g., SONET self-healing rings) are being used today to support reliable video transport by CATV network providers as well as by public carriers.

VIPs provide video applications to VIUs. It is expected that video information will be in highly compressed digital formats for more efficient use of storage and transport facilities. For MPEG2 (Motion Picture Image Coding Expert Group Level 2) video, the bandwidth need ranges from 2 Mbps to 6 Mbps for most applications. To provide real-time access to VoD applications, a VIP may use an ATM switch. The VIP interfaces with the VDT backbone network across the VNI. The VNI requires asymmetric flow of data across a SONET–ATM interface. Video applications, data, and signaling/control are carried downstream across the VNI to VIUs; signaling/control and data are carried upstream from VIUs. Some interfaces of VoD system components have been defined in [20].

The UNI environment consists of television receivers, VCRs, personal computers, and so forth. To receive video applications, new VDT-specific CPE, such as set-top units, may be needed. The access network used to deliver VDT video applications to VIUs can be based on different distribution architectures, including Asymmetric Digital Subscriber Line (ADSL), fiber/coax bus, cablelike fiber-rich bus, Fiber-In-The-Loop (FITL), and terrestrial radio. The use of a particular distribution architecture depends on economics, the geographic area, and demand for video applications.

3.4.3.2 Video on Demand (VoD)

Among VDT services, Video on Demand (VoD) service has been specified in both ITU-T [20] and the ATM Forum [21]. The VoD service allows subscribers to order movies whenever they want and control movie processing in a VCR-like environment (e.g., pauses, rewind, stop, fast forward and play). Figure 3-37 depicts an example of VoD operations based on the communication flow model specified in ITU-T Rec. I.375 [20]. Note that the communication flow specified in ITU-T [20] is a generic flow model. A more updated and detailed communications protocol, DSM-CC (Digital Storage Media–Command and Control), for VoD, which is specified by Digital Audio Visual Council (DAVIC) can be found in References [22–23].

In the access network shown in Figure 3-37, it is assumed that the video signal will be carried by a fiber-coax bus and the signaling and control messages will be carried by a FITL system. Each VIU will have a preestablished signaling VC (i.e., VCC x in the figure) to the IP (i.e., video gateway) through the SCP. When the VIU is ready to use the VoD service, this VC (i.e., VCC x) will allow the VIU to signal the IP for a menu of VIPs, which will be sent to the VIU over the same VC.

Readers who are interested in the large-scale distributed VoD system design, including movie placement and replication, and fault-tolerant system access protocols may refer to Wu *et al.* [24] for design details.

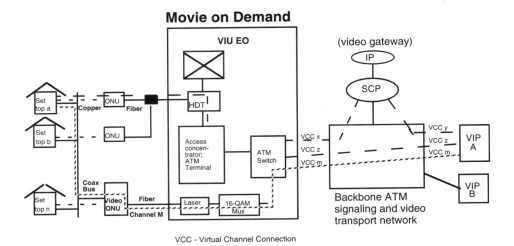

Figure 3-37: An example of VoD operations and configuration.

When the VIU selects a VIP there are two possibilities, based on where the movie titles associated with this selected VIP are stored. According to ITU-T Rec. I.375, the list of movie titles for that VIP may be stored in the IP or the VIP site, depending on the mutual arrangement between the network provider and the VIP. If the list of movie titles is stored in the video server of the VIP site, the SCP returns the address of that VIP to the local switch so that it can establish the video and control connection to that VIP. This VIP then downloads the movie titles to the subscriber's Set-Top Box (STB) with optional features (e.g., movie preview). Once the subscriber returns the selection of the movie title to the VIP's video server, the video server searches the directory that stores address information of the selected video server, if the VIP's video server function is distributed within the network.

If the list of movie titles is stored in the IP site, the IP will download a list of movie titles associated with the selected VIP via VCC x. Once the subscriber returns the selected movie title to the IP, the IP searches the directory that stores the address information of the selected video server. This IP then notifies the SCP of the destination address of the selected video server. The SCP passes that information back to the End Office's (EO's) ATM switch. Based on the addressing information, the ATM switch establishes a control VC (VCC z) and a video VC (VCC m in the figure) from the VIU EO to the selected video server. The EO's ATM switch also assigns Channel M in the fiber-coax bus and connects VCC m to Channel M. Because there will be more households than channels served on the fiber-coax bus, the VIU requesting a particular movie must be told on which channel the movie is being carried so that it can tune itself to the right channel. Thus, the SCP must notify the VIU that the movie will be carried on Channel M.

When the movie is played, there are two channels between the VIU and the VIP. These are a video channel (VCC m and Channel M in the figure) and the control channel (VCC z). The video channel is used to transport the movie content, whereas the control channel is used to exercise VCR-like controls for the movie.

At the conclusion of the movie, the VIP notifies the SCP that the movie is over via VCC y (see the figure). The SCP then stops the billing process and takes down two VCs (i.e., VCC z and VCC m).

3.4.3.3 VDT Transport Network Platform

VDT network design needs to be closely related to video service requirements. There are two standards and requirements for video service performance requirements for the AAL (connection) and ATM layers. These include ITU-T's Integrated Video Service (IVS) Baseline Document [25] and Bellcore GR-2901-CORE [26]. VDT services can be supported through the SONET/STM or SONET/ATM transport platform. Criteria that may be used to compare video transport using SONET or ATM transport technology include transport node complexity, bandwidth efficiency and quality of services. In what follows, we briefly discuss criteria and comparisons of transport node complexity and bandwidth efficiency.

Transport Node Complexity

To support the multiple rate requirement for VDT applications, a SONET/STM transport node may require three different types of equipment: a W-DCS, possibly a B-DCS, and a packet switch. The W-DCS is used to cross-connect VT1.5s for 1.5 Mbps video, and VT6s (6 Mbps) for 4 Mbps video. A B-DCS is needed when STS-3c signals for High Definition Television (HDTV) and other uncompressed video signals are transported. The packet switch is needed for control and signaling, which may be an STP in today's SS7 network or an independent system (e.g., X.25, TCP/IP) used for VDT transport only. Signaling in this case is performed in an out-of-band manner and cannot support point-to-multipoint signaling, unless the existing SS7 protocol is modified.

In contrast, an ATM transport node requires only a single VPX for provisioned video VPs, signaling, and control. In the ATM VP-based transport network, signals with different rates may be assigned to different VPs and are routed through the same physical transport network. Also, the current available ATM VPX or ATM switch inherently handles multirate signals in parallel. The ATM signaling network may be designed for a public network or an enterprise network (we will discuss ATM signaling networks in Chapter 4). Although the initial VDT services could be supported through a SONET/STM transport system, the stringent design and performance requirements of VDT services may make ATM a transport technology of choice because it can support high-speed transport with distributed message processing and flexible bandwidth allocation for different bit-rate traffic.

For the movie-on-demand application, a small-scale video server may be asked to support 1000 concurrent streams and to store 500 movie contents on-line with a further 2000 contents in off-line tape archives. For acceptable quality of service, the video server needs an output bandwidth of 3 Mbps. It also needs to respond in less than a second to requests by the user for such transactions as movie database searches and movie ordering [27]. These requirements can be converted into several key design characteristics. The server's output data rate will be 400 Mbps, and its storage will be 1.5 Terabytes (TB) of on-line disk capacity and 6 TB of off-line archival tape. ATM is a better transport technology because it can support those tremendous output data rates. In particular, the

ATM signaling network has the advantage of being able to provide signaling bandwidth on demand with flexible bandwidth allocation to support a variety of bit rates. The bandwidth allocation for each ATM signaling connection is 173 cells per second (about 66 Kbps) or multiples of that where the multiple is up to 23 (i.e., about 1.5 Mbps) [28], depending on the application or service supported. This bandwidth-on-demand feature makes the ATM network attractive for supporting services having unpredictable or unexpected traffic patterns. In addition, it is expected that the ATM signaling network may offer more cost-effective point-to-point and point-to-multipoint signaling links than its X.25/TCP/SS7 counterparts. This is because, in the ATM-based transport network, signaling messages and service data are carried through the same physical transport network, and the cost of adding relatively low bandwidth signaling traffic to the physical path carrying high-bandwidth broadband services (e.g., video) may not be significant.

Details on architectural design and evolution issues for ATM signaling transport networks will be discussed in Chapter 4.

Bandwidth Efficiency

Comparison of different transport technologies depends on the video service architecture being considered. There are two possible video service architectures: one with a continuous mode and the other with a burst mode. The service architecture with a continuous mode, which is being tried by the CATV industry, dedicates STS-3c links from the VIP to serving COs, and video information is transmitted continuously over these links from the VIP to the selectors. In the serving CO, the incoming STS-3c signal is demultiplexed into the constituent video channels (e.g., 1.5 Mbps). Individual video channels are selected, in accordance with the VIU choice, and delivered to the individual VIU via the access network.

In this continuous-mode architecture, no buffers are required in the serving CO, and even the selectors and VDT service control units can be implemented in the VIP node. Thus, essentially, in this service architecture the VDT backbone network may amount to pure transport. The transport system supporting this continuous mode service architecture can be implemented in a provisioned manner initially, and then upgraded to the switched manner when signaling standards become available and costs can be justified.

For the provisioned transport network, if all VIUs associated with one video selector of the serving CO request services continuously, the STS-3c link may be fully utilized. In this scenario, the STM transport, using W-DCS, may be more economical than its ATM counterpart for carrying this STS-3c channel. However, that STS-3c capacity may not be shared with other STS-3c paths even if only a small portion of the cluster users request video services. Thus, utilization of the STS-3c capacity would be low if STM transport is used. This scenario would worsen considering that STS-3c paths have to be dedicated from the VIP to several clusters.

Consider for example, Figure 3-38, in which an STS-3c is provisioned from the VIP to each of the serving COs (CO-4 and CO-5). Thus, using STM transport, each CO can serve a maximum of 84 VIUs (assuming 1.5 Mbps video channels) simultaneously out of that VIP. The rest of the link (OC-12/48) capacity is provisioned similarly from other VIPs. If 42 VIUs in CO4 and 126 VIUs in CO-5 request service at the same time, all from the same VIP, then 42 users out of CO-5 will be denied service, even though sufficient capacity is available on the links connecting the VIP with CO-5.

Consider, on the other hand, the same OC-12/48 facilities, but with ATM transport (with circuit emulation adaptation) deployed in this VDT backbone. In that case, VPs are provisioned from the VIPs to each of the serving COs. However, each VP capacity can support, depending on usage in all the clusters and VIPs, up to 336 (STS-12 payload) VIUs simultaneously out of each serving CO. In the previous example, no customer would be denied service. The ATM transport allows sharing of transport capacity among different COs served by any of the VIPs and among different VIPs.

For switched, nonprovisioned applications, STM transport requires a SONET Bandwidth Manager (SSBN)–like system to allocate bandwidth on demand via DCS reconfiguration. According to the FCC ("green book"), a VIU expects to receive the services within 3 seconds after it makes the request. SSBN may not meet this goal, as its connection setup time is targeted for 30 seconds, due to slow DCS reconfiguration time (order of a few hundred milliseconds for a physical path reconfiguration) and complex protocol stack processing. In contrast, ATM VPX can reconfigure its connections much faster than its STM counterpart by simply updating its routing table (order of 10 ms or less), and it has simpler OS interface protocol processing. Thus ATM VPXs may support this video connection setup objective of 3 seconds.

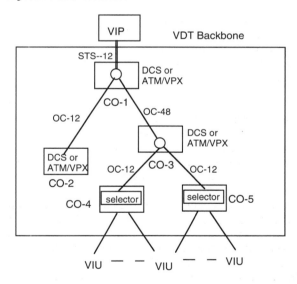

Figure 3-38: An example for analysis of transport bandwidth efficiency.

Unlike the first service architecture, which distributes a 2-hour movie to the video selector within the serving CO, and to the customers, continuously, the VIP in the second service architecture sends the movie at fast burst mode, so as to increase the flexibility of accommodating VDT customers and services [29]. Each burst contains, for example, a

20-second segment of the requested movie [29]. The video selector, equipped with buffers to store the 20-second bursts, distributes the movie continuously to the viewer.

In this service architecture, an STS-3c channel may be allocated for each VIP–Selector pair, for example, from the VIP to CO-5 in Figure 3-38. In this scenario, it takes approximately 0.2 seconds to distribute a video burst of 20 seconds real-time length. This 0.2-second burst may be transmitted anytime within the 20-second period, thus enabling more flexible sharing of transport capacity. STM transport is not considered for supporting this service mode, since the resultant bandwidth utilization would be approximately 1 percent all the time. In contrast, if the ATM transport system is used, the capacity corresponding to the 19.8 seconds idle time on the CO-1 to CO-5 VP can be used flexibly by other users in CO-5, as well as by users on the VIP to CO-4 VPs. This bandwidth utilization difference would have a significant impact on the number of VIUs that could be supported by the same transport network, especially for multicast applications. Note that the gain in transmission efficiency for ATM networks supporting the "burst" video service mode comes at the expense of needing a relatively complex path scheduling protocol to ensure no two or more incoming video streams will arrive at the same output port of the ATM switch simultaneously.

3.5 Broadband Transport Network Evolution

3.5.1 Network Evolution Strategies

SONET/STM networks were originally designed to support private line services and to provide a high-speed transport mechanism for switched services. The network connections supporting these services are either semipermanent or permanent and are established and torn down using a service provisioning process. The network works well for the aforementioned services because those functions that are needed do not require dynamic bandwidth control capability. However, due to the introduction of new broadband services such as Frame Relay Services (FRSs), SMDS, and Cell Relay Services (CRSs) that introduce bursty traffic, the SONET/STM network infrastructure may not be efficient enough to accommodate these bandwidth-on-demand requirements. ATM transport technology may alleviate the problem of bandwidth inefficiency. Also, SONET/STM network protection is through simple and cost-effective self-healing capability. One example is the SONET self-healing ring. The question here is, when the network moves to ATM transport technology in order to accommodate bandwidth-on-demand requirements, will the SONET layer be sufficient to provide protection for new broadband services that may have very different delay and cell loss requirements? Figure 3-39 depicts such a challenging new networking environment.

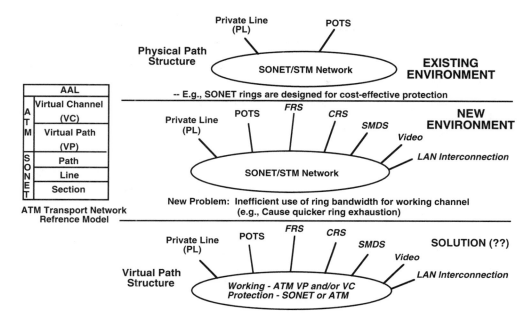

Figure 3-39: A challenging new networking environment.

To achieve the goal of building a B-ISDN transport network infrastructure, the target network may evolve from the present SONET/STM network infrastructure or it may be built from scratch by skipping the STM world, as shown in Figure 3-40. The latter approach is revolutionary. The evolution path means that the present SONET/STM equipment either will serve as the physical layer transport (e.g., STS-Nc transport) within the SONET/ATM network without any change or SONET/ATM equipment may be upgraded to accommodate ATM capabilities. Whether the evolution or the revolution path should be used depends on the existing network infrastructure, equipment availability, the new service implementation time frame, the network evolution plan, network transport, and operations costs.

Network evolution or revolution depends on whether or not the considered network architecture has been implemented and/or the needed equipment is available. For example, SONET self-healing rings have been widely deployed in LEC networks that provide very reliable high-speed transport for both narrowband and broadband services. In this case, it may be more attractive if the SONET ADM can be upgraded to accommodate burst-type broadband data services and still keep the simple, cost-effective SONET layer protection scheme intact. On the other hand, SONET DCSs and ATM VPXs appropriate for mesh core networks may become commercially available during 1997–98. In this case, the company needs to decide whether to use just one type of equipment (SONET DCSs or ATM VPXs) or use both in the same network. For example, NTT is implementing an ATM VP transport network throughout Japan. If both SONET DCSs and ATM VPXs are used in the same network, the SONET DCS may perform STS path cross-connects, whereas the ATM VPX cross-connects VPs, which are accommodated within SONET STS paths. Which approach should be used depends on the network provider's network evolution plan, economics, operations, and service implementation time frame.

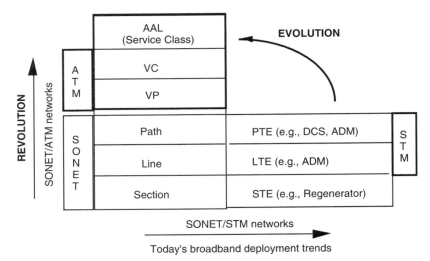

Figure 3-40: Broadband network revolution or evolution.

To answer the question of whether the network should evolve from the present STM infrastructure or proceed directly to the future B-ISDN network infrastructure requires good understanding of the relative roles of SONET/ATM and SONET/STM technology in broadband transport. In addition, identifying potential limitations of using ATM technology in broadband transport is necessary to understand appropriate application areas for different transport technologies. This information will provide insights for a cost-effective transition from today's SONET/STM network infrastructure to the future B-ISDN infrastructure. Some cost-effective ATM evolutionary scenarios can be found in Wu [30]. However, a cost-effective network evolution plan varies from one network provider to another and is a great challenge for network planners who are seeking a cost-effective and implementable plan.

3.5.2 Interworking between B-ISDN Transport and Other Networks

During the network evolution process, interworking arrangements between B-ISDN networks and other existing networks must be made to ensure existing services as well as emerging services can be supported on an end-to-end basis over a large communication area. The interworking of SONET/ATM networks with 64 Kbps–based N-ISDN, Frame Relay networks, and LAN interconnected networks are particularly crucial to the success of ATM deployment, as services supported by these networks have been widely available and popular around the world. Examples of LAN interconnections include TCP-IP over ATM and ATM network emulation. In the remainder of this section, we will discuss principles and some sample arrangements for network interworking based on ITU-T Rec. I.580 [31].

Figure 3-41 depicts three possible interworking scenarios specified in ITU-T Rec. I.580. Scenario I is an interconnection scenario between a B-ISDN and another network. Scenario II is a network concatenation interworking scenario, but the interfaces and services are the same as those currently provided by another network. In Scenario III, the service capabilities provided between broadband user access points are limited by other network capabilities. In these three scenarios, network interworking is accomplished through a Interworking Function (IWF) system.

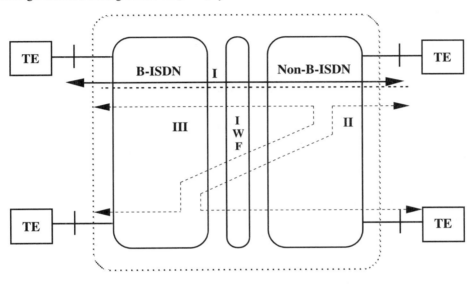

IWF: Interworking Function
TE: Terminal Equipment

Figure 3-41: Interworking scenarios specified in ITU-T Rec. I.580.

Figure 3-42 depicts a case where B-ISDN is interposed between Frame Relaying networks. In this case, the Frame Relay service is offered to two FRS customers via a core ATM network. This scenario is likely to be one of the initial network interworking applications. The equivalent B-ISDN service for interworking with FRS is Class C (message mode, unassured option without flow control) with AAL5 specified by both ITU-T and the ATM Forum. Other network interworking configurations can be found in Reference [31].

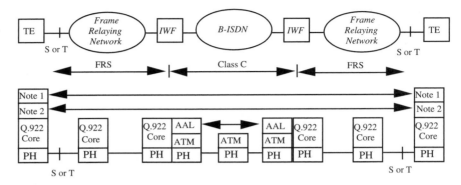

Note 1 : User-specified Upper Layers   PH: Physical Layer
Note 2 : User-specified additional Layer 2 protocols

Figure 3-42: Frame Relay network interworking with B-ISDN.

3.6 ATM Development and Deployment

Figure 3-43 depicts the history of B-ISDN standardization and B-ISDN research trends. Since the middle of the 1980s, ITU-T has played a leading role in standardizing B-ISDN.

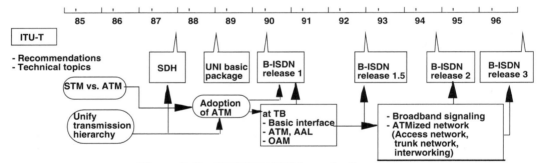

Figure 3-43: ITU-T B-ISDN standard progress.

Initially intending to investigate the framework for B-ISDN, ITU-T did not plan to discuss the network structure owned by each network provider even if the UNI for broadband service had to be standardized. However, after discussing some technical topics for broadband services, ITU-T recognized that broadband services could not be established unless the transmission network structure itself, namely the digital hierarchy, was standardized. At that time, there were three different digital hierarchies around the world: Japanese (1.5 /6.3 /32 /100 /400 Mbps), North American (1.5 /6.3 /45 Mbps), and European (2 /8 /34 /139 Mbps). In 1986, a framework to unify these three digital hierarchies was proposed to ITU-T. As a result, ITU-T studied a new hierarchy and standardized a unique worldwide digital hierarchy, SDH in 1988. On the other hand, in 1986 France proposed a new multiplexing method, called ATM, that uses fixed-length cells. After this mechanism (i.e., ATM) was improved through many technical meetings, ITU-T agreed to use ATM as the only transfer mode for B-ISDN. The basic requirements for ATM including cell length, cell header functions, and OAM functions were summarized as a total package in 1989, and these were standardized in 13 recommendations in 1990. ITU-T specified the ATM Layer, AAL, OAM principles and traffic control in more detail in 1993. Some specific B-ISDN recommendations can be found in Chapter 1 (also see Figure 1-15).

Most ATM network deployment around the world is primarily driven by state-owned Postal Telephone and Telegraphs (PTTs) in various countries based on their national policies. However, ATM network deployment in the United States is primarily driven by economics rather than national policy. Furthermore, ATM network development in the United States is driven not only by the telecommunications industry, but also the computer networking industry (sometimes referred to as the Internet community or IP community). The creation of the ATM Forum dealing with issues of concern to both public and private networks is one example of this ATM development trend. This unique trend (especially as compared with the trend around the world) is the result of the computer networking industry's crucial role, which matches that of its telecommunications counterparts in economic and social impact due to the popularity of LANs and the Internet in the United States. In fact, the number and size of the ATM networks deployed today by the IP community in the United States are larger than those deployed by telecommunications network carriers, and the ATM switches designed for enterprise networks dominate the U.S. ATM switch market.

Today, all major public carriers have small-scale ATM networks that are being deployed or are already operational. For private ATM networks, [32] has reported five real and operational ATM networks, which include the Hollywood-based Cinesite's ATM network, Chrysler Corporation's ATM network, the Indiana University Library of Music ATM network, the Naval Surface Warfare Center's ATM network, and the Pinellas County (Florida) Sheriffs Office's ATM network. Primary applications for these networks include client-server applications, graphics processing on high-end platforms, the movement of large image-based files, real-time multimedia services, LAN replacement, disaster recovery, and applications across the LAN/WAN boundary at high speeds. Multimedia traffic types include CAD images and full-motion video. In particular, the Navy network, called NEWNET, is a nationwide ATM-based backbone network that connects to 47 Navy sites and is fully operational. This network has allowed the consolidation of many dedicated point-to-point connections, resulting in significant cost savings (more than 60 percent cost savings, representing millions of dollars). In addition, this ATM network infrastructure has also provided support for new applications that were not previously possible (e.g., Navy ship engineering drawings transfer, telemedicine) [32].

In addition to the Navy's ATM network, Advanced Technology Demonstration Network (ATDNet) is one of the largest ATM network testbeds in the United States [33]. It was established by the Defense Advanced Research Project Agency (DARPA) in 1994. The ATDNet is a major collaborative effort of the U.S. government, Bell Atlantic, and Bellcore to stimulate high-speed ATM and SONET infrastructure. It provides a testbed for hands-on experiments in next-generation communication protocols, advanced network services, bandwidth-demanding applications, and advanced network management. Today, the network has grown from the original seven primary sites and tens of hosts to hundreds of hosts from almost 20 organizations and agencies.

In Japan, NTT has already conducted two basic ATM experiments and is now developing the first commercial ATM network. High-speed data transmission services supported by ATM technology were introduced in 1994. Figure 3-44 depicts NTT's ATM experimental progress measured against ITU-T's standardization progress.

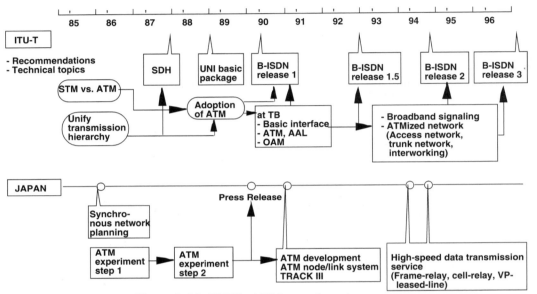

Figure 3-44: NTT's ATM experiments.

In addition to ITU-T's activities, NTT has also contributed to and participated in B-ISDN-related activities at the ATM Forum, the Internet Engineering Task Force (IETF), and the Network Management Forum. In 1994, NTT developed an experimental ATM network, as depicted in Figure 3-45, to evaluate network performances based on the standardized results. The experiments covered all of Japan, and 10 cities had ATM terminal equipment. These experiments enable NTT not only to establish network design and management technologies for B-ISDN, but also to develop and create new applications needed in the new information age by coordinating networks, user facilities, and software. Target applications in these ATM experiments are high-speed computer communications covering high-speed inter-LAN communication, large-capacity file transfer, and multimedia services. These applications are supported on the broadband backbone network using a combination of ATM and optical fiber technologies. Tests of CATV video transmission, also known as video on demand, also have been conducted.

A pan-European ATM pilot project involving at least 14 European PTTs began in early 1994. The purpose of this pilot project is to check international interpretability issues that are specified in ITU, European Telecommunications Standards Institute (ETSI), and

European Institute for Research and Strategic Studies in Telecommunications (EURESCOM). Benchmark services of this project include Frame Relay service, SMDS, and CBR services. The VP Cross-Connect System (VPX) used in the project is a 16 x 16 ATM switch with 256 VPs per port. The access line has at least 34 Mbps for the Plesiochronous Digital Hierarchy (PDH) link and the STM-1 rate if SDH links are available. In this project, each PTT manages its own VP network. Management interworking among PTTs is not within the scope of the project.

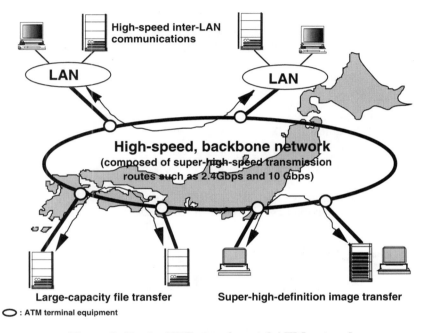

Figure 3-45: An NTT experimental ATM network.

3.7 Summary and Remarks

We have reviewed ATM technology and associated potential transport network architectures. Network evolution and interworking issues were also covered in this chapter. The target broadband network architecture and associated network evolution vary from one network provider to another depending on economics, network operation strategies, and the service evolution plan.

The broadband network architectures discussed in this chapter constitute a transport platform to support network control, signaling, and network operation functions. The messages associated with signaling, control, and operations may be transported along with service data through the same physical transport network but with different QoS requirements. A great challenge here is how to add these signaling, control, and operations capabilities to the transport platform. Currently, there are two competing models for control, signaling, and traffic management over the ATM network: one is from the ATM Forum, and the other is from the Internet Engineering Task Force (IETF). The model used in the ATM Forum treats the ATM network as an intelligent network supporting needed QoS for a variety of services with few control features above the ATM Layer. This ATM Forum model reflects the traditional telecommunications network design philosophy that makes the network more intelligent and the end system simpler.

In contrast, the IETF model enhances the current Internet Protocol (IP) model to accommodate signaling, control, and operations features in order to provide guaranteed QoS for a variety of services (primarily for multimedia services) at the IP layer (i.e., the network layer). In this IEFT model, the ATM network is treated as a high-speed pipe and may not contribute enough intelligence. This model is consistent with the traditional Internet design philosophy that makes end systems more intelligent and the network simpler. Interworking these two models is an ongoing effort in both the telecommunications and the information networking industries; research on this challenging area is just beginning. We will discuss some of these issues in Chapters 4 (ATM Signaling Networks), Chapter 5 (ATM Network Traffic Management), and Chapters 6 and 7, both of which deal with ATM network integrity.

References

[1] ITU-T Recommendation I.321, "BISDN Reference Model," 1992.
[2] ITU-T Recommendation I.362, "B-ISDN ATM Adaptation Layer (AAL) Functional Description," Study Group XVIII, Geneva, June 1992.
[3] ITU-T Recommendation I.363, "B-ISDN ATM Adaptation Layer (AAL) Specifications," Study Group XVIII, Geneva, June 1992.
[4] The ATM Forum, *Traffic Management Specification Version 4.0*, April 1996.
[5] Bellcore GR-1110-CORE, "Broadband Switching Systems (BSS) Generic Requirements," Issue 1, October 1996.
[6] J. Y. Hui, *Switching And Traffic Theory for Integrated Broadband Networks*, Kluwer Academic Publishers, Dordrecht, The Netherlands, 1990.
[7] K. Sato, H. Ueda, and N. Yoshikai, "The Role of Virtual Path Crossconnection," *IEEE Mag. Lightwave Telecommunications Systems*, Vol. 2, No. 3, pp. 44–54, August 1991.
[8] ITU-T Recommendation I.311, "B-ISDN General Network Aspects," January 1993.
[9] K. Y. Eng, M. A. Pashan, R. A. Spanke, M. J. Karol, and G. D. Martin, "A High-Performance Prototype 2.5 Gb/s ATM Switch for Broadband Applications," *Proc. IEEE GLOBECOM*, pp. 111–117, December, 1992.
[10] K. Yamaguchi, K. Sakai, J. Shiohama, S. Unagami, H. Oka, W. Naito, H. Yamashita, and K. Yamazaki, "ATM Transport Systems Based on Flexible, Non-Stop Architecture," *Proc. IEEE GLOBECOM*, pp. 1468–1475, November 1993.
[11] J. Anderson, B. T. Doshi, S. Dravida, and P. Harshavardhana, "Fast Restoration of ATM Networks," *IEEE J. Selected Areas in Commun.*, pp. 128–138, January 1994.
[12] T.-H. Wu, *Fiber Network Service Survivability* Artech House, May 1992.
[13] The ATM Forum, *Private Network-Network Interface Specification Version 1.0 (PNNI 1.0)*, March 1996.

[14] K. Sato, S. Ohta, and I. Tokizawa, "Broadband ATM Network Architecture based on Virtual Paths," *IEEE Trans. on Commun.*, Vol. 38, No. 8, pp. 1212–1222, August 1990.

[15] T. Aoyama, I. Tokizawa, and K. Sato, "Introduction Strategy and Technologies for ATM VP-Based Broadband Networks," *IEEE J. Selected Areas in Communications*, Vol. 10, pp. 1434–1447, December 1992.

[16] T. Aoyama, I. Tokizawa, and K. Sato, "ATM VP-Based Broadband Networks for Multimedia Services," *IEEE Commun. Mag.*, pp. 30–39, April 1993.

[17] T.-H. Wu, J. Bartone, and V. Kaminisky, "A Feasibility Study of ATM Virtual Path Cross-Connect Systems in LATA Transport Networks," *Proc. IEEE GLOBECOM*, pp. 1421–1427, November 1994.

[18] Y. Inoue, N. Terada, M. Kawarasaki, K. Sano, and K. Ikuta, "Granulated Broadband Network Applicable to B-ISDN and PSTN services," *IEEE J. Selected Areas in Commun.* Vol. 10, No. 9, pp.1474–1488, 1992.

[19] M. Yoshida and H. Okazaki, "A Study on ATM Network Planning Based on Evaluation of Design Items," *Conference Records of Networks '92*, pp.147–152, Japan, October 1992.

[20] ITU-T Draft Recommendation I.375, "Network Capabilities to Support Multimedia Services," Geneva, 14–25, November 1994.

[21] The ATM Forum/95-0012R5, "Audio Visual Multimedia Services: Video on Demand Implementation Agreement 1.0," August 1995.

[22] ISO/IEC 13818-6, "Digital Storage Media Command and Control," DIS, July 1996.

[23] Digital Audio Visual Council, *DAVIC 1.0 Specification*, January 1996.

[24] T-H. Wu, I. Korpeoglu, and B. Cheng, "Distributed Interactive Video System Design and Analysis," *IEEE Commun. Mag.*, pp. 100–109, March 1997.

[25] ITU-T Document, "Integrated Video Service (IVS) Baseline Document," Study Group 13, Geneva, March 1994.

[26] Bellcore GR-2901-CORE, "Video Transport Over Asynchronous Transfer Mode (ATM) Generic Requirements," Issue 1, May 1995.

[27] G. H. Petit and D. Deloddere, "A Video-On-Demand Network Architecture Optimizing Bandwidth and Buffer Storage Resources," *Proc. ISS'95*, pp. 319–323, Berlin, Germany, April 1995.

[28] Bellcore GR-1111-CORE, "Broadband Access Signaling Generic Requirements," Issue 2, October 1996.

[29] A. D. Gelman, H. Kobrinski, L. S. Smoot, S. B. Weinstein, M. Fortier, and D. Lemay, "A Stored-and-Forward Architecture for Video on Demand Service," *Proc. IEEE ICC'91*, pp. 842–846, Denver, 1991.

[30] T.-H. Wu, "Cost-Effective Network Evolution," *IEEE Commun. Mag.*, pp. 64-73, September 1993.

[31] ITU-T Recommendation I.580, "General Arrangements for Interworking between B-ISDN and 64 Kbps Based ISDN," March 1994.

[32] *THE ATM REPORT*, Broadband Publishing Corporation, Vol. 4, No. 2, May 22, 1996.

[33] S. Bajaj, N. Cheung, G. Hayward, and Y. Tsai, "High Speed ATM/SONET Infrastructure Research in ATDNet," *IEEE Network*, pp. 18–29, July/August, 1996.

Chapter 4

ATM Signaling Networks

4.1 Introduction

Signaling is a family of protocols used for call and connection setup. The purpose of the signaling function is to establish the connection on a demand basis to allow network resources to be efficiently used. This feature is particularly needed for large scale networks. The signaling feature is provided through a switching system, which has call processing and routing capability. Figure 4-1 depicts a switching system architecture reference model [1].

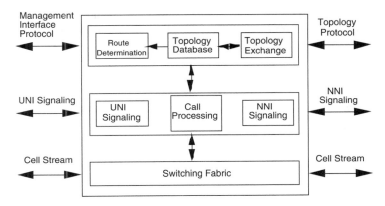

Figure 4-1: Switching system architecture reference model [1].

As shown in Figure 4-1, a switching system consists of three major functional components: switching fabric, call processing, and routing decision. The routing decision component uses up-to-date topology information to determine the inquiry and converts it to the appropriate destination format that can e best route based on the most current network status and available resources. The call processing module accepts the user's be used by the routing decision module. The inquiry of the destination address may be through the information embedded in the user inquiry message (e.g., direct phone call) or through some kind of database like SCP (e.g., 1-800 calls). The lower level of system module (i.e., switching fabric) is used to perform the physical routing function to establish a connection between the source and destination switches.

Signaling protocols are designed based on physical transport network interfaces. For example, today's signaling protocols that support the switched telephone service are based on the SS7 signaling transport network interface, defined in both ITU-T [2] and ANSI [3]. For emerging ATM networks, ATM signaling is a set of protocols used for call/connection setup over ATM interfaces. These ATM interfaces depend on the involved networks are public or private (e.g., enterprise) networks.

In emerging broadband networks, the network signaling function conceptually can be provided at the ATM Layer (OSI link layer), the Network Layer (OSI), or the Internet Protocol (IP) Layer (IETF), as depicted in Figure 4-2. ITU-T has been active since early 1990 in standardizing the signaling capability and operations for public broadband telecommunication networks. The broadband signaling activities involved in ITU-T include signaling capabilities at the ATM Layer as well as the Network Layer [i.e., Message Transfer Part Level 3 (MTP-3), which has been used in standard SS7 networks for narrowband telecommunication networks] [4]. Although private networks are not part of the ITU-T charter, due to the popularity of these networks (e.g., LANs) in the United States, the ATM Forum has taken a leading role in standardizing signaling for private ATM networks, called PNNI [1], which provides the signaling capability only at the ATM Layer. Recently, in response to the demand to carry multimedia applications through the IP-based Internet, the Internet Engineering Task Force (IETF) approved a signaling standard, called Resource Reservation Protocol (RSVP), in 1996 [5] that would be added to the host and routers to provide needed real-time QoS requirements for multimedia and other real-time video services. Note that the RSVP may or may not run on the ATM transport network.

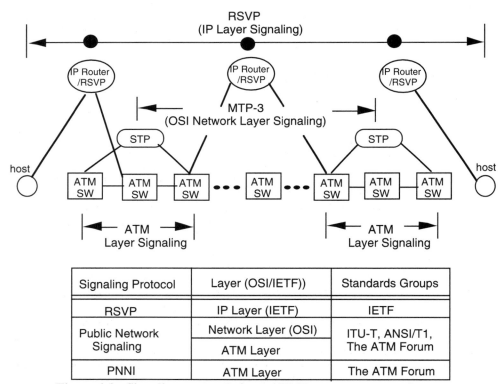

Signaling Protocol	Layer (OSI/IETF))	Standards Groups
RSVP	IP Layer (IETF)	IETF
Public Network Signaling	Network Layer (OSI)	ITU-T, ANSI/T1, The ATM Forum
	ATM Layer	
PNNI	ATM Layer	The ATM Forum

Figure 4-2: Signaling protocols from ATM layer to transport layer.

Although conceptually the signaling capability can be provided through different layers as described in the foregoing, it would be expensive and operationally difficult for a network to support signaling capabilities at all these layers. Thus, which layer(s) will be used for various kinds of applications and networks may depend on both market timing and economics. This chapter discusses the basic concepts and operations for each layer signaling capability from a transport perspective. This may provide some background for interested readers who are searching for signaling solutions for their own research and/or networks. Since this book centers on ATM transport, we will focus more on signaling transport capability at both the ATM Layer and the Network Layer. However, we still devote one section to a discussion of the RSVP and operations and differences between the RSVP and the ATM signaling.

We will review the existing signaling network architectures and processes used to support both public and private networks. Then the signaling requirements of emerging services and their effects on the signaling network architectures will be examined. After that, both public and private broadband signaling network architectures will be discussed and analyzed. The RSVP will then be introduced, and differences between the RSVP and its ATM signaling counterpart will also be discussed. These sections are followed by signaling network evolution. This chapter concludes with a summary of new challenges for broadband signaling network design. We hope that these discussions can form a necessary background for readers and stimulate research in and development of broadband signaling networks.

4.2 Existing Signaling Networks

4.2.1 SS7 for Public Networks

Figure 4-3 depicts an architecture model used in today's SS7 networks [6]. The SS7 architecture comprises three major components: Service Switching Point (SSP), Signal Transfer Point (STP), and Service Control Point (SCP). All nodes in the network with Common Channel Signaling (CCS) capability are SSPs that are interconnected by signaling links. Nodes that serve as intermediate signaling message transport switches are called STPs. SCPs are the SSPs that provide database access to support transaction-based services. Signaling links are the transmission facilities that convey the signaling messages between two SSPs.

The STP is engineered on a paired basis to enhance reliability. The SS7 network is physically separated from the POTS transport network, because the SS7 network is a packet-switched network and the POTS network is a circuit-switched network. The associated signaling mode is referred to as point-to-point signaling transport; in the quasi-associated mode, the SSP pair and SSP–SCP pair communications must be through the STP. Although the SS7 network architecture can support both the associated and quasi-associated signaling modes, today's SS7 networks in the United States implement only the quasi-associated mode due to economic considerations. However, those design assumptions used to optimize the SS7 network in early 1980 may not fully apply to the new ATM network environment (e.g., the service and signaling messages may be carried on the same physical ATM network with overlay end-to-end connections).

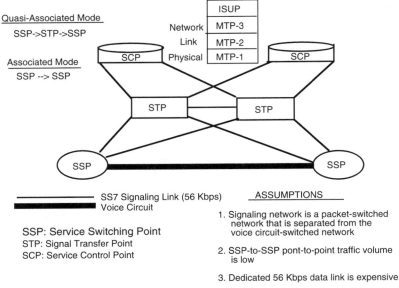

Figure 4-3: The architecture model for today's SS7 networks.

4.2.2 Connection Setup for Packet-Switched Networks

The connection setup process in current value-added packet switched networks (e.g., X.25) can be summarized as follows. The call setup is processed through the call control packet using a datagram form to reach its destination. When the receiving node receives the call control packet, which includes information regarding an end-to-end routing path, it verifies the packet and returns an acknowledgment (ACK) message through the path established by the call control packet, if the receiving node decides to accept the call. The transmitting node starts to send data packets through the established virtual circuit after it receives the ACK packet from the receiving end.

4.3 Broadband Signaling Requirements and Design Impacts

4.3.1 Impact of Service on Signaling Evolution

There is a wide range of factors driving service involved in signaling transport network evolution; these include existing revenue producing services such as cellular services to new emerging broadband and multimedia services. Some of these services and their potential impact on broadband signaling transport are discussed here [7,8].

4.3.1.1 Voice and Data Network Services

One of the earliest and most prolific of SS7 based services was "freephone," or 1-800 service. In countries where freephone is available, it is estimated that 10 percent of all toll revenues can be attributed to this service. Worldwide revenues in 1994 generated by freephone service are estimated to be $17B.

The advent of ISDN and CLASS (Custom Local Area Signaling Service) has also contributed significantly to the revenues earned by SS7. As these services become more universally deployed and as innovative services are added, total services revenues including freephone are expected to increase by at least 10–15 percent per year to reach more than $80B worldwide by the year 2000.

There are currently 36 standard services defined internationally that require the use of the Telephone User Part (TUP), ISUP, and Transaction Capabilities Application Part (TCAP) components of SS7. Typical examples include call forwarding, calling line (caller) identification, name display, and call completion.

4.3.1.2 Intelligent Network (IN)

The advent of the Intelligent Network (IN) with its attendant service creation environment will require additional signaling functionality and will substantially increase network signaling traffic. Initial views of IN have, however, focused on Class A services (single-ended, single point of control), which use existing SS7 functionality. Even these relatively simple services require longer messages, increased signaling density, and more service capability. Increased message length has been taken into account in national and international standards via segmentation and reassembly of the messages.

Delay in the signaling network is not a current concern. However, IN will affect signaling response times to varying degrees, depending on service characteristics such as the message length, messages per call, and the supporting service network architecture. Delay analysis using queuing information from [8] indicates that link delay for long IN messages may be up to 10 times the delay for current messages. Thus, 64 Kbps signaling links may lead to unacceptable network response times, especially for services requiring multilink network paths or SCP–SCP interactions. A study [9] on the increased signaling density due to IN also indicates that it will be necessary to increase signaling link speeds and the processing capacity of the STPs.

4.3.1.3 Mobility Management

The term *mobility management* covers both terminal and personal mobility. Mobile services are currently classified as cellular, Universal Personal Telecommunication (UPT), Personal Communication Service (PCS), and Future Public Land Mobile Telecommunications Systems (FPLMTSs).

Personal mobility allows users to access any network terminal to request and use services according to their registered service profile. Establishing the service profile, authenticating its use, and providing remote access to the specified services all require additional signaling capabilities and activity.

Mobility management and the provision of mobile-based supplementary services will require SS7 enhancements, increased database access, and increased signaling network capacity. Reference [10] suggests that the increase in database access for mobility management will require additional SCPs with additional signaling link capacity and increased processing.

Signaling link loading also depends on the interval used for location update messages; loading increases substantially with a reduction in the update interval. One approach is to introduce flow control for update messages, which will then vary the update interval depending on link availability. This control capability certainly will add more overhead into the signaling transport network.

4.3.1.4 Broadband Services

Broadband services require additional signaling capability to support the wide range of new services, such as video on demand, distance learning, and high-speed data transfer. Broadband networks will also support, or be interactive with, services on existing 64 Kbps–based networks.

The need to interwork broadband and existing services, such as database access, high-speed data, video conferencing, and video phone, places key demands on the evolution of the network signaling system.

Additionally, services on ATM networks require signaling to specify QoS and traffic parameters that depend on bit rate, burstiness, and transport media.

Broadband services include the requirement to establish a call and subsequently to add, change, or subtract connections. This requires separation of the call and connection control. Call and connection separation used with lookahead capability will allow more service flexibility and more efficient call setup. For example, networks can determine logistics before setup, including access resource, feature availability, terminal and access compatibility, and redirection location, if required. These functions all require additional signaling capability.

4.3.1.5 Multimedia Services

Multimedia services, both broadband and narrowband, have potentially significant new requirements for signaling. By their nature, many multimedia services will require multiple connections to support one call; for example, a mixed voice and image call will require two connections for the single multimedia call. This also requires separation of call and connection control, and places additional burdens on network signaling.

Taking account of the foregoing service and network requirements, evolving signaling networks must include [7]
- Increased signaling link speeds and processing capability
- Increased services functionality, e.g., version identification, mediation, billing, mobility management, quality of service, traffic descriptors, message flow control, lookahead that would require bandwidth-on-demand capability
- Separate call control and connection control

- Reduced OAM costs (including provisioning) for services and signaling that would require flexibility of connection establishment

In Section 4.4, we will discuss some alternative signaling transport network architectures that may provide capabilities to meet these requirements.

4.3.2 Impact of Traffic on Signaling Transport

Studies regarding the impact of the introduction of new services on today's signaling networks have been conducted and reported in [8,11–14]. These studies evaluated the signaling network capacity and delay arising from the penetration of PCS and Advanced Intelligent Network (AIN) services. The results in [8] have suggested, from a delay perspective, that use of today's 56 Kbps links may lead to unacceptable network response times, particularly for services requiring long network paths and extensive user–network or SCP–SCP interactions. Higher-speed signaling links are needed to alleviate this signaling delay performance concern. For example, as shown in Figure 4-4 [8], the total network response time improves by approximately 30 percent when 56 Kbps links are replaced by 1.5 Mbps links under the same load and network configuration. The results also suggested that a significant increase in the speed of SCP A-links would be needed, and the number of SCPs also increases significantly due to the capacity limitation of the incoming SCP A-linksets (which, therefore, limit the traffic entering the SCP). These potential limitations would eventually result in increased network complexity and cost due to additional investments in new SCP, reconfiguration of network components (nodes and links), and additional resources needed to maintain and administer the network [8].

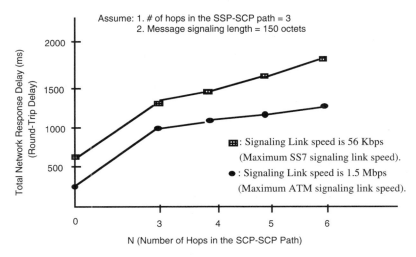

Figure 4-4: Mean network response times vs. number of hops in SCP-SCP path [8].

It is expected that the network capacity requirement for broadband signaling will be much higher than that for its narrowband counterpart due to its more intelligent nature and the need to support both point-to-point and point-to-multipoint connections. It has also been suggested that existing SS7 network physical interfaces may not be able to support the stringent end-to-end signaling time goals (20–100 ms) advocated by potential users of the broadband Switched Virtual Connection (SVC) services.

As already discussed, PCS is expected to make extensive use of Common Channel Signaling (CCS) to support user/terminal mobility. In a study reported in [12], the per-call signaling load for a PCS call in a dense inner city environment was compared with an ISDN call and a GSM (Global System for Mobile communications) cellular call. It was concluded that the load on the signaling network would be 4 to 11 times greater for cellular than for ISDN and 3 to 4 times greater for PCS than for GSM cellular. Similar results are reported in [13]. In addition, a study reported in [14] shows PCS may generate an enormous database transactions rate that would significantly affect the design of SCP system architecture and capacity. These results obviously will have an impact on the design of high-capacity signaling networks. Some other important teletraffic issues that should be considered on signaling network design can be found in [11].

4.4 Broadband Signaling Transport for Public Networks

The signaling capabilities needed for public telecommunication networks may be provided through the ATM Layer, the Network Layer using STPs, or a combination of both. The standards groups responsible for broadband signaling for public networks include ITU-T and ANSI/T1.

4.4.1 ATM Role in Broadband Signaling

The broadband signaling transport platform is an ATM-based network due to flexible establishment of a connection and the ease of bandwidth allocation. Table 4-1 summarizes major differences between the existing SS7 network and the emerging ATM signaling transport network. In general, the ATM signaling transport network has advantages over its SS7 counterpart in higher speed, ease of bandwidth allocation, and the flexibility of connection establishment. For example, the ATM signaling transport network may offer bandwidths of 64 Kbps to 1.5 Mbps on a demand basis [15], compared with a dedicated 64 Kbps bandwidth for its SS7 counterpart.

Two major components of the changing signaling requirements for today's SS7 signaling networks and emerging ATM-based broadband signaling transport networks are (1) the evolution of the signaling user parts, and (2) the evolution of signaling transport in a broadband environment. The evolution of signaling user parts (i.e., B-ISUP) has made significant progress, with most parts already standardized [16,17]. Work on broadband signaling transport in the ATM environment has made some progress, and some results have been reported in ITU-T Recs. I.311 and Q.2010) [4,18].

Table 4-1: Comparison between SS7 and the ATM signaling network.

System Attributes	SS7 [*]	ATM Signaling Network
Connection establishment	dedicated	flexible
Bit rate	up to 56 Kbps	up to 1.5 Mbps [#]
Bandwidth on demand	no	yes

* SS7 is a packet-switched network with delay-sensitive requirement.
\# Bandwidth allocation for each ATM signaling connection is 173 cells
     per second (approximately 66 Kbps), or multiples of that where the
     multiple is up to 23 (i.e., approximately 1.5 Mbps).

The issues of broadband signaling transport that need to be addressed include signaling transport architectures and protocols that may be used in an ATM environment to provide reliable signaling transport, while also making efficient use of the broadband capabilities of ATM networks in support of new, vastly expanded signaling applications. A number of suggestions have been made for possible broadband signaling transport architectures, ranging from (1) retention of MTP to (2) adoption of a fully associated signaling mode using signaling Permanent Virtual Channels and a small part of MTP Level 3 (MTP-3), and finally to (3) a fully distributed signaling transport architecture supporting only the associated mode. One of the major issues that make the signaling transport standards progress slower than that of B-ISUP is whether the target ATM-based signaling transport network should evolve from existing SS7 networks or be designed based on unique ATM characteristics. If the ATM signaling transport network evolves from the existing SS7 network, its signaling capabilities need to be provided through both the ATM Layer and the Network Layer using the MTP-3 protocol. If the ATM signaling transport network does not follow the evolution path from SS7, it may be possible to provide required signaling capability only at the ATM Layer. These different perspectives form a platform for alternative broadband signaling transport network architectures that we will discuss in Section 4.4.4.

To utilize ATM transport capability cost-effectively, services and control/signaling messages are carried by the same physical network, as depicted in Figure 4-5 (see Rec. I.311 [18]). In Figure 4-5, the ATM control and signaling transport network and the ATM service transport network use the same physical network, but logically they are separated from each other by using VCs or VPs with different QoS requirements. Several potential benefits to be gained by using an ATM VP service transport backbone network to interwork with the signaling/control network have been described in ITU-T Rec. I.311. These include possible simplification of existing protocols for control and management transport, better performance mainly by reducing the control and management message delay (i.e., high-speed transport), and reliability enhancement via possible self-healing capability at the VP level.

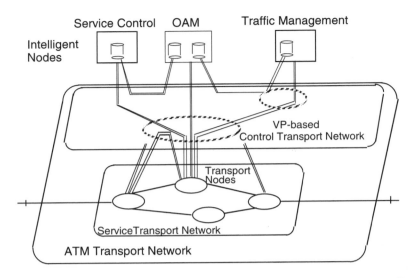

Figure 4-5. An integrated ATM service and control transport network.

4.4.2 Broadband Signaling Protocols

4.4.2.1 Broadband Signaling Protocol Stack

The release of broadband signaling standards in ITU-T is associated with the three-phase release for B-ISDN service features and network capabilities. Signaling requirements for Release 1 B-ISDN are based on existing call control procedures. The Release 1 signaling architecture has been specified in ITU-T Rec. Q.2010 [4]. The Private Network–Network-Interface (PNNI) signaling protocol stack shown in Figure 4- 6 was specified in ITU-T Rec. Q.2010 in December 1993. This broadband signaling protocol stack supports both the associated mode and the quasi-associated mode, if applicable. However, the actual implementation of this signaling protocol stack for ATM-based signaling transport networks may vary from one network provider to another, depending on the signaling network evolution plan, the services to be supported, and the implementation time frame. Services for Release 1 are broadband connection-oriented bearer service (Classes A and X) and broadband connectionless bearer services (Class D). These services are specified mainly by the peak traffic parameter (bit rate and cell rate). Services for Release 2 and Classes B and C provide Variable-Bit-Rate services. Resource allocation based on statistical multiplexing is also provided in Release 2. Multimedia or distributed services are included in Release 3, although the technical issues associated with this phase are not yet clear. Since functions provided by the Signaling Adaptation ATM Layer (SAAL) are equivalent to those provided by the current SS7 link layer, the existing network layer of SS7 can be used on ATM transport network, if the network provider chooses to develop the broadband signaling transport network from SS7. We discuss the NNI signaling protocol stack depicted in Figure 4-6. More details follow in the remainder of this section.

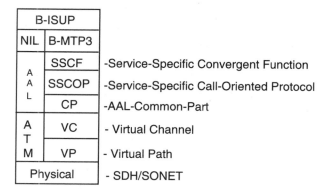

Figure 4-6: The NNI signaling protocol stack.

4.4.2.2 Signaling AAL (SAAL)

The SAAL builds on the service offered by the ATM Layer in order to offer a Data Link Layer service that is equivalent to that offered by the current Data Link protocols used for signaling, namely, MTP-2 for network signaling and Q.921 for user-to-network signaling. The SAAL is divided into a service-specific part and a common part. The service-specific part implements most Data Link functions, for example, connection establishment, release, and reset, and retransmission of errored Service Data Units (SDUs) for assured service. The primary function of the common part is the segmentation (if necessary) of AAL SDUs into ATM cell payloads at the transmitter and the reassembly of the SDUs from ATM cell payloads and detection of any errors at the receiver.

The service-specific part comprises a Service-Specific Connection-Oriented Protocol (SSCOP) and a Service-Specific Coordination Function (SSCF). In fact, the use of "service-specific" for the SSCOP may be misleading, since a goal of this sublayer is to avoid a special protocol that is used only by signaling. The SSCOP is designed not only for signaling, but also for connection-oriented end-user applications. The use of the same basic protocol for user applications and network signaling may reduce implementation and operation costs. Of course, network signaling may require some special features, such as security and reliability, that are not used by all end-user applications. In this case, those special features should be provided with modular extensions to or options within the basic protocol. To the extent possible, special features should be located in the SSCF, which customizes the service provided by the SSCOP to create the service required by a particular AAL user. In many cases the SSCF may be null or a simple mapping function between the primitives used by the AAL user and the primitives supported by the SSCOP. To provide some of the services expected by MTP-3, however, more complex functions may be required in the SSCF.

The basic operation of the SSCOP protocol was agreed at a ITU's interim meeting in September 1992. During the previous few meetings, two fundamentally different proposals had been discussed: one, initially favored by delegates from Japan and several European countries, was to modify the Q.921 protocol to operate over the service provided by the common part of the AAL; the other, proposed by the United States, was to adopt a new link layer protocol, much more suited to links with a high bandwidth delay product. At the September (1992) meeting, the U.S. proposal was accepted with some minor modifications to better accommodate low-speed and lightly utilized links. Meetings in February and May 1993 further refined the protocol, and the specification was approved as a Recommendation in December 1993.

For a detailed description of the SSCOP, interested readers should consult the draft Recommendation [19], but some of the important features of the protocol are described here. The protocol is viable for end-user applications at very high speeds, which may also accommodate satellite links. The transmitter and the receiver are decoupled to permit parallelism in implementations. Two timers operate at the transmitter, none at the receiver. Multiple polls can be sent during one round trip delay to provide prompt error recovery even on long-delay links, such as those using satellites, but multiple retransmissions are suppressed through a poll numbering scheme. The processing per user data unit is minimized (e.g., no piggybacking of acknowledgments) to allow very high speed data transfer.

The common part is so named because the same functions are expected to be common to a variety of AAL protocols. The common part is itself sublayered into a Segmentation and Reassembly (SAR) sublayer and a Common Part Convergence Sublayer (CPCS) (not shown in Figure 4-6). This common part can be used by various service-specific sublayers or even directly by an AAL user that needs no additional functions from the AAL. The common part protocol specifications were developed in ITU-T SG13 (formerly CCITT SG XVIII) as part of the infrastructure for B-ISDN. The two candidate common part protocols for an asynchronous, message oriented application such as signaling are the Type 3/4 (AAL-3/4) common part, specified in the 1992 Rec. I.363, and the Type 5 (AAL-5) common part, which specification was frozen in the January 1993 SG XVIII (now SG13) meeting. Although either common part, may meet the need of signaling, AAL-5 is simpler to implement and has been approved for ATM signaling by ITU-T SG13 as well as by the ATM Forum.

Although, it is generally expected that the SAAL for the NNI should provide the same layer service as does today's MTP-2, the key issue here is whether these MTP-2 functions can be efficiently and/or cost-effectively implemented in SSCOP or whether they can be implemented at the ATM Layer. These MTP-2 specific functions include proving, error monitoring, and retrieval, which provide a required level of reliability for an existing SS7 signaling network. *Proving* is part of the signaling link establishment procedures for SS7 networks. It essentially consists of sending some "dummy" Protocol Data Units (PDUs) over the link for a time interval and verifying that PDUs will get through on the link with an acceptable success ratio in both directions before MTP-3 is notified that the link is in service, and signaling traffic is placed on the link. It reduces the probability that signaling traffic will be placed on a link that has an unacceptable delay, throughput, or residual error rate, and will likely fail in short order. *Error monitoring* continuously monitors the ratio of corrupted PDUs that are received and declares the link failed if the ratio becomes unacceptably high. Another MTP-2 specific function is *retrieval*. When a link is declared failed, usually there are messages in the retransmission buffers at the two ends of the link. The transmitting end has no idea which, if any, of these messages were received successfully by the other end of the link before it failed. As part of the changeover procedure, which diverts traffic from the failed link to alternative links, if there still exists a communication path between the nodes at the two ends of the link, the nodes exchange messages that say what the sequence number is of the last in-sequence message received before the link failed. Messages subsequent to the last one received correctly at the other end are then retrieved from the retransmission buffer and rerouted to alternative links ahead of any traffic that arrived after the failure.

Making a rational decision about how best to fill the needs within an ATM network that are satisfied by the retrieval, error monitoring, and proving procedures in today's networks is difficult, because there are no data on types and frequencies of failures in ATM networks. However, some preliminary analysis reported in [20] has suggested error monitoring may not be appropriate at the SAAL due to the high overhead involved. This error monitoring function is now not included in the SSCOP and may be handled by the ATM Layer, which already has had that function. In the May 1993 ITU-T SG11 meeting, it was agreed that SSCOP should support retrieval, and the frozen draft specification includes all the necessary procedures to perform retrieval. In addition, the Go-Back-N retransmission method used in MTP-2 has been replaced by a multiple selective rejected-based method in the SSCOP, since the latter retransmission method is more suited to high-speed transmission.

4.4.2.3 Signaling Network Layer (MTP-3)

Once the SAAL is available, one can begin building a signaling network where the links consist of ATM Virtual Channel Connections over which the SAAL is run. Since SAAL provides the same layer service as MTP-2, MTP-3 and the Signaling Connection Control Part (SCCP) can be used unchanged to provide the Network Layer functions. High-capacity transmission systems, ATM multiplexing, and AAL protocols that require less processing than today's MTP-2 should bring down the cost per link, making direct signaling links between end nodes (associated mode signaling) economical for high-volume signaling relations. There will still be a need for nodes with the MTP relay function, that is, STPs, to support low-volume signaling relations. The STPs also can provide backup routes for the signaling relations using direct links and can be convenient points for gateway screening and for interworking to the pre-ATM SS7 network. An STP could terminate some ATM links and some STM links and relay messages from any link to any other. SCCP relay functions will also still be needed to provide functions such as global title translation and intermediate signaling network identification. Because of the administrative overhead involved in keeping the frequently changing global title translation tables up to date, retaining the tables in a relatively few nodes in the network is preferable to distributing them to every node. Although SCCP and MTP relay functions do not have to be located in the same physical nodes, they probably will be in most cases, since this minimizes the number of nodes through which any particular message must be relayed.

Over the past year, no changes specific to broadband transport have been made in MTP-3 or SCCP. Thus, it is reasonably expected that the initial signaling networks over the ATM platform will use today's SS7 MTP-3 essentially unaltered over signaling links using SAAL over ATM Virtual Channel Connections. Associated mode signaling may be economically viable for high-volume signaling relations, but the STP function may still be needed for low volume signaling relations. It will be logical to colocate with the STP function, functions such as interworking to existing SS7 network, gateway screening, and global title translation. For the longer term, feasibility should be investigated for enhancing end-to-end recovery procedures and simplifying network management procedures at Network Layer relays (STPs).

4.4.3 Broadband Signaling Transport Network Architectures

The signaling protocol stack described in Figure 4-6 can be implemented in three different ways, all of which form a class of ATM-based signaling transport architectures, as shown in Figure 4-7 [21]. These possible architectures are classified based on the signaling message routing principle: (1) the quasi-associated mode only,(2) the associated mode only, and (3) a hybrid of the quasi-associated mode and the associated mode. For architecture options (1) and (3), the MTP-1 and MTP-2 functions that reside within today's STP will be replaced by ATM Layer and AAL functions in the ATM-based signaling transport network. To differentiate broadband signaling network components from today's SS7 network components, the B-STP, the B-SSP, and the B-SCP used in Figure 4-7 are defined as an ATM-based Broadband STP, SSP, and SCP, respectively.

4.4.3.1 Quasi-associated Mode Only

Figure 4-8 depicts two examples of the quasi-associated signaling transport architecture using ATM VP transport and ATM VC transport (see Section 3.4.1). Figure 4-9 depicts an example of the protocol stacks of network elements involved in the VP signaling transport path shown in Figure 4-8(a). This architecture follows today's SS7 network concept with enhanced STP and SCP capabilities for broadband signaling needs. In this architecture, B-SSPs communicate with each other, and the B-SSP communicates with the B-SCP through the B-STP, as do today's SS7 networks. Note that in this architecture, the B-STP is an ATM-based switching system with an MTP-3 controller that supports signaling traffic only.

The quasi-associated signaling transport architecture uses ATM transport as a one-to-one replacement of today's dedicated copper data transport. This architecture requires a significant capacity upgrade of today's SS7 components, which could make these components difficult and expensive to build. Due to economics and reliability considerations, two or more B-SCPs are needed to accommodate the requirements of both narrowband and broadband network intelligence. The increased number of B-SCPs means more A-links between the B-STP and the B-SCP are needed. This could make it more difficult both in terms of costs and operations considerations for any signaling transport network to be developed into a large-scale one.

	Broadband Signaling Transfer Mode		
	Quasi-associated Mode	Associated Mode	Hybrid of Associated/Quasi-associated Mode
Network Configuration	- Signaling Messages between two SSPs are transported via B-STPs (Use B-STPs).	- Signaling Messages between two SSPs are transported without higher-layer termination (No B-STPs).	- No B-STPs for Associated mode. - Use B-STPs for Quasi-associated mode.
	- Signaling and Service messages are carried via physically separated networks. - One-to-one replacement of today's dedicated copper data transport.	- Both Signaling and Service messages are carried in the same physical network, but logically separated.	- Same Physical Network for Associated mode. - Separate Physical Network for Quasi-associated mode.
Signaling Message Routing	- Network Layer Routing (MTP-3)	- Network Layer Routing (e.g. MTP-3) - ATM Layer Routing (without MTP-3 functions in ATM switches.)	- Network Layer Routing or ATM Layer Routing for Associated Mode - Network Layer Routing for Quasi-Associated Mode
Hierarchical Database Access	- B-SSP - B-STP - B-SCP.	- B-SSP - B-SCP. - Database Management (Distributed/ Centralized)	- B-SSP - B-SCP (for Associated mode) - B-SSP - B-STP - B-SCP. (for Quasi-Associated mode)

Figure 4-7: A class of ATM signaling transport architectures [21].

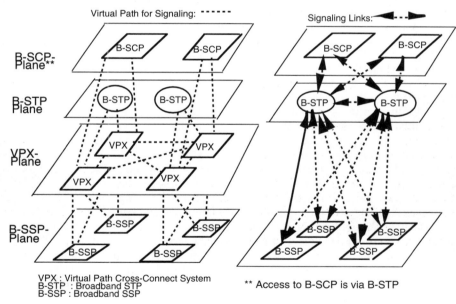

(a) VP Connectivity for Signaling Transport (b) VC Connectivity for Signaling Transport

Figure 4-8: Examples of ATM-based quasi-associated signaling transport network architecture.

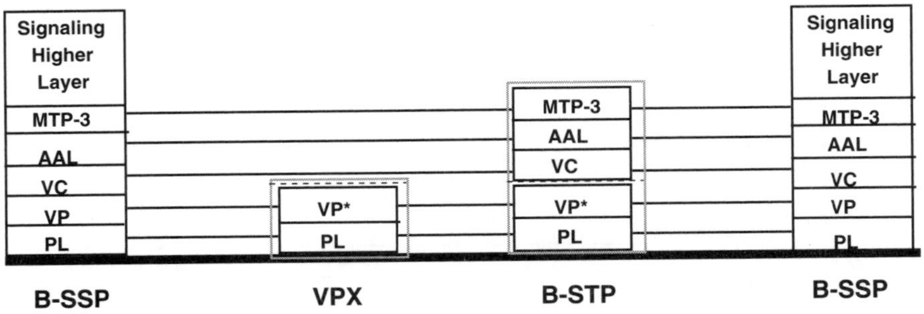

VPX : Virtual Path Cross-Connect
B-SSP : Broadband SSP
B-STP: Broadband STP
* : VP links are connected between VPXs or between B-SSP and VPX, whereas VP
connections are terminated at the B-STPs or the B-SSP.

Figure 4-9: Protocol stacks of network elements involved in VP signaling transport shown in Figure 4-8(a).

4.4.3.2 Associated Mode Only

Figure 4-10 depicts an example of the associated signaling transport architecture using ATM/VP transport and associated protocol stacks involved in the signaling transport path. In this figure, messages related to a particular signaling relation between two adjacent B-SSPs are conveyed over a linkset, directly interconnecting those B-SSPs by Virtual Paths. This architecture essentially distributes the STP function into each signaling node; signaling message routing is then performed on a distributed basis. For that associated mode, it should include "Network Layer Routing" (e.g., MTP-3) or "ATM Layer Routing." The option of Network Layer Routing has been used in Europe and proposed by most major organizations. This option does not use B-STP; rather, it implements the "simplified" MTP-3 in each ATM switch. This supports management functions such as changeover of signaling traffic from a failed signaling link to an alternate signaling link. The approach is a natural choice from a signaling evolution perspective. Another option is used to ATM Layer Routing, which is primarily taken from the network integration perspective. Whether Network Layer Routing or ATM Layer Routing should be used remains to be determined.

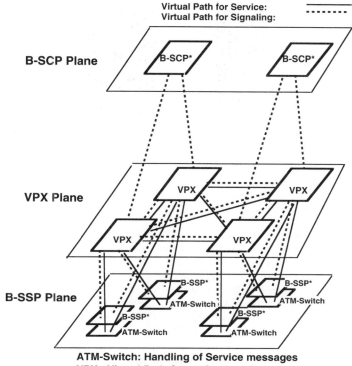

Virtual Path for Service: ―――――
Virtual Path for Signaling: ....

B-SCP Plane

VPX Plane

B-SSP Plane

ATM-Switch: Handling of Service messages
VPX : Virtual Path Cross-Connect System
* : a part of STP function(MTP-3) may be included.

(a) An Example of ATM-Based Associated Signaling Transport.

Signaling Higher Layer				Signaling Higher Layer
MTP-3				MTP-3
AAL				AAL
VC				VC
VP	VP	VP		VP
PL	PL	PL		PL

B-SSP   **VPX**   **VPX**   **B-SSP**

VPX : Virtual Path Cross-Connect
B-SSP : Broadband SSP; PL: Physical Layer

(b) Protocol Stacks of Network Elements Involved in Signaling Transport.

Figure 4-10: ATM-based associated signaling transport.

In the associated mode architecture, each ATM signaling node can directly access the B-SCPs, which may be either centralized or distributed. Note that the example in Figure 4-

10 shows only a centralized SCP placement configuration with protection (i.e., duplication). As discussed in Section 4.3, the existing SS7 network may require more than two SCPs and more physical connections between STPs and SCPs to support new services (e.g., PCS and AIN) requiring access to SCP. This could impose a potential limitation on SS7 network growth. This potential network scalability limitation on existing SS7 networks may be alleviated when ATM-based signaling transport is introduced. By partitioning and replicating the database, one can meet high throughput and reliability requirements for a large-scale signaling network. This distributed B-SCP placement scenario can be more cost-effective in the ATM-based signaling transport network than in its SS7 counterpart, because fewer dedicated signaling links for B-SCP access are required. One example of the distributed database architecture and its performance analysis in the integrated signaling and control ATM network can be found in [10].

4.4.3.3 Hybrid Signaling Transport

Hybrid signaling transport architecture supports both the quasi-associated mode (through B-STPs) and the associated mode in the same ATM control transport network. This architecture is similar to that of the SS7, except that point-to-point signaling links are also supported. If the signaling messages use the associated mode, the provision of point-to-point signaling links may provide flexible and direct linking connectivity to access B-SCPs. The hybrid signaling transport architecture has been studied in ITU-T [4,18]. The rationale for using hybrid network architecture is that the associated mode transport will be used most of the time, whereas the B-STPs may be used in situations in which the traffic volume is low between two signaling points or between the B-SSP and B-SCP [4,18]. The concept of using B-STPs is based on the basic understanding of today's SS7 network design (i.e., the direct point-to-point dedicated data link is not cost-effective when the traffic volume on that link is low). If this design philosophy is still applicable in a new ATM network environment, the hybrid architecture may be a target broadband signaling network.

4.4.3.4 Signaling Transport Architecture Comparisons

This section compares alternative signaling transport architectures described in the previous three subsections based on criteria that include economics, performance, evolvability, and scalability. The criterion of economics and performance is used to evaluate which architecture may best utilize the inherent ATM technology potential to support both narrowband and broadband signaling requirements. Thus, the architecture that is chosen is likely to be a longer-term solution. The best architecture according to this criterion may not necessarily be the best architecture when other factors, such as network evolvability, are considered. However, it is important to study this criterion, as the best technology-oriented architecture may be preferable if it has an acceptable capital gap compared with other architecture alternatives. Thus, the analysis using this criterion serves as a basic model for further architecture comparisons.

A primary benefit of quasi-associated architecture is that it can reuse most of the operations systems designed for today's SS7 networks. This architecture simply uses ATM signaling links as a one-to-one replacement of today's dedicated copper signaling link. However, the B-STP may require a significant capacity upgrade; in addition, the engineering rules should also be reexamined due to the expected increase in signaling volumes and the different performance and reliability requirements for broadband services. This significant capacity upgrade requirement could make the B-STP very difficult or expensive to build or could result in a totally new system that may not be upgradable from the existing STP. This results in a higher cost for operations and maintenance, compared with other alternatives.

Conceptually, the quasi-associated architecture may not be economical from operations or equipment costs points of view, when compared with the associated signaling transport architecture that can support both signaling and services in the same physical network. This is because any two signaling points in ATM networks have already carried ATM connections for services, and the cost of adding additional signaling connections on the same connection paths would be minimal (compared with the stand-alone B-STP). The bandwidth requirement for these signaling channels is relatively small compared with that of broadband services (e.g., video). This is particularly true when the ATM broadband network is designed to support Video Dial Tone services. Thus, the questions that should be answered for supporting the quasi-associated architecture include (1) Can the stand-alone B-STP system and operations costs justify the costs of concentrating low-volume signaling pair traffic? and (2) Could the use of the B-STP improve any aspect of signaling performance, compared with the case of no B-STPs? These two questions require detailed quantitative studies, and results may vary from one network to another. We will discuss some research results for these quantitative studies in Section 4.4.4.

To realize the associated signaling transport architecture, many control and management mechanisms need to be redesigned and redeveloped. Due to significant capital and the longer development cycle that will be necessary to implement a distributed associated broadband signaling transport architecture, a detailed analysis should be performed to explore and quantify potential strengths, in terms of operations and transport economics, over its quasi-associated counterpart. Section 4.4.4 shows the preliminary economics and availability analysis for these architectures.

As mentioned earlier, PCS and other new services (including broadband services) would certainly exhaust today's SCPs if the centralized SCP placement strategy remains unchanged. Thus, the only economical and flexible way to accommodate new database access needs for these broadband services is to distribute the SCP functions throughout the network. In that case, if the B-STP is used (e.g., for the hybrid signaling transport architecture), more A-links, which physically connect the B-STP to the SCP, are needed; this would make it fairly difficult for quasi-associated signaling transport network architecture to grow to large signaling transport network size as compared to its associated signaling transport counterpart, because the latter does not require physical connectivity from B-STPs to B-SCPs.

On the other hand, scalability involves the following issue. One of the drawbacks of the associated mode is the requirement of fully meshed connections in SS7 networks and fully meshed Virtual Connections in ATM networks. For example, when a single B-SSP is added in the associated architecture, new signaling links from/to every other B-SSP or B-SCP must be designed at the same time. On the other hand, in the quasi-associated architecture, signaling link design is easier than in the associated architecture, because the architecture is adequate for connecting new signaling links to B-STPs. Thus, the associated architecture is not readily scalable to large networks. To develop flexible signaling networks, special servers are necessary to manage VP/VC connections. Intelligent operation using ATM technology is expected to meet this requirement.

4.4.4 Signaling Network Architecture Analysis

To better understand trade-offs between the two possible target broadband signaling transport architectures described in Sections 4.4.3.1 and 4.4.3.2 (i.e., quasi-associated mode vs. associated mode), detailed quantitative analysis in terms of economics and performance is needed. This analysis depends on the services and the network being considered. The following summarizes some results reported in [22] that compare the quasi-associated mode architecture with the associated mode architecture with VP transport in terms of cost, reliability, and survivability. We will discuss cost and reliability comparisons in the rest of this subsection, and the survivability comparison will be discussed in the next subsection (Section 4.4.5).

4.4.4.1 Network Model

The network models considered here are based on NTT's existing signaling transport network in Japan. The existing trunk network in NTT is divided into seven blocks. In each block, some nodes are defined as Block Nodes, and four disjoined routes are defined through seven blocks. Every transit-switch node belongs to two Block Nodes in the same block. In this network model, VPXs are located at Block Nodes and VPs between SSP systems are connected through these VPXs. Between the blocks, the VPs are routed on the four disjoined routes. Figure 4-11 depicts typical configurations between two blocks for the quasi-associated signaling mode architecture and the associated signaling mode architecture that will be analyzed in this case study. In Figure 4-11, the first architecture model [called the Type-1 model, see Figure 4-11(a)] is for the quasi-associated mode architecture, and the other three models are for the associated mode architecture. The second architecture model [called the Type-2/S model, see Figure 4-11(b)] provides only one VP routing path between each pair of SSPs within two blocks, and the third architecture model [called the Type-2/4 model, see Figure 4-11(c)] provides four disjoint VP routing paths between each pair of SSPs within two blocks. The last architecture model [called the Type-2/4C model, see Figure 4-11(d)] is the same as the Type-2/4 model, except that the VPX in this model has copy and selection functions. In this model, the originating SSP copies ATM cells and sends them to two originating VPXs. The originating VPX then copies the received cells and sends them to two disjoint destination VPXs, which select one of two copied cells before passing it to the destination SSP. The destination SSP then selects only one cell from two identical cells sent from two destination VPXs. Note that, although SCP nodes are also connected to Block Nodes, they have been omitted from Figure 4-11.

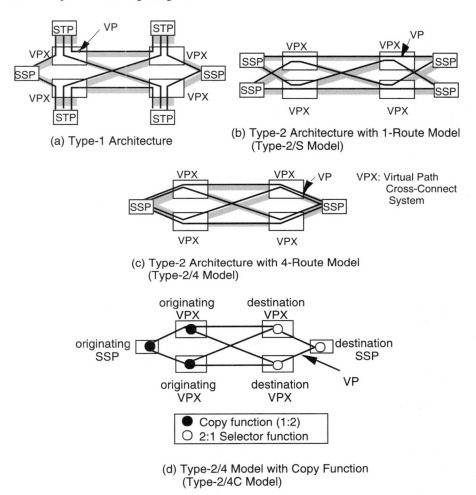

(a) Type-1 Architecture

(b) Type-2 Architecture with 1-Route Model
(Type-2/S Model)

(c) Type-2 Architecture with 4-Route Model
(Type-2/4 Model)

VPX: Virtual Path
Cross-Connect
System

(d) Type-2/4 Model with Copy Function
(Type-2/4C Model)

Figure 4-11: Four architecture models for case study.

In this case study, there are 84 SSPs and 34 STPs. The number of VPXs needed can be computed based on the block model previously described. To simplify the study, SCP nodes are assumed to be colocated with the same nodes with STPs. This assumption does not affect the cost study results, because the result of the cost evaluation is not greatly influenced by the SCP location. In this case, we assume that the quantity of signals between SSPs was 1.15 Kbytes per call and that the signaling quantity from an SSP to an SCP was 1.82 Kbytes per call, which is approximately 10 times that of the signaling traffic in the existing NTT network. We also assumed that 20 percent of all calls need to be processed by SCPs. The VP capacity was calculated by summing these traffic quantities. In the Type-1 architecture, we assumed that signals were carried from an originating SSP to two SCPs in the block of the destination SSP. In the Type-2 architecture, we assumed that signals were first carried from an originating SSP to two

SCPs in the same block, and then from these SCPs to two SCPs in the block of the destination SSP.

4.4.4.2 Cost and Reliability Analysis

Figure 4-12 summarizes the cost results reported in [22]. As shown in the figure, the Type-2 architecture has a cost advantage of approximately 38–52 percent over the Type-1 architecture. However, this figure also shows that using STPs would make VP transport more efficient than if STPs are not used. The cost savings for the Type-2 architecture occurs mostly because expensive STPs are not necessary (those in which STP cost is higher than transport cost, i.e., fiber, VPXs). This study suggests that the cost ratio of the STP to the transport system would affect the result of an architecture analysis.

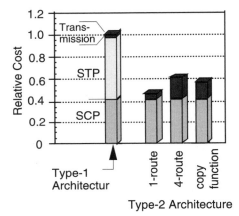

Figure 4-12: Results of cost analysis [22].

The Type-2/S architecture, also reported on in Figure 4-12, is approximately 23 percent less expensive than the Type-2/4 architecture, but with much less reliability (average unavailability of 3.5×10^{-4} vs. 2.5×10^{-9}, see Figure 4-13). The copy function of VPXs reduces the network cost by 20 percent, as compared with the network that does not have a copy function (i.e., the Type-2/S model). This cost savings from the copy function is expected to increase when the network becomes larger.

Figure 4-13 summarizes the reliability comparison results in [22], where unavailability is defined here as the ratio of the system downtime to the system downtime plus the operating time. Figures 4-13(a) and (b) depict distribution histograms of unavailability for the Type-1 and Type-2/4 architectures. In this study, there is a total of 1900 SSP pairs in the model network. In Figure 4-13, 10 percent of the number of SSP node pairs means that there are 190 SSP pairs being considered. In the Type-1 architecture, the average unavailability is 3.3×10^{-8} with the worst at 1.5×10^{-7}. In this case, the unavailability of STP nodes is the major contributing factor to unavailability. In the four–route model of the Type-2 architecture (i.e., the Type-2/4 model), the average unavailability is 2.5×10^{-9}, with the worst at 3.3×10^{-8}. The unavailability for more than 80 percent of SSP node pairs is under 1.0×10^{-9}. The unavailability in the Type-2/4 architecture is much smaller than that in the Type-1 architecture due to the elimination of STP nodes. On the other hand, in the one-route model of the Type-2 architecture, the average unavailability is 3.5×10^{-4}, with the worst at 1.0×10^{-3} (not shown in the figure). This observation suggests that the one-route model of the Type-2 architecture (i.e., Type-2/S model) may be the least expensive among architecture alternatives, but with the worst reliability.

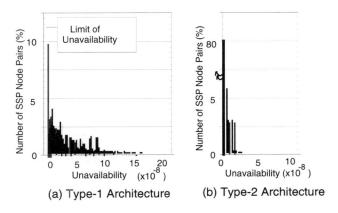

Figure 4-13: System reliability analysis results [22].

4.4.5 ATM Signaling Network Survivability

4.4.5.1 Architecture Modeling

The signaling network survivability can be achieved by implementing network rerouting capability. For an ATM-based broadband signaling transport network, the network rerouting function can be performed at the SONET layer (e.g., SONET Self-Healing Rings), the ATM Layer (e.g., ATM/VP and/or VC self-healing networks), the Network Layer (MTP-3), or a combination of these three layers, as depicted in Figure 4-14. Thus, it is necessary to minimize the functional redundancy across layers to reduce both the network and operation costs.

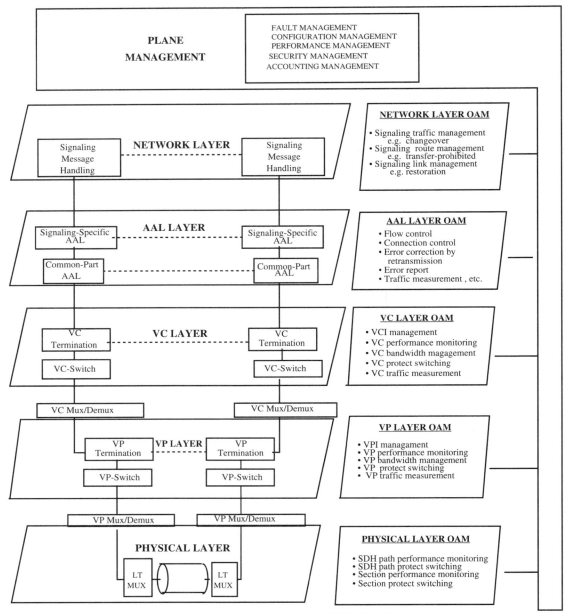

Figure 4-14. Rerouting capabilities across layers in ATM signaling transport networks.

The multilayer functional capabilities shown in Figure 4-14 suggest a need for modeling the interworking scenario among ATM transport, control systems, and a network management system across layers, as depicted in Figure 4-15 [21]. In this model, the network control systems of interest include ATM self-healing [22–25] and protection switching systems [26], which are implemented at the VP layer, and a two-level ATM congestion control system (based on ITU-T standards), which may be implemented at the VP or VC layer. The signaling routing controller may be implemented by using the existing or simplified MTP-3 protocol at the Network Layer or a new distributed signaling routing protocol at the ATM Layer (which may be integrated with the ATM Layer's rerouting protocol). The link retransmission function for signaling message protection is implemented at the AAL layer using AAL Type 5 protocol, as defined in ITU-T SG11. At each layer, these control systems are managed and triggered by layer management, which is then coordinated by the system (or plane) management module. The layer management system includes fault management for the network failure scenario and performance management for the network congestion scenario. The fault management system at each layer is used to trigger either protection switching at the SONET layer, the self-healing or protection switching scheme at the VP layer, and/or the rerouting scheme at the Network Layer. The system (or plane) management collects the failure information, interprets the failure messages, identifies and isolates the failure location(s), coordinates the timing of generating next higher layer AIS messages or other protection switching messages [26], and takes necessary actions to recover the network from failures.

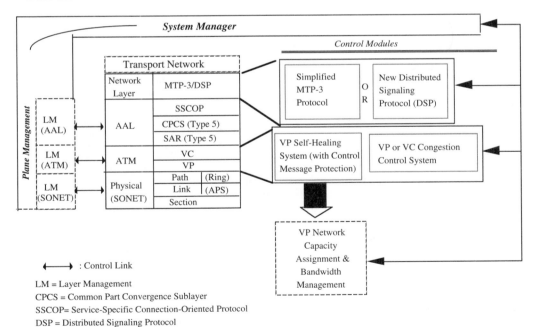

: Control Link

LM = Layer Management
CPCS = Common Part Convergence Sublayer
SSCOP= Service-Specific Connection-Oriented Protocol
DSP = Distributed Signaling Protocol

Figure 4-15: A layer interworking model for survivable ATM signaling transport networks [21].

In the following subsection, we will discuss a case study of survivable ATM signaling transport architectures based on the concept shown in Figure 4-15 and reported in [27].

4.4.5.2 A Case Study for Survivable Signaling Network Architecture Analysis

This section quantitatively analyzes two rerouting options for achieving the desired level of network survivability for ATM signaling transport networks. These two rerouting options are ATM VP rerouting and the Network Layer's MTP-3 rerouting. Before we discuss the quantitative analysis, it is helpful if the functional differences between MTP-3 rerouting and VP rerouting are summarized, as done in Table 4-2.

Table 4-2: A comparison between rerouting schemes.

System Attribute	MTP-3 Rerouting	VP Rerouting
Functional layer	Network Layer	ATM VP Layer
Objects processed	signaling messages	Virtual Path
Routing information	signaling link selection/originating & destination point codes	Virtual Path Identifier
Network element	SSP/STP	SSP/VPX
Way to decide alternate paths	preplanned	preplanned or dynamic
Role of rerouting	restoration against node and link failures	restoration against node/link failures and possible rerouting for congestion
Lost signaling message recovery	more efficient (check before reroute)	less efficient (one round-trip delay)

As shown in the table, both Network Layer rerouting and VP rerouting are autonomous rerouting mechanisms triggered by failure indication message (e.g., AIS) or some other control message [26]. The major difference between the MTP-3 and the VP rerouting mechanisms is that MTP-3 rerouting is used for failure recovery only and is performed in a connectionless manner (although it can be preplanned), whereas VP rerouting may be able to handle both failure and congestion situations in a connection-oriented manner and on a preplanned or an on-demand basis. Furthermore, in general, the VP rerouting scheme may reroute ATM cells faster than the MTP-3 Network Layer rerouting scheme, but some signaling messages will be lost during network failures. To recover these lost messages, SSCOP retransmission functions [19] require one round-trip process of control messages between SSP nodes. Thus, the VP rerouting system may not recover signaling messages efficiently. On the other hand, MTP-3 rerouting checks the sequence numbers of lost messages before rerouting messages. This checking function may decrease restoration time and allows the MTP-3 rerouting scheme to retransmit lost messages more efficiently. Therefore, without simulations it is not clear which rerouting mechanism can restore the failed signaling network faster. In the following, we briefly discussed a simulator model and results that were reported in [27].

Figure 4-16 depicts the structure of a network simulator developed for quantifying the differences of two signaling rerouting schemes.

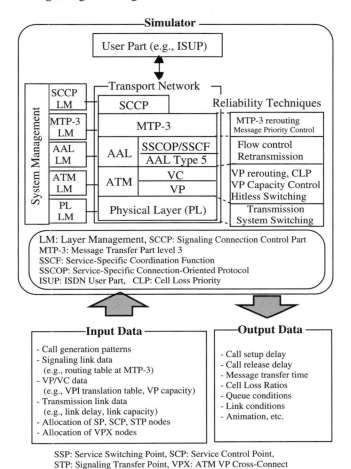

Figure 4-16: The structure of a reliable signaling network simulator [27] (©1995 IEEE).

The simulator shown in Figure 4-16 is applicable to both associated and quasi-associated modes. The simulator is implemented based on the commercially available OPNET simulation platform [28]. This simulator can be constructed on a hardware that supports upper versions of SUN OS 4.1 (Solaris 1.0). The user module processes call setup or call release messages based on ISDN User Part (ISUP) procedures [15]. Furthermore, this module has capabilities for end-to-end information transfer and Service Control Point (SCP) node access (e.g., database access for location registration). The control modules at each layer are managed and controlled by layer management at each layer, which is coordinated by a system-management module. The layer-management modules monitor failure and congestion conditions, and they manage reliability techniques at each layer (e.g., the ATM layer-management module manages VP rerouting by changing the VPI translation table). The system-management module is an important module in the coordination of multiple reliability techniques. It is used to collect failure information, coordinate the timing of generating the next higher layer AIS messages, and order the actions to be taken. More details on the simulator can be found in [27].

Figure 4-17 depicts a network model used for the simulation study in [27]. In this simulation study, the CCS network architecture is assumed to be based on the associated modes. Call and connection control information are used because they are the main type of information transferred in signaling networks. Call setup and call release scenarios are based on the ISUP scenarios. SSPs are assumed to generate call setup or call release messages based on Poisson distributions. The processing times of the SSP nodes are assumed to be the same as those of existing transit-switching nodes in SS7 networks, except in the ATM-related layers. The following results indicate the behavior of MTP-3 rerouting and VP rerouting in the case of a transmission link failure.

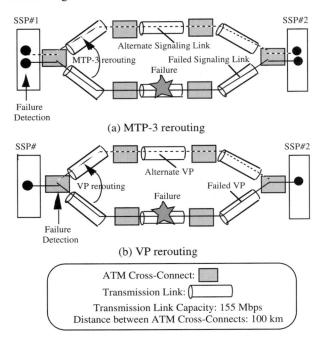

Figure 4-17: Network model for reliable signaling network simulator [27]
(©1995 IEEE).

Figure 4-18 depicts call setup delays for different traffic loads for two rerouting schemes. In this evaluation, it is assumed that SSP#1 and SSP#2 generate call setup messages at the same average interarrival time, and failure detection times at VPXs and SSPs are the same. This result suggests that the performance of the two rerouting mechanisms is similar in the case of low traffic loads, because few messages are transferred before completing rerouting. However, in the case of heavy traffic loads (over 2 calls/second), the setup delay time of VP rerouting is better than that of MTP-3 rerouting. The traffic load in the current NTT signaling network is likely to be in the low-traffic-load area, but if future traffic loads are assumed to be 10 times larger than current loads due to emerging broadband services and other new services, VP rerouting will have a great advantage over MTP-3 rerouting. The foregoing evaluation applies to a VP rerouting mechanism that reroutes cells via alternate VPs as soon as VPXs detect a failure. This situation resembles that of the VP 1+1 protection switching mechanism. The other VP rerouting mechanisms require longer rerouting times after failure detection in order to reserve alternate VP capacities or establish alternate VP connections. If VP rerouting times are longer, it takes more time to recover lost messages.

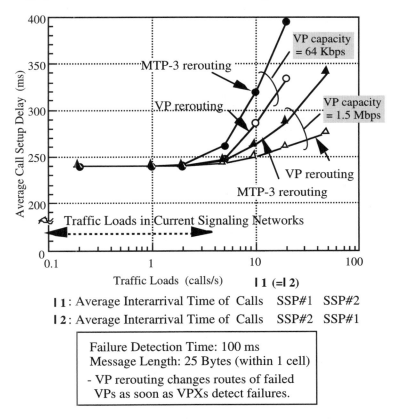

Figure 4-18: Call setup delay vs. traffic loads [27]
(©1995 IEEE).

Figure 4-19 depicts a relationship between call setup delay and rerouting time for two alternative rerouting schemes. This result suggests that, when the VP rerouting time is 200 ms, the performance of VP rerouting becomes worse than that of MTP-3 rerouting. If VP rerouting is applied in broadband CCS networks, the VP rerouting time must be 100 ms or less. To meet this requirement, the VP 1+1 protection switching mechanism and preassigned VP rerouting mechanisms are likely to be potential candidates for survivable broadband signaling networks.

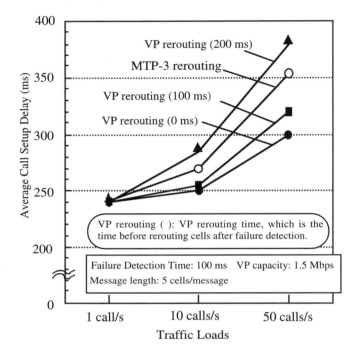

Figure 4-19: Call setup delay vs. rerouting time [27]
(©1995 IEEE).

4.4.6 Public Signaling Network Evolution

4.4.6.1 General Considerations

Several approaches may be used to implement additional network signaling capabilities. These approaches range from a smooth evolution from existing SS7 implementations to overlay by a new signaling architecture. An intelligent signaling network evolution will take advantage of the existing substantial investments in SS7 signaling networks and equipment in tandem with signaling architectures that can then maximize ATM transport benefits.

Functions provided by the Signaling ATM Adaptation Layer (SAAL) are equivalent in many ways to those provided by the current SS7 link layer. However, issues of transmission efficiency (e.g., signaling frames do not necessarily occupy full ATM cells) and the requirements of the MTP-3 in an ATM environment need further analysis.

Associated mode signaling should be economically viable for high-volume signaling situations, but the STP function will still be needed for low-volume signaling requirements. The STP function will also be needed for access to existing services on SCPs, for interworking between existing and broadband services, and for access to existing databases for broadband services. Additional STP functions will be provided as needed for improved gateway screening, for the expansion of global title translation, and for increased processing for the Intelligent Network (IN) and mobility management services. Additional signaling functions, coupled with the emergence of high-speed services, will require increased link signaling speeds.

The impact of ATM on signaling transport will also depend on the services to be implemented and the time frame considered. Some studies have suggested a variety of approaches, which include a simplified MTP-3 protocol, a new SCP planning strategy, distributed signaling, and new signaling engineering rules. New engineering rules will utilize ATM technology to its best advantage to provide reliable, high-speed signaling transport.

An interim step, however, is likely to include the use of T1 or E1 rate (1536 or 2048 Kbps) signaling links to, or between, STPs as an initial upgrade. This step provides the increase in link speed required for network services, IN, mobility management, and access to existing databases for support of new broadband services. ATM interfaces will then be added to STPs to accommodate further speed and capacity increases as local switches are upgraded for high-speed services or additional high-capacity traffic.

One alternative, which will postpone the need to increase signaling link speeds when using ATM networks, is to reduce the loading on STPs by making use of mesh connected ATM transport to provide more associated mode signaling between SSPs connected via the ATM transport network. Indeed, the associated mode could perhaps be implemented less expensively on ATM transport than on today's SS7 networks because of the flexible bandwidth allocation and the inherent logical mesh topology of the ATM transport network.

4.4.6.2 Evolution Scenarios and Interworking Architectures

Signaling transport evolution will depend on economics, the network signaling requirements summarized in Section 4.3, planning time frames, and progress in standards for broadband signaling. Figure 4-20 depicts one possible network evolution architecture, which is designed to interwork with today's SS7 network [7]. It requires few changes to current signaling networks and will take advantage of ATM transport where possible. It uses the ATM transport system to interconnect existing SS7 components; for example, connecting SSP-(a) to the STP in Figure 4-20. This enhanced SS7 network architecture preserves today's signaling access hierarchy with improved STP and SCP processing capabilities.

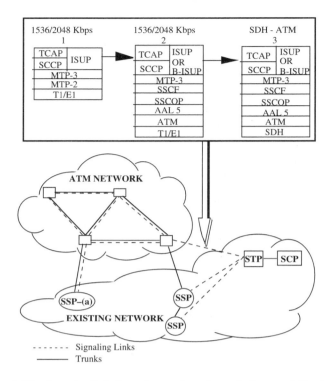

Figure 4-20: A potential signaling network evolution architecture [7].

Reference [7] proposed the three-phase evolution scenario for the network architecture depicted in Figure 4-20. Phase 1 provides existing SS7 essentially unaltered using MTP-2 and MTP-3, SCCP, TCAP, and ISUP (with B-ISUP where necessary), over high-speed links at T1 or E1 rates.

Phase 2 will replace MTP-2 and -3 with ATM and Signaling AAL. The Physical Layer remains as T1 or E1, with the octets mapped into ATM cells instead of into frames. There are some transmission penalties associated with this approach because signaling messages may not fully occupy ATM cells. For example, in the worst case, where all the signaling messages comprise 45 octets, the effective signaling rate on a 1.544 Mbps signaling link will be approximately 0.8 Mbps.

Phase 3 replaces the T1/E1 transmission path with an SDH transport, providing signaling link rates as required up to 150 or 600 Mbps.

To realize the cost-effective target signaling transport networks, interworking between today's SS7 networks and the broadband signaling networks is a critical evolution issue. If the quasi-associated signaling network becomes a target transport architecture, it will be easy to interwork with today's SS7 networks. This is because the quasi-associated network architecture can easily evolve from today's SS7 network by upgrading its current STP to ATM-based B-STP and supporting existing POTS and some other new services.

On the other hand, the associated signaling network architecture may not evolve easily from today's SS7 networks. In this case, however, the associated broadband signaling transport network may interwork with today's SS7 networks as shown in Figure 4-21. Note that this figure describes only NNI signaling. The interworking scenario using IWF1 (Interworking Function 1) has been specified in ITU-T Rec. I.580 [29]. In this case, the IWF1 will translate both today's signaling messages and broadband signaling messages. For instance, the IWF1 translates between B-ISUP and Narrowband ISUP (N-ISUP) messages. This covers the generation, termination, and protocol conversion of signaling messages. The interworking scenario using IWF2 in the figure may be an option for network migration, but it is not yet specified in the current version of ITU-T Rec. I.580. This interworking configuration would allow the signaling transport network to support broadband services first and then integrate voice services later when operations and transport costs can be justified.

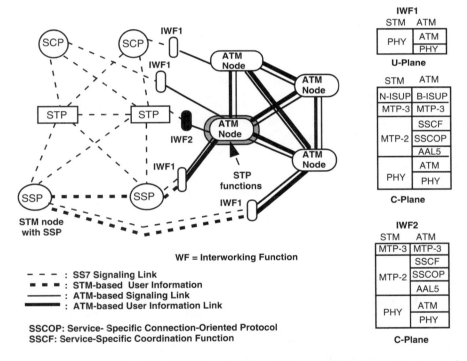

Figure 4-21: Interworking between the ATM-based signaling transport network and the SS7 network.

It is also possible to evolve from today's networks using the associated mode from a service evolution perspective. In this scenario, POTS can still be carried through the STP-like structure (which can be upgraded later to ATM-based STPs if costs can be justified; this corresponds to the hybrid architecture, as described in Section 4.4.3.3; other services may be carried through the new associated mode. This service mapping strategy would minimize the change of existing STPs for POTS, while also taking advantage of ATM network flexibility and economical connectivity to support new services requiring direct access to service modules (e.g., B-SCPs). For POTS, this architecture avoids the expensive capacity upgrade of existing SS7 system components (e.g., STP) and reduces Operations and Maintenance costs due to the possible use of a single ATM network management and operations system. For PCS and AIN, this architecture offers a flexible, fast scheme to access B-SCPs and AIN service modules directly. Also, the flexible ATM connectivity may allow this architecture to accommodate the message volume increase easily, and to implement the distributed B-SCP architecture when the capacity of the SCP in today's SS7 network becomes exhausted. Thus, from the evolution perspective, the interworking architecture with today's SS7 networks will be used to support POTS and new broadband services.

4.5 Signaling for Broadband Enterprise Networks

The signaling transport network architectures discussed in the foregoing section are primarily for a public broadband network. However, due to the popularity of LANs and private networks in the United States, the ATM Forum was created to address the signaling needs for those private networks that may be upgraded using ATM technology. The public network is typically defined as a network offering telecommunication services. The telephony networks operated under Bell Regional Telephone Companies and PTTs are examples of public networks. A private network is defined as either a dedicated network or an enterprise network, which consists of LANs and their interconnections. For example, the IP network own by any LEC is considered to be a private network. Table 4-3 summarizes major differences between the public networks and enterprise networks. These major differences are in network size and stability. In public networks, the interoffice networks are typically on the order of several hundred nodes with careful and long planning for node additions and deletion. In contrast, the enterprise network can easily extend to thousands, even tens of thousands, of nodes with frequent node deletion and addition. Other major differences involve nodal complexity and associated network control and management complexity.

Table 4-3: Differences between public networks and private networks.

Attributes	Public Networks	Enterprise, Private Networks
Network size	few hundred nodes	> tens of thousands of nodes
Node addition/deletion	not frequent	frequent
Nodal complexity	complex	simpler
Network control and management	complex	simpler

The comparison in Table 4-3 implies that the ATM switch for public networks has far more stringent reliability requirements than its enterprise counterpart in terms of redundancy and automatic operations. Thus, the ATM switches designed for public networks are expected to be much more expensive than those designed for enterprise networks. In general, the ATM switching system designed for enterprise networks may not be able to support the stringent QoS requirements needed to support public networks.

However, due to the earlier availability of an ATM signaling system for private networks (e.g., PNNI) and the slow progress of ATM signaling in ITU-T for public networks, many efforts are taking place to investigate the feasibility of using the ATM signaling system designed for private networks to support early deployed ATM public networks. Whether the enterprise ATM signaling system would be able to meet the public network QoS requirements remains to be determined. This trend creates a big challenge, as shown in Figure 4-22, for network engineers and planners who are concerned about evolution of broadband signaling systems in their networks. Thus, the purpose of this section is to review the progress of signaling transport architectures for both public and enterprise networks and explore several evolution scenarios.

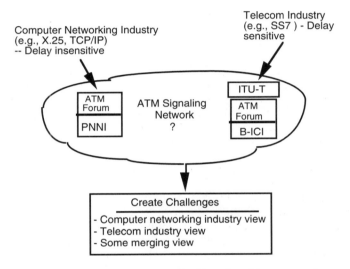

Figure 4-22: Challenges of ATM signaling networks.

In the rest of this section, we will discuss the general concept and operations of Private Network–Node Interface or Private Network-to-Network Interface (PNNI) specified by the ATM Forum for private networks and the interworking of private networks with public networks.

4.5.1 Private Network-to-Network Interface (PNNI) Specifications

PNNI is defined by the ATM Forum as a signaling standard for private networks [1]. This Implementation Agreement specifies the procedures to dynamically establish, maintain, and clear ATM connections at the private network-to-network interface between two ATM networks or the network–node interface between two ATM nodes. Figure 4-23 depicts an ATM Forum signaling interface model that interfaces both public and private networks. In addition to the PNNI, the B-ICI (B-ISDN InterCarrier Interface) is the network-to-network interface between two public networks or switches [30].

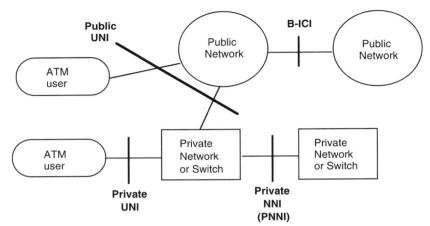

Public UNI: the user-network interface between an ATM user and a public ATM network
Private UNI: the user-network interface between an ATM user and a private ATM network
B-ICI: the network-to-network interface between two public networks or switches (B-ICI stands
     for B-ISDN InterCarrier interface)
Private NNI (PNNI): the network-to-network interface between two private networks or switches

Figure 4-23: A signaling interface model (ATM Forum).

4.5.1.1 PNNI Signaling Protocols

The PNNI solution can be based on both short-term and long-term perspectives. The short-term PNNI solution can be provided by Interim Interswitch Signaling Protocol (IISP). The purpose of the IISP is to enable Switched Virtual Channel (SVC) interoperability in a small, static environment. It uses existing standardized signaling (i.e., Q.2931 and Q.2100) between switches. However, it requires manual configuration of static topology and resource tables.

The complete long-term PNNI solution can be offered by PNNI Phase I protocol, as specified by the ATM Forum [1]. The PNNI Phase I protocol enables extremely scalable, dynamic, multivendor ATM networks. This PNNI Phase I specification includes two categories of protocols. The first protocol is an ATM routing protocol, which is defined for distributing topology information between switches and clusters of switches (used to compute paths through the network). Such a protocol uses a hierarchical mechanism to ensure network scalability, and it has the ability to configure itself automatically in networks in which the address structure reflects the topology. The topology and routing in this protocol are based on the link state routing technique. In other words, the routing decision is based on the link utilization or other link-related performance measure (e.g., delay) at the time the routing decision is made.

The second protocol is defined for signaling the flows of which are used to establish point-to-point/multipoint connections across the ATM network. This signaling protocol is based on the ATM Forum UNI signaling (i.e., Q.2931), but it includes additional features supporting source routing, crankback, and alternate routing of call setup requests in case of connection setup failure. The finite state machine definition and the specification of the symmetrical interactions of PNNI are derived from the Frame Relay NNI signaling defined in an ITU-T draft Recommendation.

4.5.1.2 ATM Addresses

There are two types of ATM addresses defined in the ATM Forum: one for private networks, and the other for public networks. ATM addresses for public networks use private ATM address structure (as in the PNNI) or E.164 addresses (supplied by the network). In contrast, ATM addresses for private networks use private ATM address structure as defined in the ATM Forum. Figure 4-24 depicts the three possible ATM private address formats defined in the ATM Forum [1].

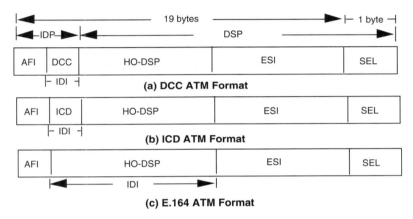

IDP= Initial Domain Part; DSP= Domain Specific Part; DCC= Data Country Code
ICD= International Code Designator; ESI= End System Identifier; SEL= Selector
AFI=Authority & Format Identifier; IDI= Initial Domain Identifier

Figure 4-24: Private ATM address structure and formats (ATM Forum).

As depicted in Figure 4-24, private ATM addresses are 20 bytes, but PNNI routing operates only on the first 19 bytes. The private ATM address structure is designed for private networks, but it may also be used in public networks.

The PNNI addressing principles specified in the ATM Forum include the following:

- It operates in a topologically hierarchical environment.
- The structure of the hierarchy is defined by the peer group IDs used in the routing domain.
- Address assignment has a hierarchy that should generally correspond to the topological hierarchy for proper scaling.

4.5.2 PNNI Routing Protocol

PNNI routing uses and extends Internet routing concepts from Open Shortest Path First, (OSPF, IETF-RFC1583), OSI IS (Intermediate System)-IS (IETF-RFC1142/1195), and

Inter Domain Policy Routing (IDRP, IETF-RFC1478). It includes three major principles: source routing, link state routing, and hierarchical routing.

4.5.2.1 Source Routing

In source routing, the source node chooses a complete end-to-end path to the destination. After the end-to-end routing path is selected during the call setup process, the source node adds the full path information to the signaling message. When the transit nodes receive the signaling message, they simply forward the message based on routing information specified in the signaling message. Figure 4-25 depicts the concept of source routing. In this example, when the source node (Node N1) receives the call setup request message, it computes the end-to-end path, N1 Æ N2 Æ N4 Æ N5, and adds this path information to the signaling message. This signaling message will then be sent to Node N2, which forwards the signaling message to Node N4, based on the routing information presented in the signaling message. The same process is repeated for transit nodes until the signaling message reaches its destination (i.e., Node 5). Note that the PNNI uses a hierarchical version of this concept.

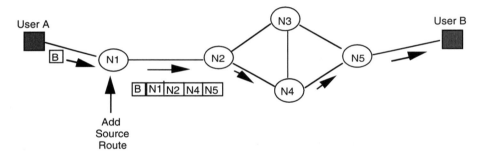

Figure 4-25: The concept of source routing.

4.5.2.2 Link State Routing

The operation of link state routing includes three processes: discovery of neighbors and link status, synchronization of link state databases, and flooding of link state update messages. In link state routing, each node periodically (1) exchanges "Hello" packets with directly neighboring nodes, (2) constructs a "Link State Update" (LSU) that lists links to direct neighbors, and (3) floods LSUs to all other nodes. This operation allows each node to perform source routing based on a complete network topology. Figure 4-26 depicts the concept of link state routing. A hierarchical link state algorithm is the key to PNNI routing.

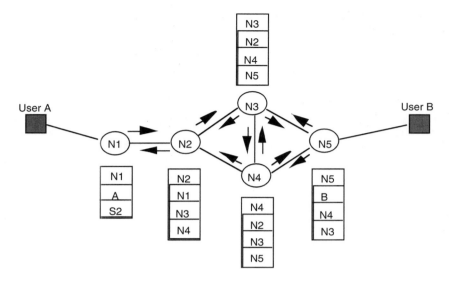

Figure 4-26: Concept of link state routing.

The QoS-related link state parameters, such as available bandwidth, need to be advertised in LSUs to allow better paths to be chosen and make call admission more efficient. Link state parameters required in the PNNI Phase I protocol include maximum cell transfer delay, maximum cell delay variation, maximum cell loss ratio, administrative weight, and available cell rate. A list of optional link parameters has also been specified in [1].

The end-to-end path computed by the source node may not reflect the most current network status, due to potential delay and database synchronization in the link status update process. Thus, two types of connection admission controls are needed here: Generic Connection Admission Control (GCAC) and Actual Connection Admission Control (ACAC). The GCAC is run by a switch choosing a source route, which determines if a path route can probably support the call. The ACAC is run by each switch (including the source switch) along the chosen path, which determines whether or not the switch can actually support the call.

4.5.2.3 Hierarchical Routing

The PNNI uses hierarchical routing for scalable large networks based on hierarchical topology information. The hierarchical topology construction consists of three operations: election of Peer Group Leaders (PGLs), summarization of state information, and construction of the routing hierarchy. We use the following example to explain such a hierarchical topology construction process.

PNNI routing applies to a private network of lowest-level nodes. Figure 4-27(a) depicts such a network consisting of 26 interconnected lowest-level nodes, shown as small circles. Data passes through lowest-level nodes to other lowest-level nodes and to end systems. End systems are points of origin and termination of connections. End systems are not shown in the figure. For purposes of route determination, end systems are identified by the 19 most significant octets of private ATM addresses. The selector octet (i.e., the last octet) is not used for purposes of PNNI route determination but may be used by end systems.

Each arc in Figure 4-27(a) represents a "physical" link attaching two switching systems. A port is the attachment point of a link to a lowest-level node within a switching system. Physical links are duplex (traffic may be carried in either direction). However, physical link characteristics may be different in each direction, either because the capacities are different or because existing traffic loads differ. Each physical link is therefore identified by two sets of parameters, one for each direction. Such a set consists of a transmitting port identifier plus the node ID of the lowest-level node containing that port. Note that PNNI port IDs may be different from equipment-specific port identifiers.

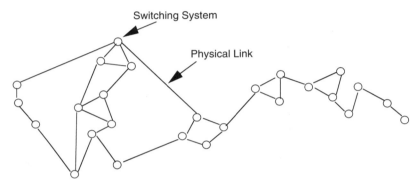

(a) Example of Physical Network Topology.

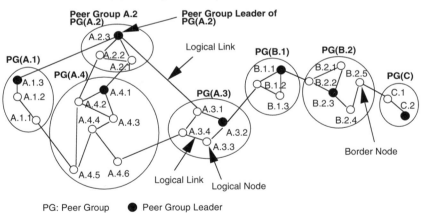

(b) Bottom Level Peer Grouping.

Figure 4-27: PNNI hierarchy example (a) a private physical ATM network topology. (b) Partially configured PNNI hierarchy (bottom-level peer grouping).

If the PNNI protocol supported only the flat network representation depicted in Figure 4-27(a), then each lowest-level node would have to maintain the entire topology, including information for every physical link in the network and reachability information for every node in the network. Although feasible for small networks, this would create enormous overhead for large networks. The PNNI routing hierarchy is designed to reduce this overhead while providing for efficient routing.

The PNNI hierarchy begins at the bottom level, where the lowest-level nodes are organized into peer groups. A "logical node" in the context of the bottom hierarchy level is a lowest-level node. For simplicity, logical nodes are often denoted as "nodes." A Peer Group (PG) is a collection of logical nodes, each of which exchanges information with other members of the group, such that all members maintain an identical view of the group. Logical nodes are uniquely and unambiguously identified by "logical node IDs."

In the example shown in Figure 4-27(b), the network is organized into seven bottom-level Peer Groups: A.1, A.2, A.3, A.4, B.1, B.2, and C. To avoid later confusion between nodes and Peer Groups, Peer Groups are represented in the figure by PG( ). For example, PG(A.3) denotes Peer Group A.3. Here, a "border" node has at least one link that crosses the Peer Group boundary. Hence, neighboring nodes with different Peer Group IDs are border nodes. In the presence of certain errors or failures, PGs can partition, leading to the formation of multiple PGs with the same Peer Group ID.

A "logical group node" is an abstraction of a PG for the purpose of representing that PG in the next PNNI routing hierarchy level. For example, in Figure 4-28, logical group node A.2 represents Peer Group A.2 in the next higher level Peer Group A. Figure 4-28 shows one way in which the PGs in Figure 4-27(b) can be organized into the next level of PG hierarchy.

The functions of the logical group node and the Peer Group Leader of its child PG are closely related. In PNNI specification, the functions of these two nodes are considered to be executed in the same system. Therefore, the interface between these two components is not specified.

The function of a logical group node includes aggregating and summarizing information about its child PG and flooding that information into its own PGs. A logical group node also passes information received from its PG to the PGL of its child PG for flooding. A logical group node does not participate in PNNI signaling.

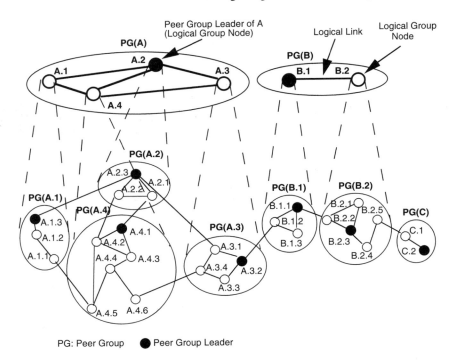

Figure 4-28: Partially configured PNNI hierarchy (bottom two hierarchical levels).

The PNNI routing hierarchy example shown in Figure 4-28 is still incomplete where the higher-level PGs do not exhibit connectivity. Figure 4-29 depicts one possible completion of the hierarchy. Completion is achieved by creating ever-higher levels of PGs until the entire network is encompassed in a single highest-level or top PG. In this example, this is achieved by configuring one more PG containing logical group nodes A, B, and C. Node A represents Peer Group A, which in turn represents Peer Groups A.1, A.2, A.3, A.4, and so on. Another possible completion would be if Peer Groups B and C were aggregated into a Peer Group BC, which was then aggregated with Peer Group A to form the top PG.

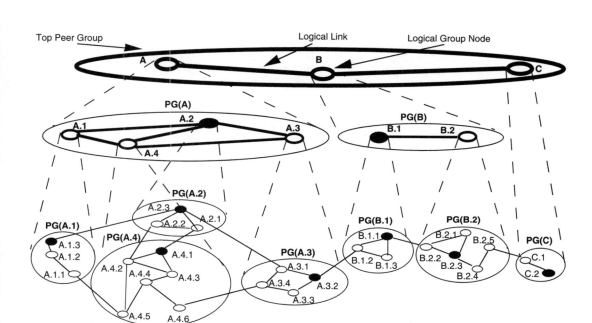

Figure 4-29: A complete PNNI hierarchical configured network (an example).

Figure 4-30 depicts a single switch's view (e.g., A.1.1) of the global topology.

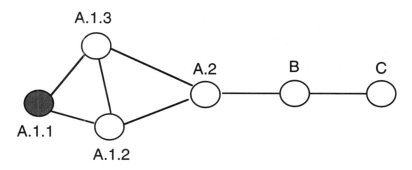

Global topology as seen by Node A.1.1

Figure 4-30: A global topology viewed by a single switch.

The hierarchical topological structure consisting of the topology database for each node in the same PG is identical, and the PGL (one per PG) performs the function of representing the PG in the next hierarchy level.

The path selection in PNNI routing is based on the following principles:

- The path is selected by the source node (i.e., source routing) on an end-to-end basis.
- The routing path based on the most current link information (i.e., link state routing) must be able to support required QoS.
- Source routing is used for all calls, and it is not necessary for each node in the same Peer Group to use an identical algorithm for routing path computation.
- Multilevel hierarchical routing is allowed, but path selection algorithms for intragroup and intergroup cases may be different.

Note that because the PNNI specifies only the routing principles (i.e., source routing and link state routing) and its hierarchical structure, the algorithm for path computation is considered to be vendor-dependent and thus is not specified in the ATM Forum. This is another factor in helping to ensure signaling interoperability.

4.5.3 PNNI Signaling

PNNI signaling makes use of the information (e.g., route calculation) gathered by PNNI routing. Specifically, it uses the route computations derived from the reachability, connectivity, and resource information dynamically maintained by PNNI routing. These routes are computed as needed from the node's view of the current topology. Key concepts of PNNI signaling are to complete source routing across each level of hierarchy and reroute a signaling message around the failed area via crankback messages. In the rest of this section, we discuss the PNNI signaling protocol stack and then the aforementioned additional features needed to meet PNNI signaling requirements.

4.5.3.1 PNNI Signaling Protocol Model

Figure 4-31 depicts the relationship between the signaling procedures and the underlying services. As shown in the figure, the signaling layer contains two entities: the PNNI Call Control and the PNNI Protocol Control. The PNNI Call Control serves the upper layers for functions such as resource allocation and routing. The PNNI Protocol Control entity provides services to the PNNI Call Control and processes the actual signaling finite state machines (incoming and outgoing calls) using symmetric procedures. The PNNI signaling symmetric procedure uses the services of the UNI signaling SAAL with Type 5 (AAL-5) CPCS and SAR sublayers as specified in ITU-T Rec. I.363.

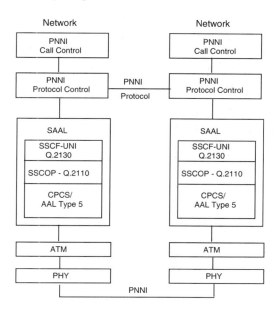

Figure 4-31: PNNI signaling protocol stack.

4.5.3.2 PNNI Control Features

To meet the requirements, PNNI signaling uses the four following additional features, which are beyond those defined for UNI signaling (i.e., Q.2931): Designated Transit Lists (DTLs); crankback and alternate routing; associated signaling; and support of soft PVPCs/PVCCs (Permanent VPCs/VCCs). DTLs are used to carry hierarchically complete source routes. Crankback and alternate routing allow for attempting alternate paths to cope with inaccurate information. Associated signaling is used for PNNI operation over virtual paths. A soft PVPC/PVCC is one where the establishment within the network is done by signaling, as compared with a PVPC/PVCC that is established by provisioning.

Designated Transit Lists (DTLs)

In processing a call, PNNI signaling may request a route from PNNI routing, and it specifies these routes as DTL stacks. A DTL is a complete path across a Peer Group, consisting of a sequence of Node IDs and, optionally, Port IDs traveling the PG. The switch specifying the source route in each PG may or may not be the PGL. A hierarchically complete source route represents a route across a PNNI routing domain that includes each hierarchy routing level between the current level and the lowest visible level in which the source and destination are reachable. This is expressed as a sequence of DTLs ordered from the lowest to highest PG level and organized as a stack. The DTL at the top of the stack is the DTL corresponding to the lowest-level PG. The DTL in each PG is then appended to Q.2931 SETUP and ADD_PARTY messages in a hierarchical manner.

Crankback and Alternate Routing

When creating a DTL, a node uses the currently available information on resources and connectivity. That information may be inaccurate for a number of reasons. Among these reasons: the link failed, or some nodes along with the source route in the DTL became congested after the DTL was produced. In that case, the call is "cranked back" to the ingress node of that Peer Group. This ingress switch may choose to generate an alternate source route, as depicted in Figure 4-32, if it has the provisioning of alternate routing or cranks back the call to the ingress switch of the next higher level PG along the source route. In order to set up an alternate route, any received higher-level DTL must be used and the blocked node(s) or link(s) should be avoided. The "crankback" process is repeated until the alternate path is found or the source node of the complete end-to-end path is reached.

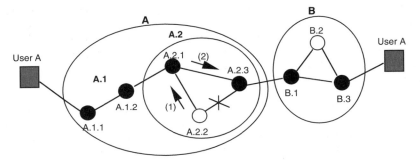

The ingress node of Peer Group A.2 = Node A.2.1
Original Path in Peer Group A.2: A.2.1 -> A.2.2 -> A.2.3
Alternate Path in Peer Group A.2: A.2.1 -> A.2.3

(1) Call is cranked back to Node A.2.1 from Node A.2.2
(2) Node 2.1 computes alternate path: A.2.1 -> A.2.2 -> A.2.3

Figure 4-32: An example of call crankback and alternate routing.

The "cranked back" signaling feature of PNNI can also be used to detect network failure and inform the source switch to perform the traditional network restoration function for ATM service transport networks, which traditionally use the OAM cells to convey the failure information that triggers the network restoration process. We will discuss an example of using cranked back signaling of PNNI for network failure detection and propagation to trigger network restoration for ATM-based video transport networks in Chapter 6.

4.6 IP Layer Signaling: RSVP

While ITU-T and the ATM Forum are making progress in ATM signaling standards and implementation agreements for emerging broadband services including multimedia, the Internet Engineering Task Force (IETF), on the other hand, has approved a resource ReSerVation Protocol (RSVP) used at the IP Layer to provide signaling and QoS support for integrated services like the ones supported by the ITU-T and ATM Forum standards.

Given that both IP and ATM will provide similar signaling capability to support similar services, the question here is whether we really need both. If we do not need both sets of signaling capability for the same network, choosing one of them is a complex issue and the outcome could be determined by market timing and economics. If the market and/or applications determine that both are needed, then the question is how to make these signaling protocols at different layers interwork together. The purpose of this section is to briefly discuss the operations of RSVP and the difference between ATM signaling and RSVP. Details of RSVP and ATM signaling support for IP over ATM can be found in [5,31,32].

4.6.1 Resource Reservation Protocol (RSVP)

Currently, the Internet architecture delivers services based on the point-to-point connection environment with the "best effort" model. This model also provides reliable data transport through the Transport Control Protocol (TCP). This is adequate for traditional data applications [e.g., File Transfer Protocol (FTP) and Telenet], where correct data delivery is more important than the delay. However, emerging broadband services, such as multimedia and video teleconferencing, require point-to-multipoint or multipoint-to-multipoint connections with real-time QoS support, which the current Internet architecture cannot support.

The RSVP is designed to work on the point-to-multipoint or multipoint connection environment and provide real-time QoS needed to support multimedia and teleconferencing applications. The RSVP is used by applications to request specific QoS from the network. The RSVP is not a routing protocol; rather, it is an Internet control protocol that establishes and maintains resource reservations over a distribution tree, independent of how it was created. In the following, we briefly discuss the key features and operations of the RSVP [5,31].

The RSVP uses the receiver-based model; that is, each receiver in the distribution tree for point-to-multipoint connection determines the QoS needed for the path bandwidth reservation from the source node to that receiving node. This is different from the traditional model, where the source node of the distribution tree determines QoS and bandwidth reservation for each receiver; that would make the network scalability difficult. For example, when adding or deleting a leaf, that leaf will have to inform the source node. This would require a lot of overhead and tree reconfiguration, which becomes difficult when the network becomes very large. In addition, the source-based model can offer only one type of QoS to all receiving nodes; this may not work well for an emerging network-of-networks environment in which one network may have a much lower bandwidth than the other. In addition, path messages carrying the routing information are separate from reservation messages. The path messages are generated by the routing protocol and sent by the sender node, whereas reservation messages are generated by the RSVP and sent by the receiving nodes.

The operation of the RSVP can be summarized as follows. Each sender implementing the RSVP periodically sends path messages in the network. These are forwarded by the switches/routers using existing routing tables. These messages contain the sender's filter, its TSpec, and something called ADspec. The TSpec defines the traffic characteristics of the data stream the sender will generate, and it is used by traffic control to prevent over reservation and other failure situations. The ADspec is advertising information, which is used by local traffic controls at the switch; this switch generates an updated ADspec (which the switch can support).

When each receiver receives a path message, it sends an RSVP reservation request as a Resv message upstream toward the senders via the reverse path of the path message. In fact, path messages are used solely to do this reverse forwarding for reservation messages. Each intermediate switch stores the relevant reservation information. Both path messages and reservation messages carry a time-out value that is used by the switches to set the intermediate timers. Whenever the timer expires, the corresponding state is deleted. In this way, resource blocking is prevented in case the receiver fails to send explicit tear-down messages or the underlying routing changes. Thus, it is the responsibility of the senders and receivers to update state information periodically in the network.

4.6.2 RSVP vs. ATM Signaling

The RSVP is an IP signaling protocol that may run on any kind of underlying transport network. The ATM network is expected to become the transport choice for IP networks due to its enormous high-speed capability and scalable architecture. It is expected to support a variety of services with different QoS requirements. The implementation of the RSVP on the IP-over-ATM may depend on how the connectionless IP multicast service is mapped to the connection-oriented ATM services [33]. Given that the RSVP runs on the IP-over-ATM transport network, several issues need to be resolved before it can be practically realized. For example, one issue is the QoS mapping between the IP layer and the ATM Layer, particularly if the ATM Forum model is used here. When the ATM Forum model is considered, the IP service is carried through AAL-5, which would be dropped when the network becomes congested or failed, because it is designated as low-priority traffic in the ATM Forum model. If the ATM Forum model is not used, then a new AAL type will have to be created to ensure its bandwidth guarantee while also maintaining bandwidth reservation flexibility. Another issue is how to carry multiple RSVP flows on a VC so that large VCs could be setup between IP routers in an ATM network. Detailed discussions of these issues can be found in [32,34].

Table 4-4 summarizes differences between the ATM signaling protocol and the RSVP. In general, ATM signaling operates at the OSI link layer (i.e., ATM Layer) and/or the Network Layer (i.e., MTP-3) with a single QoS supported for multicast connections; the RSVP runs at the IP Layer with heterogeneous QoS supported for multicast connections. In contrast to the source-oriented reservation model with no QoS renegotiation during the connection that underlies ATM signaling, the RSVP uses the receiver-oriented reservation model and allows QoS to be renegotiated during the connection by using a timer. To support dynamic QoS and the receiver-oriented reservation model, the network overhead and the end system using the RSVP will be much higher and more complex, respectively, than their ATM counterparts, because the RSVP system needs periodically to advertise QoS that can be supported by the source and intermediate routers/switches. Readers who are interested in Integrated Service IP-over-ATM networks with the RSVP may refer to [32,34] for more details.

Table 4-4: Differences between ATM signaling and the RSVP.

System Attribute	ATM Signaling Protocol	RSVP
Protocol layer	link or network layer (OSI)	IP layer (IETF)
Multicast connection setup model	source-oriented	receiver-oriented
Nodes responsible for path and reservation messages	source node responsible for path and reservation messages	Source node: path messages; Receiving node: reservation messages
QoS supported	uniform (one)	heterogeneous
Resource reservation state	stay as long as the connection exists (static QoS)	time-out model (allow dynamic QoS)
Advertise QoS support by source/switches	no	yes

4.7 Summary and Challenges

We have reviewed several ATM-based broadband signaling transport network architectures for broadband public telecommunication networks, private networks, and integrated services IP-based networks. Some network-dependent quantitative signaling transport network architecture analyses were also discussed to provide technical insights for understanding the differences in alternative architectures.

The real challenge in the broadband signaling area is that there exist three very different broadband signaling systems specified by three major standards groups, ITU-T, the ATM Forum, and the IETF. These three broadband signaling systems were originally designed for three major types of networks: broadband public networks, broadband private networks, and broadband IP networks. However, their features and capabilities could potentially support the same emerging broadband services such as multimedia. The implementation complexity and operation effects would be different, since these three competing signaling systems support broadband services at different layers (ATM Layer, Network Layer, and IP Layer). In this race, the ITU-T-based broadband signaling system faces the challenge of system complexity and slow standardization and implementation. The challenge faced by the PNNI is whether it can support broadband signaling QoS and reliability requirements as competently as telecommunications signaling systems already do. The challenges for the RSVP include performance verification of needed QoS for broadband services and acceptance outside the IP world. Theoretically, we may not need three signaling networks supporting the same broadband services. In reality, they may coexist for a long period of time before they are merged into a single system, which may depend on market timing and costs. Thus, another big challenge generated by this heterogeneous broadband signaling network environment is how these networks can be interworked to provide an end-to-end signaling connection in the network of networks.

References

[1] The ATM Forum, *Private Network-Network Interface Specification Version 1.0 (PNNI 1.0)*, March 1996.

[2] ITU-T Recommendation Q.700, "Introduction to CCITT Signaling System No.7," 1993.

[3] ANSI Standard T1.111, "Signaling System 7 (SS7) Protocol - Message Transfer part (MTP)," American National Standards Institute, New York, 1988.

[4] ITU-T Draft Recommendation Q.2010, "Broadband Integrated Service Digital Network Overview - Signaling Capability Set 1, Release 1," 1995.

[5] R. Braden, L. Zhang, S. Berson, S. Herzog, and S. Jamin, "Resource ReSerVation Protocol (RSVP) Functional Specification," IETF Draft Version 1, March 1996.

[6] A.R. Modarressi and R.A. Skoog, "Signaling System No.7: A Tutorial," *IEEE Commun. Mag.*, pp. 19–35, July 1990.

[7] J. Luetchford, N. Yoshikai, and T-H. Wu, "Network Common Channel Signaling Evolution," *Proc. International Switching Symposium (ISS)*, Vol.2, pp.234-238, April 1995.

[8] Bellcore SR-NWT-002897, "Alternatives for Signaling Link Evolution," Issue 1, February 1994.

[9] Y. Inoue, and Y. Harada, "Evaluation and Integration of the Signaling and OAM&P transport Network," *ITC 7th seminar*, October 1992.

[10] L. Demounem, and H. Arai, "A Performance Evaluation of an Integrated Control & OAM information Transport Network with Distributed Database Architecture," *IEICE Trans. Commun.*, Vol. E-75-B, No.12, pp.1315-1326, 1992.

[11] E. H. Lipper, And M. P. Rumsewicz, "Teletraffic Considerations for Widespread Deployment of PCS," *IEEE Network*, Vol. 8, No. 5, pp. 40–49, September/October, 1994.

[12] K. Meier-Hellstern, E. Alonso, and D. O'Neill, "The Use of SS7 and GSM to Support High Density Personal Communications," *Wireless Communications: Future Directions,* J. M. Holtzman, and D. J. Goodman, (eds.), Kluwer Academic Publishers, 1993.

[13] G. Columdo, *et al.*, "Mobility Control Load in Future Personal Communications Networks," *Proceedings of the 2nd International Conference on Universal Personal Communications*, Ottawa, Canada, October 12-15, 1993.

[14] C. N. Lo, S. Mohan, and R. S. Wolff, "An Estimate of Network Database Transaction Volume to Support Voice and Data Personal Communications Services," *Proceedings of the 8th ITC Specialist Seminar on Universal Personal Telecommunications*, Santa Margherita Ligure, Genova, Italy, Oct. 12-14, 1992.

[15] Bellcore GR-1111-CORE, "Broadband Access Signaling Generic Requirements," Issue 2, October 1996.

[16] ITU-T Recommendation 2761 "Broadband integrated services digital network (B-ISDN) . Functional description of the B-ISDN user part (B-ISUP) of Signaling System No.7," 1994.

[17] ITU-T Recommendation 2762, "Broadband integrated services digital network (B-ISDN) . General functions of messages and signals of the B-ISDN user part (B-ISUP) of Signaling System No.7," 1994.

[18] ITU-T Recommendation I.311, "B-ISDN General Network Aspects," Geneva, 1992.

[19] ITU-T Recommendations Q.2110, "B-ISDN - ATM Adaptation Layer - Service Specific Connection Oriented Protocol (SSCOP)," 1993.

[20] R. Franz, K. D. Gradischnig, M. N. Huber, and R. Stiefel, "ATM-Based Signaling Network Topics on Reliability and Performance," *IEEE J. Selected Areas in Commun.*, Vol. 12, No. 3, pp. 517-525, April 1994.

[21] T.-H. Wu, N. Yoshikai and H. Fujji, "ATM Signaling Transport Network Architectures and Analysis," *IEEE Commun. Mag.*, pp. 90–99, December 1995.

[22] N. Yoshikai, T. Kawata, H. Fujii and T.-H. Wu, "Proposal of ATM based Signaling Transport Network Architectures," *Conference Record, 7-th Joint Japan and Korea Technical Conference (JC-CNSS)*, Taejon, Korea, July 1994.

[23] T.-H. Wu, *Fiber Network Services Survivability*, Artech House, May 1992.

[24] N. Yoshikai, and T-H. Wu, "Control Protocol and Its Performance Analysis for Distributed ATM Virtual Path Self-Healing Networks," *IEEE J. Selected Areas in Commun.*, pp. 1020-1030, August 1994.

[25] R. Kawamura, K. Sato, and I. Tokizawa, "Self-Healing ATM Networks Based on Virtual Path Concept," *IEEE J. Selected Areas in Commun.*, Vol. 12, No. 1, pp. 120–127, January 1994.

[26] Y. Fujita, T-H. Wu and H. Fowler, "ATM VP Protection Switching and Applications," *Proc. International Switching Symposium (ISS)*, pp. 234–238, April 1995.

[27] H. Fujii, and T. Kawata, "Simulation Study of Reliability Techniques for ATM-based Common Channel Signaling Networks," *Proc. IEEE GLOBECOM,* pp. 708–712, Singapore, November 1995.

[28] OPNET Modeling Manual, Release 2.4, MIL3, Inc., Washington, DC, 1993.

[29] ITU-T Draft Recommendation I.580, "General Arrangements for Interworking between B-ISDN and 64 Kbps Based ISDN," March 1994.

[30] The ATM-Forum, *Draft B-ICI Specification Document, Version 2/0*, April 10–14, 1995.

[31] M. Perez, F. A. Mankin, G. Hoffman, A. Grossman, and A. Malis, "RFC 1755, ATM Signaling Support for IP over ATM," February 1995.

[32] M. Borden, E. Crawley, B. Davie, and S. Batsell, "RFC 1821: Integration of Real-Time Services in an IP-ATM Network Architecture," August 1995.

[33] G. Armitage, "Support for Multicast over UNI 3.0/3.1 based ATM Networks," Draft-ietf-ipatm-ipmc-12.txt, February 1996.

[34] A. Demirtjis, S. Berson, B. Edwards, M. Maher, B. Braden, and A. Mankin, "RSVP and ATM Signaling," The ATM Forum Contribution 96-0258, 1996.

Chapter 5

ATM Network Traffic Management

5.1 Introduction

The flexibility of bandwidth management and connection establishment in the ATM network has made it attractive for supporting a variety of services with different Quality of Service (QoS) requirements under a single transport platform. However, because this network is of such high speed, these powerful advantages could cause a serious problem when the network becomes congested. In an ATM transport network, a variety of traffic characteristics such as bandwidth, burstiness, delay time, and cell loss must be considered, because many calls have various traffic characteristics or quality requirements that require calls to compete for the same network resources. Thus, to make the ATM network operationally effective, traffic carried on the ATM network must be managed and controlled effectively to take advantage of ATM's unique characteristics with a minimum of problems for users and the network when the network is under stress. The control of ATM network traffic is fundamentally related to the ability of the network to provide appropriately differentiated QoS for network applications.

A primary role of Network Traffic Management (NTM) is to protect the network and the end system from congestion in order to achieve network performance objectives. An additional role is to promote the efficient use of network resources. ATM NTM includes proactive ATM network traffic control and reactive ATM network congestion control [1–3]. The ATM network traffic control is the set of actions taken by the network to avoid congested conditions. The ATM network congestion control is the set of actions taken by the ATM network to minimize intensity, spread, and duration of congestion, where these actions are triggered by congestion in one or more network elements. The key issues that need to be resolved to achieve the aforementioned design goals include:

- How quality of services is defined at the ATM Layer
- How users declare traffic characteristics that the network can use to recognize and monitor traffic for policing
- How the network measures traffic to determine if the call can be accepted and if congestion control should be triggered
- How the network avoids congestion whenever and wherever possible
- How the network reacts to network congestion to minimize its impact

Compared with present low-speed packet- and circuit-switched networks, the following unique characteristics of ATM networks have made the foregoing three traffic management functional designs much more complex:

- Various B-ISDN Variable-Bit-Rate (VBR) sources generate traffic at significantly different rates (few Kbps–hundreds of Mbps) with very different QoS requirements.
- Traffic characterization of various B-ISDN services are not well understood.

- A single source may generate multiple connections with different types of traffic patterns and characteristics.
- The high-speed transmission speed results in a large number of cells outstanding in the network.
- The high-speed transmission speed limits the available time for message processing at immediate nodes.

This chapter will review the modeling concepts and methodologies that may address these design questions.

5.2 ATM Network Traffic Management Reference Model

Figure 5-1 depicts a reference configuration for the ATM Network Traffic Management system. The functions depicted in Figure 5-1 form a framework for managing traffic and controlling congestion in ATM networks and may be used in an appropriate combination [1].

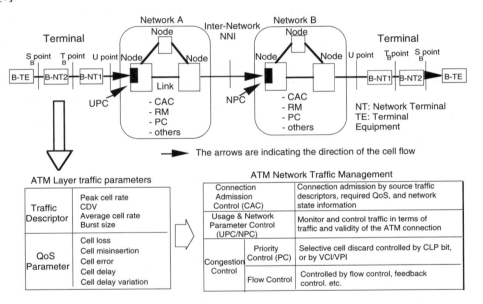

Figure 5-1: Reference configuration for ATM traffic control and congestion control (ITU-T's Rec. I.371).

As depicted in Figure 5-1, during call setup process the user declares the bandwidth and QoS needed to support the application. The network then decides whether a Virtual Channel/Virtual Path Connection request should be accepted or rejected, after evaluating the network resources available for this request. This set of actions taken by the network during the call setup process is called Connection Admission Control (CAC) (in which routing is part of these actions).

The network monitors and controls traffic from users according to the negotiated traffic parameters, and network resources are allocated so that requested QoS will be provided. The network protects network resources from malicious as well as unintentional misbehavior that could affect the QoS of other existing connections, by detecting violations of negotiated traffic parameters and taking appropriate action. The set of actions is called Usage/Network Parameter Control (UPC/NPC).

The user may generate traffic flows with different priorities, and a congested network element may selectively discard cells with low priority This action is called Priority Control. Feedback control is also an effective technique for regulating the traffic injected into ATM connections according to the current state of network resources.

Other traffic controls include Resource Management (RM) by provisioning, considered to be useful for allocating network resources (e.g., Virtual Paths) to separate traffic flows according to service characteristics and QoS requirements. Each NTM control mechanism has its response time scale at which it is most effective [1], as depicted in Figure 5-2, where the response defines how quickly the controls react. In general, congestion control functions can necessarily operate at response time scales greater than the propagation delay, whereas traffic control techniques are designed to be effective at cell transmission times to long-term resource provisioning. For example, cell discarding can react on the order of the insertion time of a call. Similarly, feedback controls can react on the time scale of round-trip propagation times. CAC or call routing may act at least between two call setup time intervals. Because traffic control and resource management functions are needed at different response time scales, no single network traffic management function is likely to suffice. Considering effectiveness and response time scale, network designers may select an appropriate combination of traffic control and congestion control mechanisms to provide required QoS at an acceptable cost.

5.3 Service Class and Quality of Service (QoS)

5.3.1 QoS and Network Performance (NP)

Quality of service determines the degree of satisfaction experienced by users of a communications service. A typical user is not concerned with how a particular service is provided, or with any of the aspects of the network's internal design. However, he/she is interested in comparing one service with another in terms of certain universal, user-oriented performance concerns that apply to any end-to-end service. Therefore, from a user's point of view, QoS is best expressed by parameters that [4]

- Focus on user-perceivable effects, rather than their causes within the network
- Do not depend, in their definition, on assumptions about the network's internal design
- Take into account all aspects of the service from the user's point of view, which can be objectively measured at the service access point
- May be assured to a user at the service access point by the service provider(s)
- Are described in network-independent terms and create a common language understood by both the user and the service provider

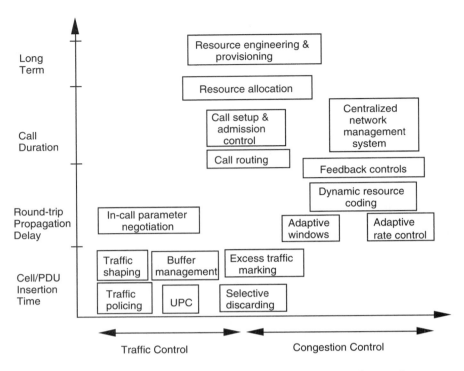

Figure. 5-2: ATM traffic management and response time scale.

In providing services to customers, a network provider is concerned with the efficiency and effectiveness of the network. Therefore, from a network provider's point of view, Network Performance (NP) is best expressed by parameters that provide information for system development, network planning, and operations and maintenance.

The user-oriented QoS parameters provide a valuable framework for network design, but they are not necessarily usable in specifying NP requirements for particular connections. Similarly, the NP parameters ultimately determine the (user-oriented) QoS, but they do not necessarily describe that quality in a way that is meaningful to users. Both types of parameters are needed, and their values must be quantitatively related if a network is to be effective in serving its users. The definition of QoS and NP parameters should make mapping of values clear in cases where there is not a simple one-to-one relationship between them. Table 5-1 shows some of the characteristics that distinguish QoS and NP.

Due to the separation of QoS and NP, a number of general points should be noted when considering the development of parameters [4]:

- The definition of QoS parameters should be clearly based on events and states observable at service access points and independent of the network processes and events that support the service.

- The definition of NP parameters should be clearly based on events and states observable at connection element boundaries, e.g., protocol-specific interface signals.
- The use of events and states in the definition of parameters should provide for measurements at the boundaries identified in the foregoing. Such measurements should be verifiable in accordance with generally accepted statistical techniques.

Table 5-1: Distinction between Quality of Service and Network Performance.

Quality of Service	Network Performance
User-oriented	Provider-oriented
Service attribute	Connection element attribute
Focus on user-observable effects	Focus on planning, development (design), operations and maintenance
Between (at) service access points	End-to-end or network connection elements capabilities

QoS describes the collective effects of service performance that affect the degree of satisfaction of a user of the service. QoS is normally defined through a set of objectively measurable parameters. The set of relevant parameters and the values of these parameters that correspond to an acceptable QoS depend heavily on the service class offered through the network providers.

The QoS, as defined in [5], is characterized by the combined aspects of performance factors applicable to all services, such as service support performance; service operability performance; service accessibility performance; service integrity performance; and other factors specific to a given service. Accessibility, integrity, and retainability of a service are usually clearly visible to a user.

The QoS needs to be mapped to network performance so that the network can be built to support required service accessibility, service integrity, and service retainability. Network performance parameter values are usually derived from QoS values visible to a user. Discussions of general aspects of QoS and network performance can be found in ITU-T Rec. I.350 [4], including the definition of primary parameters and derived parameters. A primary parameter is determined on the basis of direct observations of events at service access points or connection segment boundaries. A derived parameter is determined on the basis of observed values of one or more relevant primary parameters.

In the context of this chapter, QoS visible to a user is referred to as QoS at the AAL; end-to-end network performance is referred to as QoS at the ATM Layer. These QoS requirements at both the AAL and the ATM Layer depend greatly on service classes offered by the network providers.

5.3.2 ATM Service Class

The ATM Forum's Traffic Management Version 4.0 [2] explicitly describes the following ATM service classes: Constant Bit Rate (CBR), Variable Bit Rate (VBR), Unspecified Bit Rate (UBR), and Available Bit Rate (ABR), where VBR supports both Real-Time (RT) and Non-Real-Time (NRT) traffic.

CBR is a service category for the constant-bit-rate-type traffic in which cell intervals are almost constant. CBR offers consistent delay predictability of circuit-switched services or leased line services with guaranteed bandwidth allocation. CBR traffic (e.g., voice traffic supported by the conventional telephone network) requires the Peak Cell Rate bandwidth allocation to an ATM connection.

VBR is a service category for the variable-bit-rate-type traffic in which the cell inter-interval is not constant. VBR traffic (e.g., video traffic coded by Differential Coding) does not always require the allocation of the Peak Cell Rate (PCR) bandwidth. The required bandwidth is allocated on the basis of the Peak Cell Rate and the Sustainable Cell Rate (SCR) [or Average Cell Rate (ACR)].

UBR is a service category in which the Cell Loss Ratio (CLR) and Cell Transfer Delay (CTD) of each call are not guaranteed. The use of UBR does not specify the required bit rate, and cells are transported by the network whenever the network resources are available. The UBR traffic (e.g., data traffic over an ATM-based LAN) does not use any flow control scheme; however, the user may still negotiate PCR with the network (although the PCR used here is simpler than that used in CBR). The UBR traffic will be dropped when the network is congested.

ABR is a service category that is similar to CBR, except that it provides variable data rates based on whatever is available through the use of the end-to-end flow control system. ABR service is defined to operate between a Minimum Cell Rate (MCR), which can be set to zero, and a PCR. The ATM network may transfer cells with the cell loss being minimized. Thus, to some extent, the network "guarantees" the transfer of cells being sent as long as the source is sending those cells according to the feedback information received from the network. It guarantees both bandwidth usage efficiency and cell loss quality to some extent by using the end-to-end flow control to reduce the user data rate before network congestion occurs.

There is also an initial (start-up) cell rate for ABR, which can be set high or low, depending on the type of network. In a LAN, it is likely to start high, whereas in the WAN, it is likely to start low and then increase only if no congestion is detected. The toughest part of ABR service is to figure out how to control the flow of traffic into the network.

ABR may be used in the LAN and TCP/IP environments. ABR is needed because the bandwidth requirements for bursty data traffic are highly variable, and providing enough bandwidth to cover the traffic peak is expensive. Under ABR, the amount of bandwidth available changes over time. Using closed-loop flow control [6], the source can adjust its rate of transmission to match those changes. By slowing down when the network is congested and less bandwidth is available, the cell loss rate is lower and throughput is higher than if the transmission rate did not change.

Table 5-2 summarizes the preceding discussion. In general, in the case of a CBR or VBR connection, enough bandwidth to guarantee Cell Loss Ratio and Cell Transfer Delay required by the call is allocated on the basis of the notified traffic parameters. This type of bandwidth allocation is possible with voice and video traffic because both kinds of traffic are, according to the coding method employed, predictable. In the case of UBR and ABR, the CLR and CTD are not guaranteed because bandwidth is not allocated on a connection basis. However, one major difference between UBR and ABR is that the ABR has smaller CLR. Cells associated with UBR and ABR are discarded when the network is congested or the network resource is not available. However, ABR may have a higher throughput than UBR due to the use of the end-to-end flow control system.

Table 5-2: ATM network service category and attributes.

Attribute	CBR	VBR	ABR	UBR
Source Traffic Type	Predictable traffic characteristics	Predictable traffic characteristics	Unpredictable traffic characteristics	Unpredictable traffic characteristics
Application	Circuit emulation	Differential coding voice/video	LAN, IP services	LAN
QoS	Guarantee cell loss / delay*	Guarantee cell Loss / delay*	Smaller Cell Loss Ratio	No guarantee
Traffic Descriptor	PCR	PCR SCR IBT	(Maximum bandwidth, minimum bandwidth)	PCR
Resource Management	Deterministic multiplexing	Statistical multiplexing	Dynamic control	No management

IBT = Intrinsic Burst Tolerance (used to specify the bucket depth for the leaky bucket method.)
* In general, the delay requirements between CBR and non-real-time VBR are quite different.

5.3.3 Mapping between AAL QoS and ATM QoS

The QoS at the AAL needs to be mapped to the QoS at the ATM Layer so that the ATM network can be built to support required QoS visible to users. Table 5-3 shows video services and some related network requirements that have been described in ITU-T SG13's Integrated Video Service (IVS) baseline document [7]. This table concentrates on bit error and cell loss error correction techniques. CLR is an important of the QoS achievable for a video application. It determines the means, and even necessity, of providing cell loss protection for different services. It is recognized that there is degree of flexibility in this figure, because the network operators have some flexibility to dimension the network in order to provide certain CLRs if they are considered essential for some video services; the codec design can also be changed to accommodate different figures. Taking into account the impact of a range of CLRs on both the network and the codec could result in more progress. The CLRs for both priority levels need to be identified. The Experts Group in SG13 suggested that guaranteed overall CLRs for both priority levels would be essential to satisfy video QoS requirements. Guaranteed performance, at least within certain time intervals, will also be required. If the cell loss is sufficiently small, no cell loss protection may be necessary.

Table 5-3: Video service, bit rates, and Quality of Services (ITU-T SG13).

Service	Bit Rate	QoS Requirement (visualized by users)	Required BER/CLR without error handling in AAL	Required BER/CLR after single bit rate correction on cell basis on AAL
Communication				
Videophone	64 Kbps/2 Mbps CBR	30 min. error-free	$BER < 10^{-6}$ $CLR < 10^{-7}$	In user layer
Videophone	2 Mbps VBR	30 min. error-free	$BER < 3 \times 10^{-10}$ $CLR < 10^{-7}$	$BER < 1.2 \times 10^{-6}$ $CLR < 4 \times 10^{-8}$
Videoconference	5 Mbps VBR	30 min. error-free	$BER < 10^{-10}$ $CLR < 4 \times 10^{-8}$	$BER < 8 \times 10^{-7}$ $CLR < 4 \times 10^{-8}$
Video distribution				
TV Distribution	20-25 Mbps VBR	2 hrs. error-free	$BER < 3 \times 10^{-12}$ $CLR < 10^{-9}$	$BER < 1.2 \times 10^{-7}$ $CLR < 10^{-9}$
MPEG1 core	1.5 Mbps VBR	20 min. error-free	$BER < 4 \times 10^{-10}$ $CLR < 10^{-7}$	$BER < 1.4 \times 10^{-6}$ $CLR < 10^{-7}$
MPEG2 core	10 Mbps VBR	30 min. error-free	$BER < 6 \times 10^{-11}$ $CLR < 2 \times 10^{-8}$	$BER < 5.4 \times 10^{-7}$ $CLR < 2 \times 10^{-8}$

Network performance objectives at the ATM Layer are intended to help define what network ability is required to meet the requested ATM Layer QoS. It is the role of the upper layers, including the AAL, to translate this ATM Layer QoS to any specific application requested QoS. Table 5-4 shows some video performance requirements at the ATM Layer, which is specified in Bellcore GR-2901-CORE [8]. In this table, the Error Cell Block Rate (ECBR) is defined as the total number of error cell blocks observed during a specific time interval. The value of ECBR ranges from one error cell block in 1 second to one error cell block in 2 hours. An equation that may be used to estimate ECBR can be found in [8]. The Bit Error Rate (BER) is defined as the ratio of the number of bit errors to the total number of bits transmitted in a given time interval. The occurrence of bit errors in a network depends largely dependent on the physical media. The BER of an end-to-end MPEG 2 connection will be at most 1×10^{-10}.

Table 5-4. Some Video Performance Requirements at ATM Layer [8].

QoS Parameters	Performance Requirement
Error Cell Block Rate (ECBR)	1 per second–1 per 2 hours
End-to-end MPEG2 Bit Error Rate (BER)	10^{-10}
Cell Delay Variation (CDV)	at most 1 ms
Cell Loss Ratio (CLR)	max. 1 cell loss every 30 minutes
End-to-end Cell Transfer Delay (CTD)	< 1 second

Cell Delay Variation (CDV), also referred to as jitters, is defined as the variation in delay that cells belonging to the same transport stream experience while passing the encoders, multiplexers, switches, and decoders in the network. The CDV, for the full ATM end-to-end network, at the ATM Layer egress point should be at most 1 ms. Like CDV, Cell Loss Ratio (CLR) is also an ATM Layer QoS parameter, defined as the ratio of the number of cells lost (i.e., cells not delivered to the MPEG2 decoder) to the total number of incoming cells (lost cells + delivered cells). The CLR for the MPEG2 connection should be at most 1 cell loss in every 30 min interval. Cell Transfer Delay (CTD), which is also an ATM layer QoS parameter, is defined as the elapsed time from the moment that the first bit of a cell enters the ingress reference point of the ATM network to the time that the last bit of the cell leaves the egress reference point of the ATM network. The CTD for transporting MPEG2 transport streams over ATM should not exceed 1 second.

In addition to ITU-T's IVS baseline document [7], [5] provides performance guidelines and service parameters critical to the realization of a defined QoS for multimedia communications. These guidelines are based on a combination of existing industry standards and practices for digital networks, realistic assumptions about multimedia equipment and premises networks, and user expectations for typical multimedia applications.

5.4 ATM Layer QoS

For a single ATM connection, a user indicates a Quality of Service (QoS) class from those provided by the network. Because of the potential of providing different levels of performance on a connection-by-connection basis, the requested end-to-end user information transfer performance of an ATM connection is indicated to the network at connection setup time. This is part of the Traffic Contract at connection establishment. It is a commitment for the network to meet the requested QoS as long as the user complies with the Traffic Contract. If the user violates the Traffic Contract, the network need not respect the agreed QoS. As a result, the network can reject the violated connection.

ATM Layer QoS is characterized in terms of parameters that measure network performance at the ATM Layer. These performance parameters are used to measure the speed, accuracy, and dependability aspects associated with all three communication phases of an ATM connection, namely, the connection setup phase, the user information transfer phase, and the connection release phase [9]. Here an ATM connection can be either a Virtual Path Connection (VPC) or a Virtual Channel Connection (VCC). The connection endpoint of a VCC is at the ATM Layer service access point, and the connection endpoint of a VPC is within the ATM Layer where the VPC is originated or terminated.

This section discusses ATM Layer QoS parameters, negotiation, and measurements, which are primarily based on References [10–12].

5.4.1. ATM Layer QoS Parameters

A reference configuration used to illustrate the scope of the ATM Layer QoS parameters is given in Figure 5-3.

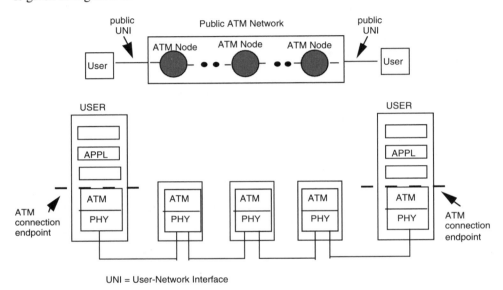

UNI = User-Network Interface

Figure 5-3: Reference configuration for ATM Layer QoS.

The performance requirements for an ATM connection between its endpoints are specified via a proper set of ATM Layer QoS parameters. These parameters are network-oriented, and they provide information on the capabilities of the network elements used by the given ATM connection. Some of the factors that could have an impact on the network performance are

- Propagation delay on transmission media
- Error characteristics of transmission media
- Switch architecture
- Processor and buffer capacity
- Traffic load
- Maximum number of nodes allowed in a route
- Resource allocation strategy
- Network failure and restoration strategy

The ATM Forum identified six ATM Layer QoS parameters associated with the user information transfer phase of an ATM connection, based on the ATM Layer cell transfer performance parameters defined in ANSI/T1 and ITU-T [2,11,12]:

- Maximum Cell Transfer Delay (MaxCTD)
- Peak-to-Peak Cell Delay Variation (Peak-to-Peak CDV)
- Cell Loss Ratio (CLR)
- Cell Misinsertion Rate (CMR)
- Severely Errored Cell Block Ratio (SECBR)
- Cell Error Ratio (CER)

Maximum CTD and Peak-to-Peak CDV are derived from a special instance of the Cell Transfer Delay (CTD) parameter defined in I.356 and T1.511. The corresponding ATM Layer QoS parameters are instances of the associated ATM Layer cell transfer performance parameters applied between the ATM connection endpoints. The remaining four parameters are special instances of the corresponding ATM Layer cell transfer performance parameters defined in ITU-T's Rec. I.356 [11] and T1.511.2 [12]. For some of these parameters, reasonable control can be exerted at the connection level and/or cell level by deploying proper network equipment, for example, with proper resource allocation, buffer management, and scheduling. Therefore, it is possible to provide different levels of objectives for this group of ATM Layer QoS parameters on a connection-by-connection basis. The parameters for which the corresponding objectives can be negotiated at connection setup time are called negotiated ATM Layer QoS parameters in this chapter. The other group of ATM Layer QoS parameters, for which there is very little control available at the connection level and/or cell level, is not subject to negotiation on a connection-by-connection basis. This latter group of ATM Layer QoS parameters is called the nonnegotiated ATM Layer QoS parameters in this chapter. Default objectives are specified for the nonnegotiated ATM Layer QoS parameters. Tables 5-5 and 5-6 summarize these two groups of ATM Layer QoS parameters.

Table 5-5: Negotiated ATM Layer QoS parameters.

Parameters	Definition	Generic Criteria
Maximum Cell Transfer Delay* (MaxCTD)	The 1 - a quantile of the CTD	speed
Peak-to-Peak Cell Delay Variation (Peak-to-Peak CDV)	The difference of the Maximum CTD and the fixed CTD	speed
Cell Loss Ratio (CLR)	The ratio of lost cells to transmitted cells	dependability

* Cell Transfer Delay (CTD): The time between when a cell starts being delivered to the originating endpoint of the ATM connection and when the corresponding cell is received at the terminating endpoint of the ATM connection.

Table 5-6. Nonnegotiated ATM Layer QoS parameters.

Parameter	Definition	Generic Criteria
Cell Misinsertion Rate (CMR)	The misinserted cells received over a given time interval	accuracy
Severely Errored Cell Block Ratio (SECBR)	The ratio of severely errored cell blocks and the transmitted cell blocks	accuracy
Cell Error Ratio (CER)	The ratio of errored cells to the sum of successfully transferred cells and errored cells	accuracy

SECBR is intended to quantify burst-type impairments. To prevent burst errors from dominating the observed CLR, CER, and CMR, cells contained in severely errored cell blocks are excluded in the calculation of CLR, CER, and CMR [10]. Further information on the ATM Layer cell transfer performance parameters used can be found in I.356 and T1.511. The various network attributes that can affect the observed values of the identified ATM Layer QoS parameters are summarized in Table 5-7 [10]. Further information on network attributes and their impact on network performance can be found in Annex B of [2].

Table 5-7: Impact of network attributes on ATM Layer QoS [10] (©McGraw–Hill 1997).

Attribute	Max CTD	Peak-to-Peak CDV	CLR	CMR	SECBR	CER
Propagation delay on transmission media	X					
Error characteristics of transmission media			X	X	X	X
Switch architecture	X	X	X			
Processor and buffer capacity	X	X	X			
Traffic load	X	X	X	X		
Maximum nodes allowed in a route	X	X	X	X	X	X
Resource allocation strategy	X	X	X			
Network failure and restoration strategy	X	X	X			

The performance perceived by the upper-layer applications or the end users is closely related to the protocols implemented between the applications and the ATM Layer, and the protocols implemented between the applications and the user. Furthermore, the performance of an ATM connection observed at the connection endpoints can be affected by the ATM Layer and the Physical Layer capabilities of the corresponding end systems.

The performance requirements for Cell Loss Ratio and Cell Transfer Delay in ATM multimedia networks are distributed roughly as depicted in Figure 5-4 [13]. Under this multimedia networking environment, the key question regarding traffic control is how many QoSs need to be defined. The approach using a single QoS is certainly simplest, but when the most stringent QoS does not apply to a large portion of traffic, the link bandwidth will be underutilized. Alternately, multiple QoSs are defined according to the performance requirements of each traffic pattern. In this case, multipriority control systems must be implemented to guarantee that high-priority traffic always goes through the network at the expense of dropped low-priority traffic when network "stress" conditions (i.e., network congestion or failure) occur.

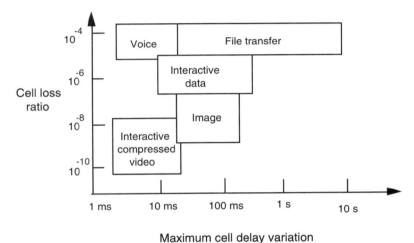

Figure 5-4: ATM multimedia traffic performance requirements [13] (©1990 IEEE).

Standards work related to the network performance associated with the connection setup phase and the connection release phase is under way in both T1 and ITU-T. An initial set of standard agreements on these topics is expected in the 1996–98 time frame. Standards work related to network performance associated with the user information transfer phase is in a more mature state. Initial T1 and ITU-T standards on ATM cell transfer performance were published in [11,12] and are currently being refined. Some of the issues that are being addressed include

*   Defining a set of standard QoS classes, including the objectives of the corresponding ATM Layer QoS parameters
*   Investigating the required adjustments to the current definitions of the ATM Layer cell transfer performance parameters when some cells of an ATM connection do not conform to the negotiated Traffic Contract

5.4.2 ATM Layer QoS Negotiation

The requested user information transfer performance of an ATM connection can be indicated, via user–network signaling, to the network by a QoS class specification at call setup time. For each supported QoS class, a different set of objectives associated with the negotiated ATM Layer QoS parameters identified in Table 5-4 is provided. Any of the ATM Layer QoS parameters can have the value "unspecified." The acceptable combination of a requested QoS class and the signaled source traffic descriptor for an ATM connection depends on the service features provided.

T1 and ITU-T are actively at work on defining, for the public B-ISDN, a set of standard QoS classes including the corresponding objective specification applied between the two public User–Network Interfaces (public UNIs, see Figure 5-3). Furthermore, T1 and ITU-T are also trying to identify a standard default objective applicable to all connections for each of the nonnegotiated ATM Layer QoS parameters identified in Table 5-6. An ATM connection may span several public networks. For those parameters that have a quantitative objective specified in the class definition, a standard allocation of the quantitative objectives among the different networks involved is included in the corresponding class definition. An end-to-end ATM connection may also span private ATM networks. Performance objectives across a private network are not subject to standardization. If the QoS class definitions developed in T1 and ITU-T are to be used, the performance impairment attributed to the private ATM network portion should also be taken into account in determining the performance of the end-to-end ATM connection.

In the ATM Forum, a new capability that accommodates negotiation of objectives for individual ATM Layer QoS parameters is being added to the signaling protocols [14,15]. Currently, only the parameters identified in Table 5-5 can be negotiated individually. This new capability also provides a simple additive accumulation mechanism for the two delay parameters, MaxCTD and Peak-to-Peak CDV, such that an estimate of the agreed end-to-end objective can be obtained at connection setup time and conveyed to the end applications. However, this capability is still under study in T1 and ITU-T.

When only QoS class specification is supported, the received individual QoS parameters are ignored and will be transported transparently by the network elements. Both QoS class and the individual QoS parameters are accommodated in [14] to facilitate compatibility with an earlier version of the ATM Forum signaling specification and the current ITU-T standards [16–18]. When both QoS class and the individual QoS parameters are received in a setup message, only the individual QoS parameters received will be acted upon in [14]. If only the QoS class is received in a setup message, for example, a request from equipment adhering to earlier ATM Forum specifications or ITU-T standards, the individual QoS parameters, as an option, may be generated based on the QoS class received in the setup message. These newly generated parameters together with the received QoS class are forwarded in the corresponding setup message during connection establishment in [14].

Further information on individual parameter negotiation procedures can be found in [14]. Note that the coding precision for the QoS parameters used in [14] should not be misconstrued as the precision at which the QoS objectives could be managed.

5.4.3 ATM Layer QoS Measurement

Within the overall network performance management framework, it is important to verify that the performance delivered to an ATM connection indeed meets the agreed objectives. Measurement procedures for the ATM Layer QoS parameters identified in Section 5.4.1 are required. These measurement procedures can be grouped into two general categories: in-service measurement methods and out-of-service measurement methods.

An in-service measurement method introduces performance–monitoring OAM cells into the transmitted user information cell stream at given intervals and derives the user information transfer performance data from the transmitted user information cell stream and the information carried in the performance monitoring OAM cells. The OAM cell that can be used to facilitate in-service measurement procedures is discussed in Section 5.6. Further information on OAM functions supporting performance measurement can be found in ITU-T Rec. I.610 [19]. A short summary of some in-service measurement methods is given in Table 5-8 [10]. Further information can be found in [11,12].

Table 5-8: In-service measurement methods [10] (©McGraw–Hill 1997).

Parameter	Measurement Method*
Maximum Cell Transfer Delay (MaxCTD)	Over a given measurement period, using the received time-stamp value carried in the TSTP subfield of the forward monitoring OAM cells to derive the CTD distribution and obtain an estimate of the delivered MaxCTD (the 1 - a quantile of CTD).
Peak-to-Peak Cell Delay Variation (Peak-to-Peak CDV)	Subtracting the minimum CTD from the maximum CTD over a given measurement period to obtain an estimate of the delivered Peak-to-Peak CDV.
Cell Loss Ratio (CLR), Cell Misinsertion Rate (CMR), Severely Errored Cell Block Ratio (SECBR)	Over a given measurement period, using the received number of cells transmitted between consecutive forward monitoring OAM cells (i.e., the received TUC_{0+1} and TUC_0 values) to derive estimates of the delivered CLR, CMR, and SECBR.
Cell Error Ratio (CER)	Over a given measurement period, using the Block Error Detection Code information carried in the $BEDC_{0+1}$ subfield of the received forward monitoring OAM cells to derive an estimate of the delivered CER.

* See Figure 5-9 (in Section 5.6) for definitions of TSTP, TUC, and BEDC.

It is assumed that the CTDs for the user information cells should be similar to the CTDs for the OAM cells on a given connection. The in-service measurement method used to estimate the delivered MaxCTD and Peak-to-Peak CDV assumes either a proper loop-back mechanism at the receiving end system or synchronized clocks at both transmitting and receiving end systems. Furthermore, OAM cells that can be transmitted on a given established user connection are limited to a small percentage of the user information cells transmitted. This may limit the use of this in-service measurement method to develop a CTD distribution. The delivered errored cell event is not used in determining whether a cell block is severely errored or not in the in-service measurement method given in Table 5-8. Furthermore, interaction between cell loss and cell misinsertion can also undercount the corresponding lost cell events and misinserted cell events. It is therefore anticipated that, to some degree, the in-service measurement method used will underestimate the delivered CLR, CMR, and SECBR. The in-service measurement method given in Table 5-8 for CER assumes that the cell block is fewer than 200 cells and the transmission medium either has almost no error or has large burst errors.

Out-of-service measurement uses a dedicated test connection. A test cell stream of known number, content, and timing is first introduced at the originating endpoint of the test connection. The test cell stream is observed over a given measurement period at the terminating endpoint of the test connection to derive the estimates of MaxCTD, Peak-to-Peak CDV, CLR, SECBR, and CER. To measure CMR, no cell will be transmitted on the test connection. Any cells received on this test connection over a given measurement period are then misinserted cells. A simple division can provide an estimate of the delivered CMR.

5.5 Traffic Characteristics and Declaration

5.5.1 Traffic Parameters and Declaration

Traffic characteristics of an ATM connection are specified by traffic parameters, which include Peak Cell Rate, Average Cell Rate, burstiness, the peak duration, and the service type (telephone, video phone, etc.). Some of these parameters depend on other parameters. For example, the burstiness measurement depends on the ACR and the PCR. Traffic parameters are grouped into Source Traffic Descriptors for exchanging information between the user and the network. In other words, a Source Traffic Descriptor is the set of traffic parameters used during the connection setup process to capture the intrinsic traffic characteristics of the connection injected from the source. Therefore, any traffic parameter to be involved in a Source Traffic Descriptor should meet the following requirements:

- The traffic parameter should be understandable by the users or their terminals.
- The traffic parameter should be useful for resource allocation for meeting network performance requirements.
- The traffic parameter should be enforceable by the UPC and NPC.

CAC and UPC/NPC have key roles in controlling traffic characteristics; the traffic control system based on these functions performs as expected only if those traffic parameters that can be definitely declared match the source traffic characteristics. However, declaring correct traffic parameters appears to be a difficult task, because most traffic parameters are statistical values. In addition, the impact of Cell Delay Variation (CDV) on traffic parameter specification should be considered when the user declares traffic characteristics. When two or more ATM connections are multiplexed, cells of a given ATM connection could be delayed as cells of another ATM connection are being inserted at the multiplexer or switches. Similarly, insertion of OAM cells for network operations and maintenance will also delay ongoing user cells. Unless the CDV is bounded at a point where the UPC/NPC function is performed, it may not be possible to design a suitable UPC/NPC mechanism and allocate network resources properly. However, controlling CDV may be difficult due to its statistical nature. Note that the CAC and UPC/NPC procedures are operator-specific. Once the connection is accepted, the values of UPC/NPC parameters are set by the network based on the policy of the network operator.

The Connection Traffic Descriptor consists of all parameters included in the Source Traffic Descriptor (e.g., Peak Cell Rate, Sustainable Cell Rate, and burst tolerance), the Cell Delay Variation Tolerance, and the conformance definition used to specify unambiguously the conforming cells of the ATM connection. The conformance definition is based on one or more applications of the Generic Cell Rate Algorithm (GCRA), which will be discussed in Section 5.7.

Figure 5-5 depicts specifications of a traffic parameter declaration. The values of the traffic parameters can be specified either explicitly or implicitly as summarized in the figure. A parameter value is explicitly specified when it is specified by the user via Q.2931 access signaling messages for SVCs or when it is specified via the Network Management System (NMS) for PVCs [during the service provisioning phase or using management system commands such as CMIP (Common Management Interface Protocol)]. A parameter value is implicitly specified when its value is assigned by the network operator using default rules, which, in turn, can depend on the information explicitly specified by the user. A default rule is the rule used by a network operator to assign a value to a traffic parameter that is not explicitly specified.

Explicitly specified parameters		Implicitly specified parameters
parameter values set at circuit setup time	parameter values specified at subscription time	parameter values set using default rules
requested by user/NMS	assigned by network operator	
SVC signaling	by subscription	network-operator default rules
PVC NMS	by subscription	network-operator default rules

Figure 5-5:   Traffic parameter declaration.

5.5.2 Peak Cell Rate (PCR)

One of the most important traffic parameters is Peak Cell Rate (PCR). The PCR is a mandatory traffic parameter to be declared explicitly or implicitly in any Source Traffic Descriptor. The PCR specifies an upper bound on the traffic flow that the user can inject into an ATM connection, which supports the CBR, VBR, or ABR service. Enforcement of PCR by the UPC/NPC allows the network operator to allocate sufficient resources to ensure that the performance objectives (e.g., Cell Loss Ratio) can be achieved.

Figure 5-6 depicts a reference configuration and equivalent terminal for the definition of the PCR. As shown in the figure, the PCR is specified at the Physical Layer Service Access Point (SAP) for an Equivalent Terminal representing the VPC/VCC. The PCR is defined as the inverse of the minimum interarrival time T between two basic events in the Equivalent Terminal, where T is the peak emission interval of the ATM connection and the basic event is the request to send an ATM Protocol Data Unit (PDU) to the Equivalent Terminal. A parameter shown in Figure 5-6 is Cell Delay Variation (CDV) tolerance (t). In addition to the PCR of an ATM connection, it is mandatory for the user to declare either explicitly or implicitly the CDV tolerance t within the relevant Traffic Contract.

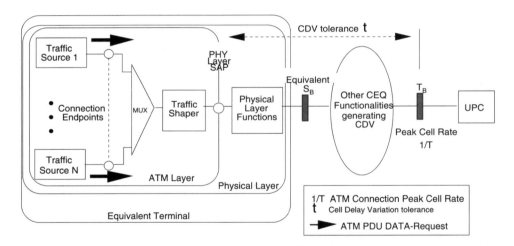

Figure 5-6: A reference configuration for defining Peak Cell Rate.

At the interworking between the ATM-based B-ISDN and another network, the PCR has a strong influence on equipment implementation, which includes Interworking Units and ATM equipment. However, it is not easy to specify the PCR because its value depends on the service/application (e.g., CBR vs. VBR, or connection-oriented service vs. connectionless service) and the interworking methods (e.g., circuit emulation and AAL types). Because the interworking between the B-ISDN and the 64 Kbps-based ISDN is the most popular interworking configuration, we discuss a method that would derive some default values for the PCR as follows. For the sake of simplicity, the following discussion assumes that the AAL Type 1 protocol is used.

The size of the available payload for AAL Type 1 is 47 bytes. In the example presented here, we use the Structured Data Transfer mode to transport cells. The Structured Data Transfer mode requires the pointer field indicating the structured boundary, which uses one extra byte per 8 SAR PDUs (i.e., cell). Thus, the payload available for the upper layer is PA = (47 × 8 − 1)/8 bytes/cell.

There are two possible scenarios here: one without the insertion of OAM cells, and the other with insertion of OAM cells. The OAM cells are inserted for both performance and fault management. For performance management, only the most frequent monitoring and reporting are considered here. In this case, 2 extra cells per 128 cells are used for this purpose, with one cell for forward monitoring and the other cell for backward reporting. For fault management, 1 cell per second is sufficient to achieve its goal.

The following discusses three scenarios for specifying the peak cell rate based on foregoing assumptions.

1. PCR with no OAM support:

$$PCR = \frac{S}{8 \times [(47 \times 8 - 1)/8]} \ [cells/s].$$

2. PCR with minimum OAM support:

In this case, the minimum OAM support means that only fault management is considered, which requires one OAM cell per second to be generated:

$$PCR = \frac{S}{8 \times [(47 \times 8 - 1)/8]} + 1 \ [cell/s],$$

where S [bit/s] is the information transfer rate.

3. PCS with maximum OAM support:

In this case, both performance and fault management OAM cells are involved:

$$PCR = \frac{S}{8 \times [(47 \times 8 - 1)/8]} \times \frac{130}{128} + 1 \ [cell/s].$$

Table 5-9 summarizes the default values for the ATM Traffic Descriptor based on these PCR formulations.

5.6 OAM Flow for Traffic Management

Performance management includes gathering and analyzing statistical data for both performance monitoring (for network maintenance) and Network Traffic Management purposes. Performance monitoring is a function that processes user information to produce maintenance information specific to the user. Although performance monitoring may not be directly related to NTM, an ATM forward monitoring OAM cell can be used to facilitate in-service traffic measurement procedures of network traffic management [10]. Thus, this section discusses a common OAM cell format and some specific OAM cell formats that are used for traffic management.

Table 5-9: Examples of default PCR values for ATM Traffic Descriptor.

Information transfer rate (Kbps)	Peak Cell Rate for user information (cells/s)	Peak Cell Rate if no OAM cells are used (cells/s)	Peak Cell Rate if one OAM cell is used (cells/s)	Peak Cell Rate with maximum OAM support (cells/s)
D:16	42.67	43	44	45
B:64	170.67	171	172	175
H0:384	1024	1024	1025	1041
H11:1536	4096	4096	4097	4161
H12:1920	5120	5120	5121	5201

5.6.1 OAM Cell Format and Types

Figure 5-7 depicts a common OAM cell format. Though the OAM cell payload format for each OAM cell type is different, there are some fields that are common to all of them. These fields include

- *OAM Cell Type:* This 4-bit field indicates the type of management function performed by the OAM cell (e.g., Performance Management, Fault Management, Activation/Deactivation, or System Management). Valid values for this field are also shown in Figure 5-7.

- *OAM Function Type:* This 4-bit field indicates the actual function performed by the OAM cell. Standardized values of this field per OAM Cell Type value are shown in Figure 5-7. For the System Management OAM cell, no processing of function-specific fields is required by the ATM network element.

5.6.2 OAM Flow at the ATM Layer

OAM functions in the network are performed on five OAM hierarchical levels (i.e., F1, F2, F3, F4, and F5) associated with the ATM and Physical layers of the protocol reference as depicted in Figure 5-8 [19]. Among them, F4 and F5 are associated with the VP and VC of the ATM Layer, respectively.

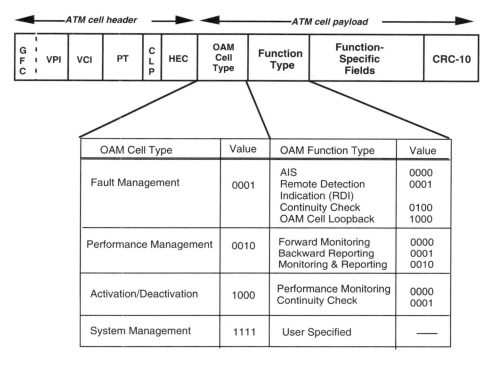

Figure 5-7: A common OAM cell format.

5.6.2.1 OAM Flow for Virtual Path Connection

The Virtual Path Connection (VPC) operations information is carried via F4 Flow OAM cells. These cells have the same VPI value as user-data cells but are identified by preassigned VCI values. Two unique VCI values are used for every VPC. VCI value 3 is used to identify the connection between ATM Layer management entities on both sides of any OAM segment, and VCI value 4 is used to identify the connection between end-to-end ATM Layer management entities.

There are two types of F4 flows, which can simultaneously exist in a VPC. These are

- *End-to-end F4 flow:* This flow, identified by a standardized VCI, is used for end-to-end VPC operations communication.
- *Segment F4 flow:* This flow, identified by a standardized VCI, is used for communicating operations information within the bounds of one VPC link or multiple interconnected VPC links; such a concatenation of VPC links is called a VPC segment.

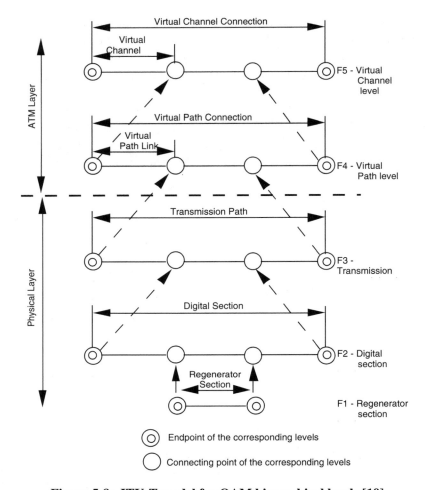

Figure 5-8: ITU-T model for OAM hierarchical levels [19].

One or more OAM segments may be defined along a VPC. Nevertheless, neither overlapped nor embedded segments can be defined. For that purpose, all intermediate connecting points between the source/sink connecting points of a segment must not be a source or sink connecting point of another segment. Note that the definition of the span of a managed segment is not necessarily fixed for the duration of a connection. In other words, the managed segment may be reconfigured as required.

End-to-end F4 flows must be terminated at the endpoints of a VPC, and segment F4 flows at the connecting points terminating a VPC segment. Intermediate connecting points along the VPC or along the VPC segment may monitor OAM cells passing through them and insert new OAM cells, but they cannot terminate the OAM flow, except when loop-backs are performed. In this case, the loop-back cell may be extracted from the OAM flow by the intermediate point where the loop-back has to be performed, and the looped cell may be extracted by the loop-back originator upon reception. The F4 flow will be initiated at or after connection setup either by the Telecommunications Management Network (TMN) or by OAM function-dependent activation procedures.

Table 5-10 shows VCI values that have been specified in ITU-T Rec. I.610 [19].

Table 5-10: VCI values on F4 flow.

VCI	Interpretation	Category
0	Unassigned cell (VPI = 0)	Nonuser cell
0	Unused (VPI > 0)	
1	Meta-signaling cell (UNI)	User cell
2	General broadcast signaling cell (UNI)	
3	Segment OAM F4 flow cell	Nonuser cell
4	End-to-end OAM F4 flow cell	
5	Point-to-point signaling cell	User cell
6	Resource management cell	Nonuser cell
7–15	Reserved for future standardized functions	
16–31	Reserved for future standardized functions	User cell
VCI > 31	Available for user data transmission	

5.6.2.2 OAM Flow for Virtual Channel Connection

The Virtual Channel Connection (VCC) operations information is carried via F5 Flow OAM cells. These cells have the same VPI/VCI values as the user-data cells but are identified by preassigned code points of the Payload Type (PT) field. Two unique PT values are used for every VCC. Table 5-11 shows PT code for OAM cells at F5 level [20]. The PT value "100" (i.e., 4) is used to identify the connection for any OAM segment; the PT value "101" (i.e., 5) is used to identify connection between end-to-end ATM network management entities.

As with the OAM F4 flow, there are two kinds of OAM cells at F5 level: end-to-end OAM cells and segment OAM cells. End-to-end OAM cells must be passed unmodified by all intermediate nodes. The contents of these cells may be monitored by any node along the path. These cells are only to be removed by the endpoint of the VPC or VCC. Segment OAM cells will be removed at the end of a segment, where a segment may be defined as a single VP or VC link across the UNI.

Table 5-11: PT code for OAM F5 flows.

PT Code	Interpretation	Category
000	User-data cell, congestion not experienced	User cells
001		
010	User-data cell, congestion experienced	
011		
100	Segment OAM F5 flow cell	Nonuser cells
101	End-to-end OAM F5 flow cell	
110	Resource management cell	
111	Reserved for future standardized functions	

5.6.3 Performance Monitoring Cells for In-Service Measurement

The two most frequently used OAM cell formats for traffic management are performance monitoring and resource management cells. We discuss the detailed performance monitoring cell format here. The format of resource management cells will be discussed in Section 5.8.3. The OAM function type for performance monitoring has been discussed in Figure 5-7. Figure 5-9 depicts the function specific field of an ATM forward monitoring OAM cell that can be used to facilitate in-service measurement procedures.

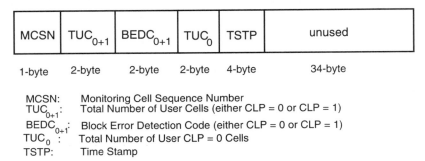

MCSN: Monitoring Cell Sequence Number
TUC_{0+1}:   Total Number of User Cells (either CLP = 0 or CLP = 1)
$BEDC_{0+1}$:   Block Error Detection Code (either CLP = 0 or CLP = 1)
TUC_0 :   Total Number of User CLP = 0 Cells
TSTP: Time Stamp

Figure 5-9: Function-specific field of forward monitoring OAM cell.

5.7 ATM Network Traffic Control

The ATM network traffic control is the set of actions taken by the network to avoid congested conditions. Its control adapts to the network condition proactively. The following functions form a framework for managing and controlling traffic in the ATM network, and they may be used in any appropriate combination:

- Connection Admission Control (CAC)

- Usage/Network Parameter Control (UPC/NPC) (traffic policing)
- Traffic Shaping
- Feedback Control
- Network Resource Management (NRM)
- Other control functions (need further study)

5.7.1 Connection Admission Control (CAC)

The CAC is defined as a set of actions taken by the network at the call setup phase (or during a call renegotiation phase) to determine whether a Virtual Channel Connection (VCC) or a Virtual Path Connection (VPC) can be accepted. A connection request is accepted only when sufficient resources are available to establish the connection through the entire network at its required QoS (and the CDV tolerance, if any) and to maintain the agreed QoS of existing connections. This applies as well to renegotiation of the connection parameters of an existing call. The CAC makes use of the information derived from the Traffic Contract to determine (1) whether the connection can be accepted, (2) traffic parameters needed by Usage Parameter Control, and (3) routing and allocation of network resources.

In a B-ISDN application environment, a call can require more than one connection (e.g., for multimedia or multiparty services such as video telephony or video conferencing). In this case, CAC procedures are performed for each VCC or VPC.

The user may negotiate the traffic characteristics of the ATM connections using the network at its connection establishment phase. These characteristics may be renegotiated during the lifetime of the connection at the request of the user. However, the network may limit the frequency of these renegotiations based on its operations planning strategy.

5.7.2 Usage Parameter Control (UPC) and Network Parameter Control (NPC)

The UPC and the NPC are defined as a set of actions taken by the network to monitor and control traffic, in terms of traffic offered and validity of the ATM connection, at the User–Network Interface (UNI) and the Network–Network Interface (NNI), respectively. The main purpose of the UPC/NPC is to protect network resources from malicious as well as unintentional misbehavior, which can affect the QoS of already established connections, by detecting violations of negotiated parameters and taking appropriate action.

The UPC may be performed on VCCs or VPCs at the point where the VP or VC links are first terminated within the network. Figure 5-10 depicts three possible locations at which the UPC function can be performed, depending on the particular configuration.

As shown in Figure 5-10, in the first configuration (i.e., Case A), the user is connected directly to the VC Connection-Related Function (CRF), such as a VC switch. In this case, the UPC is performed within the CRF(VC) on VCCs before the switching function is executed. The second configuration (i.e., Case B), connects the user to CRF(VC) via CRF(VP). In this case, the UPC is performed within the CRF(VP) on VPCs only, and within the CRF(VC) on VCCs only. For the last configuration (i.e., Case C), the user is connected to other users or to another network provider via CRF(VP). In this case, the

UPC is performed within the CRF(VP) on VPCs only. The UPC for VCCs is performed by another network provider when CRF(VC) is present.

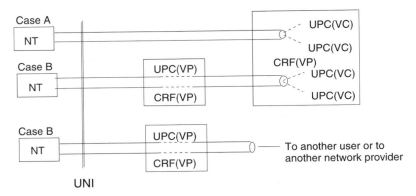

NT: Network termination   CRF(VC): Virtual Channel Connection–Related
CRF: Connection–Related Function   Function
UPC: Usage Parameter Control   CRF(VP): Virtual Path Connection–Related
                                                       Function

Figure 5-10: Location of Usage Parameter Control functions.

The locations for NPC functions is similar to those already described for the UPC, except that the UNI in Figure 5-10 is replaced by the NNI. The UPC/NPC may apply to users' VCCs/VPCs, signaling Virtual Channels. The monitoring task of the UPC/NPC is performed for VCCs (or VPCs) by checking the validity of VPI/VCI (or VPI) to see whether VPI/VCI (or VPI) values are legally assigned values, and monitoring the traffic entering the network from active VCCs (or VPCs) to ensure that traffic parameters agreed upon are not violated.

The UPC/NPC may perform its function at either the cell or connection level. At the cell level, actions of the UPC/NPC function include cell passing, cell rescheduling (when traffic shaping and the UPC are combined), cell tagging (violation tag), and cell discarding. Actions of the UPC/NPC that may be performed at the connection level, include connection release.

Cell passing and cell rescheduling are performed on cells that are identified by a UPC/NPC as conforming. Cell tagging and cell discarding are performed on cells that are identified by a UPC as nonconforming to at least one element of the Traffic Contract. Cell discarding is performed on cells that are identified by an NPC as nonconforming to at least one element of the Traffic Contact.

When an ATM connection utilizes the CLP capability on user request, network resources are allocated to CLP = 0 (high priority) and CLP = 1 (low priority) traffic flows. By controlling CLP = 0 and CLP = 0+1 (multiplexed traffic) traffic flows, allocating adequate resources, and suitably routing, a network operator may provide two QoS objectives for CLP = 0 and CLP = 0+1 cell flows. When no additional network resource has been allocated for CLP = 1 traffic flow (either on user request or due to network provisioning), CLP = 1 cells identified by the UPC as nonconforming are discarded. In this case, tagging is not applicable.

If the tagging option is used by a network operator, CLP = 0 cells identified by the UPC as nonconforming are converted to CLP = 1 cells and merged with the user-submitted CLP = 1 traffic flow before the CLP = 0+1 traffic flow enters the UPC mechanism. If the tagging option is not applied to a connection, the cells that are identified by the UPC as nonconforming to at least one element of the Traffic Contract are discarded. Tagged cells are enforced according to the traffic parameters negotiated for the aggregate CLP = 0+1 traffic. Tagged cells are admitted into the network only if they conform to the Traffic Contract.

Algorithms designed to perform the UPC/NPC function must have the following capabilities in order to ensure that user traffic complies with the agreed parameters on a real-time basis:

- Capability of detecting any illegal traffic situation
- Selectivity over the range of checked parameters (i.e., the algorithm may determine whether the user behavior is within an acceptance region)
- Rapid response time to parameter violations
- Simplicity for implementation

In addition, the accuracy of UPC/NPC should be considered in the algorithm design. For the Peak Cell Rate control, the UPC/NPC should be capable of enforcing a PCR at most 1 percent larger than the PCR used for the cell conformance evaluation. This requirement pertains to PCRs that are as low as 160 cell/s.

Algorithms for the UPC/NPC are primarily based on the Leaky Bucket method. The basic idea of the Leaky Bucket method is that a cell, before entering the network, must obtain a token from the token pool. An arrival cell will consume one token and will immediately depart from the leaky bucket if there is at least one token available in the token pool. Tokens are generated at a constant rate and placed in the token pool. There is an upper bound on the number of tokens that can be accumulated in the pool; tokens arriving at a time when the token pool is full are discarded. To police the peak rate of Virtual Connections (VPCs or VCCs), it is sufficient to set the token generation rate to the Virtual Connection bandwidth. For example, policing a 10 Mbps virtual connection would require one token generated every 42.4 ms.

Figure 5-11 depicts several methods for implementing traffic policing functions [20]. There are three types of enforcement actions that can be used with the Leaky Bucket scheme. The first, and simplest, one [see Figure 5-11 (a)] is to drop the cell if it arrives at a time when there is no token available in the token pool. Instead of dropping cells, arriving cells may be placed in a buffer if they arrive at a time when there is no token available in the pool [see Figs. 5-11(b) and (d)]. If there is a cell waiting in the buffer when a token arrives, the first cell in the buffer will immediately take that token and leave the leaky bucket. Compared with the first option, the second option has better network utilization at the expense of increased delays. The third option is that instead of discarding cells at the edge of the network, some of them are allowed to enter with the cell tagged and are discarded at the congested node within the network [see Figs. 5-

11(c)and (d)]. The third approach may result in better network utilization, but a more complex control system will be needed.

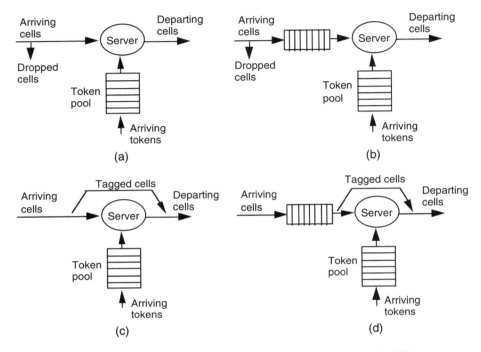

Figure 5-11: Traffic policing (Leaky Bucket method) [20]
(©ARTECH HOUSE 1994).

The Generic Cell Rate Algorithm (GCRA) [2], specified by the ATM Forum, is a Virtual Scheduling Algorithm (VSA) or a Continuous-State Leaky Bucket Algorithm (CSLBA). For each cell arrival, the GCRA determines whether the cell is conforming with the Traffic Contract of the connection, and thus the GCRA is used to provide the formal definition of traffic conformance to the Traffic Contract. The GCRA is used to define, in an operational manner, a relationship between the PCR and the CDV tolerance and the relationship between the SCR and the "Burst Tolerance." Note that the VSA and the CSLBA are equivalent versions of GCRA.

Figure 5-12 depicts a protocol flow for both the VSA scheme and the CSLBA scheme. In VSA, Theoretical Arrival Time (TAT) is initialized to the current time $t(1)$ when the first cell arrives. For subsequent cells, if the arrival time of the kth cell, $t(k)$, is actually after the current value of the TAT, then the cell is conforming and TAT is updated to the current time $t(k)$, plus the increment "I." If the arrival time of the kth cell is greater than or equal to TAT - L but less than TAT, then again the cell is conforming, and the TAT is increased by the increment "I." Finally, if the arrival time of the kth cell is less than TAT $- L$, then the cell is nonconforming and the TAT is unchanged.

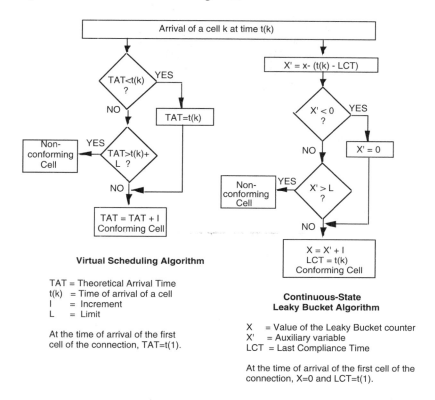

Figure 5-12: Generic Cell Rate Algorithm (GCRA).

The CSLBA scheme can be viewed as a finite-capacity bucket the real-value content of which drains out at a continuous rate of 1 unit of content per time-unit and the content of which is increased by "I" for each conforming cell. Equivalently, it can be viewed as the work load in a finite-capacity queue or as a real-value counter. If at a cell arrival the content of the bucket, X, after the arrival of the last conforming cell minus the amount the bucket has drained since then is less than or equal to the limit value, L, then the cell is conforming; otherwise, the cell is nonconforming. The capacity of the bucket (the upper bound on the counter) is $L + I$.

5.7.3 Traffic Shaping

When used at the source ATM endpoint, traffic shaping is a mechanism that alters the traffic characteristics of a stream of cells on a VCC or a VPC in a desired manner. Traffic shaping must maintain cell sequence integrity on an ATM connection. Examples of traffic shaping are Peak Cell Rate reduction, burst length limiting, and reduction of cell clumping due to CDV by suitably spacing cells in time. It is an option for the traffic shaping to be used in conjunction with suitable UPC functions, provided the additional delay remains within the acceptable QoS negotiated at call setup. It may also be used within the customer equipment or the terminal to ensure that the traffic generated by the source or at the UNI is conforming to the Traffic Contract.

The amount of bandwidth reserved for a connection is between the average rate and the peak rate. For most VBR applications, cells are generated at the peak rate during the active period and no cells are generated during the silent period. The purpose of traffic shaping is to buffer cells before they enter the network so that the departure rate is less than the peak arrival rate of cells (but still greater than the average rate). For example, as shown in Figure 5-13, the equivalent bandwidth for this LAN interconnection is decreased from 15.8 to 2.8 Mbps when the traffic shaper is used. However, the use of the traffic shaper introduces delays that may not be appropriate for delay-sensitive services or applications, such as signaling.

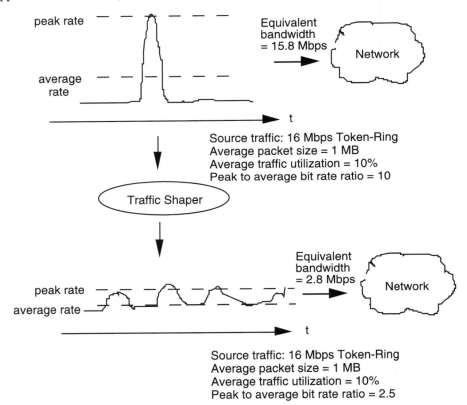

Figure 5-13: Traffic shaping concept.

There are three options for placing the traffic shaping function within the network. The first option is to reshape the traffic at the entrance of the network and allocate resources in order to respect both the CDV and the fixed nodal processing delay allocated to the network. The second option is to dimension the network to accommodate the input CDV and provide for a shaper at the output. The last option is to dimension the network both to accommodate the input CDV and comply with the output CDV without any shaping function.

Figure 5-14 illustrates the block diagram of a general-purpose ATM traffic processor, developed by Bellcore, that can be used as a traffic shaper as well as a UPC or an ATM traffic measurement tool. This research prototype would allow the network provider to conduct various traffic control experiments in the ATM network testbed. The key components of this prototype system are Bellcore's VLSI (Very Large-Scale Integration) Sequencer chip and the RISC-based (Reduced Instruction Set Computer) control microprocessor. One of the unique features of this ATM traffic processor prototype is its software controllability; for example, it can be used to implement and test different traffic control algorithms by reprogramming its microprocessor. Some of the characteristics of the ATM traffic processor are

- It performs traffic control/management on a per-Virtual-Channel and/or per-Virtual-Path basis.
- It is software controlled, so that a variety of traffic control algorithms designed for the normal and/or failure scenarios may be implemented.
- The architecture is scalable and can easily accommodate up to 4096 VCs/VPs without adding system complexity.
- Traffic parameters can be dynamically changed while the traffic processor is in use.
- It is based on readily available components and subsystems, such as Bellcore's ATM bridge [Transparent Asynchronous Transmitter/Receiver Interface (TAXI) to SONET interface] and a commercially available microcomputer board.
- It allows TAXI or SONET STS-3c interface.

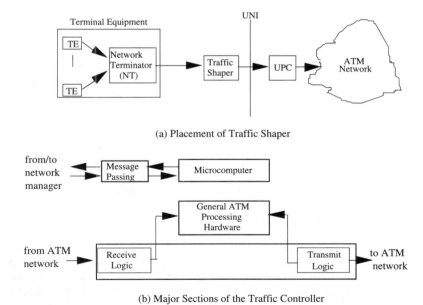

(a) Placement of Traffic Shaper

(b) Major Sections of the Traffic Controller

Figure 5-14: A Bellcore traffic shaper prototype.

5.7.4 Resource Management

Provisioning may be used to allocate network resources to separate traffic flows according to service characteristics. VPCs play a key role in network resource management. By reserving the capacity on VPCs, the processing required to establish individual VCCs is reduced. Individual VCCs can be established by making simple connection admission decisions at nodes where VPCs are terminated. Strategies for the reservation of capacity on VPCs will be determined by the trade-off between increased capacity costs and reduced control costs. The peer-to-peer network performance on a given VCC depends on the performances of the consecutive VPCs used by this VCC and how it is handled in Virtual Channel Connection–Related Functions.

Fast Resource Management (FRM) functions operate on the time scale of a round-trip propagation delay of the ATM connection. Figure 5-15 depicts the concept of FRM. One possible FRM function that has been identified is as follows. In response to a user request to send a burst, the network may allocate capacity (e.g., bandwidth, buffer space) for the duration of the burst. When a source requests an increase of its Peak Cell Rate, it has to wait until resources have been reserved in all network elements along the ATM connection before the new PCR can be used. UPC/NPC parameters would be adjusted accordingly.

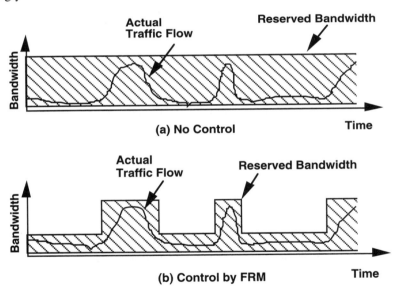

Figure 5-15: A concept for Fast Resource Management.

The VP bandwidth control scheme is the basic control feature for implementing advanced applications such as ATM protection switching and bandwidth on demand. Figure 5-16 depicts an effect of accumulating VPs using the VP bandwidth control feature. The plotted lines indicate the transitions in required bandwidth for each VP; they do not indicate the real volume of cell flow. If each VP bandwidth can be changed according to the requirements as shown in the left side of the figure, multiplexed VPs can share link resources as shown in the right side of the figure. This control technique allows the required link resources to be reduced compared with those required when the peak bandwidth is statistically assigned to each VP (see the left side of the figure). However, reduction of a required link bandwidth requires another control system. More details on bandwidth control schemes can be found in [21].

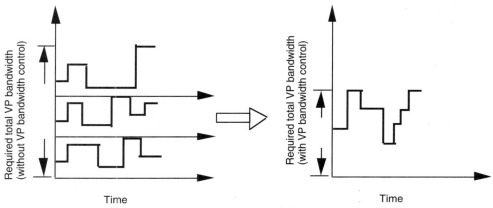

Figure 5-16: A concept of bandwidth control.

The major advantage of VP bandwidth control is reduction of the required VP bandwidth. Another advantage of VP bandwidth control is bandwidth granularity. The bandwidth of a VP can be precisely tailored to meet the demand. There is no restriction due to path hierarchy, and this would allow higher utilization of the link capacity than is possible with digital physical path bandwidth control in SONET/STM networks.

5.7.5 Feedback Control

Feedback control is defined as the set of actions taken by both the network and users to regulate the traffic submitted to ATM connections according to the state of network elements. Details of feedback control can be found in References [1,2].

5.8 ATM Network Congestion Control

The ATM network congestion control is the set of actions taken by the ATM network to minimize the intensity, spread, and duration of congestion. Network congestion is defined in ITU-T Rec. I.371 [1] as a state of network elements (e.g., switches, concentrators, cross-connects, and transmission links) in which the network cannot meet the negotiated network performance (e.g., QoS) objectives for the established connections and/or new connection requests. Note that congestion should be distinguished from the state where buffer overflow causes cell losses but still meets the negotiated QoS.

There are two possible causes of network congestion. The first is unpredictable statistical fluctuations of traffic flows in normal conditions. Network congestion can also be a result of a network under fault conditions. These fault conditions could be software faults and/or hardware failures. Software faults typically cause undesired rerouting that would exhaust some particular subset of network resources. Hardware failure can be overcome by using network restoration procedures that, in some cases, may require competition for network resources with existing unaffected connections in an ATM network.

An ATM network congestion control system includes two major subsystems: one for congestion measurement, and the other for congestion control mechanisms. A more detailed discussion for these two subsystems follows.

5.8.1 Measure of Congestion (MOC)

Measures of Congestion (MOCs) of an ATM Network Element (NE), specified in Bellcore GR-1248-CORE [22], are defined in terms of the percentage of cells discarded (Cell Loss Ratio), the percentage of ATM modules in the ATM NT that are congested, or other performance parameters. The ATM modules in an ATM NE include the ATM switching fabric, the intraswitching links, and modules associated with interfaces, such as UNI, which may buffer incoming and outgoing traffic. MOCs in the ATM module may be defined in terms of buffer occupancy (e.g., 65 percent buffer occupancy), where *buffer occupancy* = (number of cells in the buffer at a sampling time)/(the capacity of the buffer, in cells); utilization (e.g., 70 percent link utilization), where *utilization* = (number of cells actually transmitted during the sample interval)/(cell capacity of the module during the sampling interval); or the Cell Loss Ratio (e.g., 50 percent cells dropped), where the *Cell Loss Ratio* = (number of cells dropped during the sampling interval)/(number of cells received during the sampling interval). The MOC sampling period is specified to be 20 ms in Bellcore GR-1248-CORE.

Three congestion thresholds have been defined in Bellcore GR-1248-CORE, as depicted in Figure 5-17. It is expected that the threshold for "severe congestion" would be the highest congestion threshold that would need to persist for a significant time period (e.g., 30 seconds). Because merely a slight reduction in congestion below the severe congestion level should not cause the time in severe congestion to be reset, a severe congestion abatement threshold is defined. This abatement threshold may be set at any level; that is, it can be lower than other congestion thresholds. One threshold may be defined as the Explicit Forward Congestion Indication (EFCI) threshold. When the MOC is above this threshold, the EFCI bit of all cells passing through the ATM module would be set. Note that this is not required; it is acceptable for the EFCI bit to be set by other vendor-specific means, such as buffer occupancy larger than a certain percentage.

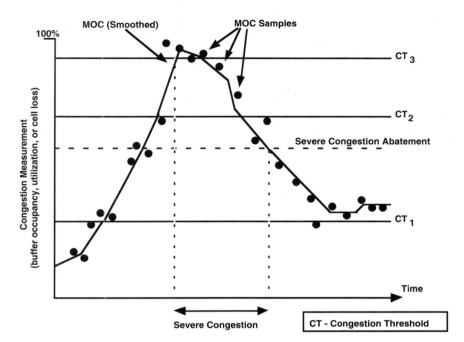

Figure 5-17: Congestion state [Bellcore GR-1248-CORE].

5.8.2 Congestion Control Functions

When network congestion occurs, there are two methods that may help the network to recover. The first is to discard low-priority cells selectively during the congestion. In this case, the network may still meet network performance objectives for both high-priority flows and aggregated flows. The second method is to use EFCI to notify the end users to lower their access rates. The EFCI is a congestion notification mechanism that may be used to assist the network in avoidance of and recovery from a congested state. Note that the EFCI may also be used for traffic control indication. A network element in an impending-congested state or a congested state may set an EFCI value in the cell header so that this indication may be examined by the destination Customer Premise Equipment (CPE) . An impending-congestion state is the state in which network equipment is operating around its maximum engineered capacity level. The end user's CPE may use this indication (i.e., EFCI) to implement protocols that would lower the cell rate of the connection during congestion or impending congestion. A Network Element that is not in a congested state or an impending congested state will not modify the EFCI value. Note that the mechanism by which an NE determines whether it is in an impending-congested state is an implementation issue and is not subject to standardization.

Other congestion control functions include rerouting mechanisms at the Network Layer and the ATM Layer. Examples of Network Layer rerouting include IP rerouting and MTP-3 rerouting as discussed in Chapter 4; examples of rerouting at the ATM Layer include ATM VP or VC rerouting as discussed in [23].

5.8.3 Congestion Control Mechanisms

There are three types of congestion control mechanisms that may be implemented in ATM networks: priority control and selective cell discard; end-to-end rate-based congestion control; and link-by-link credit-based congestion control. Among them, priority control and selective cell discard may be the simplest congestion control mechanism, and it may be used in conjunction with rate-based congestion control or credit-based congestion control.

For the priority control and selective cell discard mechanism, the user may generate different priority traffic flows by using the Cell Loss Priority bit, where CLP = 0 is used for high-priority flow, CLP = 1 is used for low-priority flow, and CLP = 0+1 is for multiplexed flow. A congested network element may selectively discard cells with low priority if necessary to protect as much as possible the network performance for high-priority cells. Network elements may selectively discard cells of the CLP = 1 flow and still meet network performance objectives on both the CLP = 0 and CLP = 0+1 flow. For a given ATM connection, the Cell Loss Ratio objective for CLP = 0 cells should be greater than or equal to the CLR objective for CLP = 1 flow. Network elements are not to change the value of the CLP bit, except when the tagging option is used.

Rate-based and credit-based congestion control mechanisms are more adaptive to network load conditions. Figure 5-18 depicts these two congestion control mechanisms, both of which have also been discussed in the ATM Forum [6].

Credit-based congestion control performs congestion control on a link-by-link basis and is based on credits allocated to the node. Rate-based congestion control adjusts the access rate based on end-to-end or segmented network status information. In a credit-based scheme, each Virtual Channel requires a credit before a data cell can be sent. When data cells are transmitted on a VC, credits are intermittently sent to the upstream node in order to maintain a continuous flow of data. In rate-based schemes, the ATM node notifies the traffic sources to adjust their rates based on feedback received from the network. Upon receiving a congestion notification (e.g., using EFCI), a traffic source slows the rate at which it transmits data to the network.

In September 1994, the ATM Forum voted for the rate-based control mechanism to be used in ATM networks that support ABR services. Detailed control algorithms can be found in [2]. In the following, we briefly discuss both rate-based congestion control and credit-based congestion control. Detailed comparisons of these two congestion control mechanisms can be found in [6].

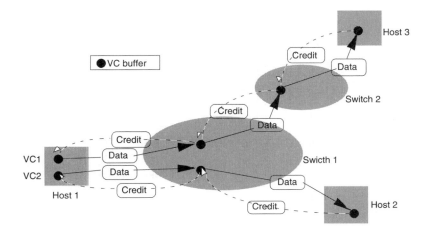

(a) Credit-Based Flow Control Applied to Each Link of a VC

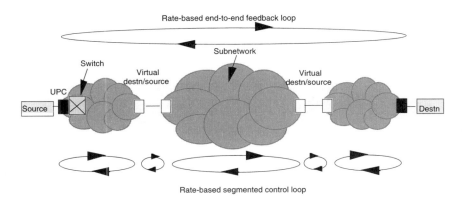

(b) Closed-Loop Flow Controlled Network

Figure 5-18: Two congestion control schemes in ATM networks.

5.8.3.1 Rate-Based Congestion Control

In the ABR service, the source adapts its rate to changing network conditions. Information about the state of the network (e.g., bandwidth availability, state of congestion, and impending congestion) is conveyed to the source through special control cells called Resource Management (RM) cells.

The RM cell uses the OAM cell format, as described in Section 5.6. Figure 5-19 depicts a RM cell format that has been specified by the ATM Forum [2]. The CLP bit in the RM cell is set to 0. The RM cells referring to a VC Connection are identified by PT value 110. VP connections are identified by PT value 110 and VCI = 6.

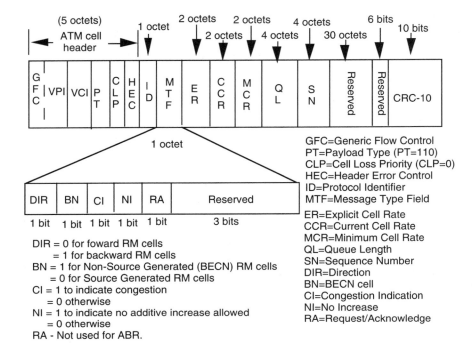

Figure 5-19: Resource Management cell format.

As specified in [2], there are four methods that may be used to control congestion at queuing points:

(a) *EFCI marking:* The switch may set the EFCI state in the data cell headers.

(b) *Relative Rate Marking:* The switch may set CI = 1 or NI = 1 in forward and/or backward RM cells.

(c) *Explicit Rate Marking:* The switch may reduce the ER field of forward and/or backward RM cells.

(d) *VS/VD Control:* The switch may segment the ABR control loop using a virtual source and destination.

In the following, we discuss a rate-based congestion control algorithm, called the Enhanced Proportional Rate Control Algorithm (EPRCA) [24], that uses the EFCI marking method. The EPRCA is based on a "positive feedback" rate control paradigm. The source increases its sending rate for a connection only when given an explicit positive indication to do so, and in absence of such a positive indication it continually decreases its sending rate. Figure 5-20 depicts a protocol flow model of the EPRCA.

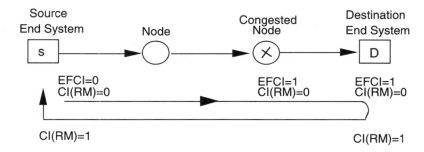

CI(RM)=Congestion Indicator in the Resource Management cell

Figure 5-20: Enhanced Proportional Rate Control Algorithm (EPRCA) [24].

A connection starts at the Initial Cell Rate (ICR) established at connection setup and defined to be in the range of the Minimum Cell Rate (MCR) to the Peak Cell Rate (PCR). When a connection is idle, the Allowed Cell Rate (ACR) is decreased to the ICR, and if it is below the ICR, its ACR remains at its current value. The ICR will be decreased with each cell sent.

For each connection, the source end system generates one forward Resource Management (RM) cell for every sending data cell with "EFCI = 0" (indicating congestion not experienced), and each of these RM cells is returned by the destination end system, creating a backward flow of RM cells, to close the feedback loop. Forward RM cells are generated with CI = 0 (CI = Congestion Indicator; 0 = no congestion), and a backward RM cell with CI = 0 received at the source for a given connection allows the connection to increase the sending rate in order both to compensate for the decreases in rate made over the last data cells and to achieve the desired overall increase in the sending rate because there is no congestion in the network. An RM cell with CI = 0 is changed to CI = 1 (i.e., congestion) by the destination if the last received data cell has EFCI = 1 (which means that congestion was experienced). It can also be changed by an intermediate network to CI = 1 in the backward RM cell if congestion is experienced by the network in the forward direction. The congestion indication that changes EFCI from "0" to "1" can be based on a cell queue depth in a switch, a threshold on an aggregate rate of cells flowing on a link, or other criteria.

5.8.3.2 Credit-Based Congestion Control

The credit-based flow control scheme, as depicted in Figure 5-21, was proposed in the ATM Forum/94-0632 [25]. This credit-based control scheme is a variant of sliding window mechanisms. It is a per-VC link-by-link flow control scheme that operates between a sender and a receiver in two adjacent nodes of an ATM network. The proposed flow control scheme is also called the *N23* scheme, and its basic operation is depicted in Figure 5-21. In this scheme, before transmitting any data cell of a VC over the ATM link, the sender needs to receive credits from the receiver of the VC. The upstream node maintains a credit balance for each outgoing VC on the link, and it can transmit a data cell associated with that VC only if that VC's credit balance is positive. The VC's credit balance decreases by one whenever the upstream node forwards a data cell of that VC to the receiver node. The downstream node is eligible to return the VC's credits back to the upstream node through out-of-band credit update cells, after forwarding every *N2* data cell of the VC, where *N2* is a design or engineer parameter. Furthermore, from time to time the upstream node sends in-band credit check cells to the downstream node to synchronize the sender and receiver "sequence number" counters and protect the VC's credit balance against cell loss due to transmission errors.

In the *N23* scheme with a static buffer allocation, the downstream node allocates at least *N2 + N3* cells to the VC, where the value of N3 is chosen so that the flow control scheme does not itself prevent the VC from sustaining its peak targeted bandwidth on the link. If RTT is the round-trip delay expressed in terms of the number of cells, and B(VC) is the target peak bandwidth of the VC in percentage of the link bandwidth, $N3 = RTT \times B(VC)$.

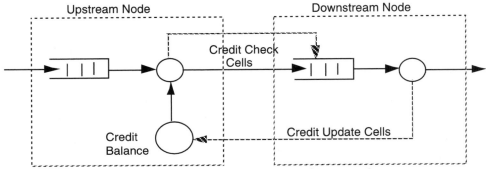

Figure 5-21: Credit-based congestion control.

5.9 Summary and Remarks

In this chapter, the technical issues and possible approaches with respect to traffic management for ATM networks have been discussed. The technology for making ATM switches is no longer in its infancy. However, how to control the network to prevent congestion and minimize effects when congestion occurs is still a major challenge for network engineers. Although several congestion control methods have been proposed, the problem of how to detect congestion still remains to be solved. The policing of flow enforcement may cause a controversial accounting problem, because the users may be reluctant to be charged for cells dropped during transmission in the network, even if they violate the negotiated Traffic Contract. Thus, how to charge users based on either cells injected from a source node or cells arrived at a destination node remains to be solved. These issues must be resolved in order to implement and operate the ATM control feature.

ITU-T, ANSI/T1, and the ATM Forum have been making significant progress on ATM Network Traffic Management standards. Some of the issues that are being addressed include the definition of a set of standard QoS classes, the objectives of the corresponding ATM Layer QoS parameters, and the required adjustments to the current definitions of the ATM Layer cell transfer performance parameters when some cells of an ATM connection do not conform to the negotiated Traffic Contract.

References

[1]   ITU-T Rec. I.371, "Traffic Control and Congestion Control in B-ISDN," March 1995.

[2]   The ATM Forum, *Traffic Management Specification Version 4.0,* April 1996.

[3]   Bellcore GR-477-CORE, "Network Traffic Management," Issue 2, December 1995.

[4]   ITU-T Recommendation I.350, "General Aspects of Quality of Service and Network Performance in Digital Networks, Including ISDNs," March 1993.

[5]   The Multimedia Communications Forum, "Multimedia Communications Quality of Service," ARCH/QoS/94-001, Rev. 2.1, June 1995.

[6]   D-H. Hsing, and F. Vakil, "A Discussion on the ATM Forum Flow Control Proposals," The ATM Forum/94-837, September 1994.

[7]   ITU-T Document, "Integrated Video Service (IVS) Baseline Document," Study Group 13, Geneva, March 1994.

[8]   Bellcore GR-2901-CORE, "Video Transport Over Asynchronous Transfer Mode (ATM) Generic Requirements," Issue 1, May 1995.

[9]   ITU-T Recommendation E.430, "Quality of Service Framework," 1993.

[10] *ATM Handbook*, Chapter 42, "ATM Layer Quality of Service (QoS)," (D-H. Hsing), McGraw-Hill Book Company, 1996.

[11] ITU-T Recommendation I.356, "B-ISDN ATM Layer Cell Transfer Performance," 1994.

[12] ANSI T1.511, "B-ISDN ATM Layer Cell Transfer Performance Parameters," 1994.

[13] G. M. Woodruff, and R. Kositpaiboon, "Multimedia Traffic Management Principles for Guaranteed ATM Network Performance," *IEEE J. Selected Area in Commun.,* Vol. 8, No. 3, pp. 437-446, March 1990.

[14] The ATM Forum, *ATM User-Network Interface (UNI) Signalling Specification Version 4.0,* April 1996.

[15] The ATM Forum, *Private Network-Network Interface Specification Version 1.0 (PNNI 1.0),* March 1996.

[16] The ATM Forum, *User-Network Interface (UNI) Specification Version 3.1,* September 1994.

[17] ITU-T Recommendation Q.2931, "B-ISDN DSS2 User-Network Interface (UNI) Layer 3 Specification for Basic Call/Connection Control," 1996.

[18] ITU-T Draft Recommendation Q.2764, "B-ISDN Signalling System No. 7 B-ISDN User Part (B-ISUP), Basic Call Procedures," September 1994.

[19] ITU-T Recommendation I.610, "B-ISDN Operation and Maintenance principles and Functions," 1993.

[20] R. O. Onvural, *Asynchronous Transfer Mode Networks: Performance Issues,* Artech House, 1994.

[21] R. Kawamura, H. Hadama, K. Sato, and I. Tokizawa, "Fast VP-Bandwidth Management with Distributed Control in ATM Networks," *IEICE Trans. Commun.,* Vol. E77-B, No. 1, pp. 5-14, January 1994.

[22] Bellcore GR-1248-CORE, "Generic Requirements for Operations of ATM Network Elements," Issue 3, August 1996.

[23] T-H. Wu, "Emerging Technologies for Fiber Network Survivability," *IEEE Commun. Mag.,* pp. 58-74, February 1995.

[24] ATM Forum/94-0438R2, "Closed-Loop Rate-Based Traffic Management," September 1994.

[25] ATM Forum/94-0632, "Credit-based FCVC Proposal for ATM Traffic Management," Revision 2, July 1994.

Chapter 6

ATM Protection Switching

6.1. Introduction

Network survivability has become increasingly critical to broadband network providers due to concerns that are intrinsically a part of any highly competitive business environment and the associated significant potential loss of uncollected operations fees. The potential loss of income is due primarily to significant network resources needed for each broadband service call, if the customer does not pay for it. For example, each Video-on-Demand (VoD) call may require a network bandwidth of 4 Mbps (using MPEG2 video compression) for a period of 2 hours. The cost and the required level of survivability are key factors that determine what kinds of network survivability systems should be implemented and deployed.

As defined in ITU-T Rec. I.311 [1], there are three types of network protection that would minimize the effects of broadband services when network failure occurs. These three network protection methods are protection switching; rerouting; and self-healing, where protection switching is performed by Network Elements automatically on a preplanned basis. Network rerouting is performed globally by the Network Management System; the self-healing method is performed dynamically without any involvement of the NMS. Among these three methods, network rerouting is generally considered as either centralized control automatic protection switching or self-healing; thus, network protection technologies developed for automatic protection switching or self-healing can also be used for network rerouting. Survivable network examples using the automatic protection switching method include Automatic Protection Switching (APS) systems and Self-Healing Rings (SHRs); survivable network examples using the self-healing method include self-healing mesh networks using SONET DCSs or ATM VPXs. This chapter will address APS systems and SHRs for SONET-based ATM networks. The self-healing mesh ATM network will be discussed in Chapter 7.

For SONET-based ATM networks, network protection methods for network survivability can be potentially implemented either at the SONET Layer, the ATM Layer, or a combination of both layers, depending on economics, required level of network survivability, and service requirements. As discussed in Chapter 2, most of the survivable SONET network architectures, including SONET APS and SONET SHRs, have been widely deployed by public carriers, private carriers (including utility companies), and CATV operators. The SONET APS is deployed for point-to-point systems of up to OC-192 rate available, and SONET SHRs are deployed at up to OC-48 line rate. These two survivable SONET network architectures may restore service within 50 ms, which will merely cause reframing for Plain Old Telephone Service (POTS). Because SONET APS and SHRs are popular among network providers, two crucial questions involving a need for additional ATM Layer protection and the interworking schemes between SONET Layer protection and ATM Layer protection must be raised and answered, if the ATM

Layer is to be considered for network protection. Please note that throughout this chapter, unless specified otherwise, the ATM network is referred to as the SONET-based ATM network.

The purpose of this chapter (and Chapter 7) is to provide technical insights that will help network planners and researchers develop their own cost-effective multilayer network protection designs. This chapter discusses limitations of SONET layer protection on emerging SONET/ATM networks, and it discusses APS and SHR system designs using both SONET and ATM technology. Some of these emerging system architectures have been implemented as commercial products, but some are still undergoing standardization. The standardization process and deployment status will also be reviewed in this chapter.

6.2 ATM Role in Broadband Network Survivability

6.2.1 The Changing Telecommunications Service Paradigm

SONET networks were originally designed to support private line services and provide a high-speed transport mechanism for switched services. The network connections supporting these services are either semipermanent or permanent and are established and torn down through the service provisioning process. The network works well for the services already mentioned, because the necessary functions do not require dynamic bandwidth control capability. However, due to the introduction of new broadband services such as Frame Relay Services, SMDS, and Cell Relay Services that introduce bursty traffic, the SONET network infrastructure may be too inefficient to accommodate these bandwidth-on-demand requirements. For example, suppose an SMDS subscriber has an Ethernet that runs at 10 Mbps. The SONET network has to dedicate an STS-1 path (51.84 Mbps) to that subscriber. Also, typically, the average Ethernet utilization is approximately 20 percent; thus, the STS-1 path utilization is only approximately 4 percent of link utilization. This example represents an emerging and changing telecommunication networking environment, as depicted in Figure 6-1. ATM transport technology may alleviate the problem of bandwidth inefficiency, as discussed in Chapter 3.

Currently, SONET network protection is provided by the simple and cost-effective self-healing capability at the SONET Layer. One example is the SONET Self-Healing Ring. The question here is, when the transport network evolves to the ATM-based transport network to accommodate bandwidth-on-demand requirements, will the SONET Layer be able to provide protection for new broadband services that may have very different delay and cell loss requirements?

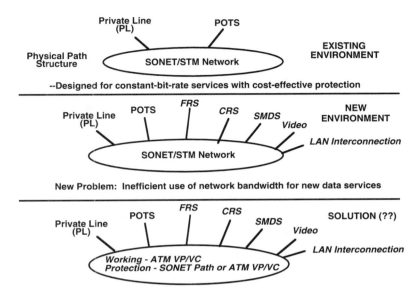

Figure 6-1: Changing service environment and assumptions.

For SONET-based ATM networks, network restoration can be performed at the Physical (SONET) Layer, the ATM Layer, or a combination of both. For Physical Layer protection, service restoration can be performed at the STS-3Nc level (where $N = 1, 4,$ or 16) or at the physical transmission link level. For ATM Layer protection, network restoration can be performed at the VC or VP (or group of VPs) level. Service restoration at the VC level may be more flexible, but slower and more expensive, than that performed at the VP level due to greater VC-based ATM network complexity. The restoration system complexity at the VP level is similar to that at the SONET STS-1/STS-3c level. Determining which layer (SONET, VP, or VC) is an appropriate restoration layer depends on the SONET-based ATM transport network architecture (see Chapter 3) and other factors such as costs and QoS.

6.2.2 Impact of ATM Technology on Broadband Network Survivability

The major differences between STM and ATM network restoration technologies are in the restoration unit and the network reconfiguration technology. SONET networks restore traffic on physical paths (e.g., STS-3c) and/or links (e.g., SONET links) through the physical network reconfiguration using Time Slot Interchange (TSI) technology. In contrast, ATM networks restore traffic carried on VPs/VCs through a logical network reconfiguration by modifying the VPI/VCI routing table. Figure 6-2 depicts these differences.

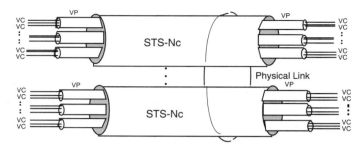

Protection switching technology	SONET/SDH Protection Switching	ATM Protection Switching
Units	STS-1/STS-Nc/Link	ATM cells accommodated on VCs/VPs
Method	Time Slot Interchange (TSI)	Routing table update
Characteristic	Physical protection switching	Logical protection switching

Figure 6-2: Differences between ATM and STM protection switching technologies.

The impact of emerging ATM technology on network restoration (as compared with SONET network restoration) may be significant due to the following factors [2–4]:

- Nonhierarchical path multiplexing (for VPs) that would simplify the survivable network design and help reduce intranode processing delays and required spare capacity

- Separation of capacity allocation and physical route assignment for VPs/VCs that would reduce required spare capacity (e.g., the capacity of protection VPs/VCs can be zero in normal conditions)

- More OAM bandwidth with allocation on demand that would reduce delays for restoration message exchanges, and much quicker detection of system degradation (i.e., soft failure) in the ATM network than in its SONET counterpart

One major difference between SONET and ATM/VP transport systems is in path structure. SONET transport uses a path structure that tightly links the physical connection and its capacity through Time Division Multiplexing (TDM) frames and their physical interfaces, and ATM/VP transport uses a logical path structure within which connections are linked to physical interfaces; however, the connection capacity can be varied depending on applications. This path structure may simplify ATM Virtual Path (VP) design and consequently increase link utilization, as compared with STM transport [2].

Unlike SONET transport, for which the capacity assignment is tied to the routing assignment, an ATM/VP route is established by setting the routing table at VP connection points between VP connection ends, and that VP capacity is not explicitly assigned at the VP connection points at VP establishment due to its logical path structure. The capacity assignment of ATM/VP transport is handled by separate management procedures such as connection admission control and usage monitoring, which are carried out at ingress VP connection endpoints. Thus, intermediate VP connection points on the VP route perform no processing for VP capacity management and so are not affected by changes in the VP capacity allocation. The independence of the route assignment from the capacity allocation makes some ATM Layer protection schemes more efficient and flexible than their STM counterparts, but this might be at the expense of network restoration speed.

The logical path structure and the separation of path and bandwidth assignments of ATM transport also result in a lower spare capacity requirement than for its SONET counterpart. Figure 6-3 shows that the required spare capacity at the ATM Layer is generally less than that at the SONET Layer. The amount of spare capacity required for the ATM network may be less than that for its SONET/STM counterpart due to its nonhierarchical path structure. For example, as shown in the figure for the STM network, working traffic is carried on two different STS-3c's with 50 percent utilization that would require one STS-3c reserved for protection, providing 1:2 path protection for this link. For the ATM/VP network, the required spare capacity carried on VPs can be engineered as just the capacity of half an STS-3c. The savings of spare capacity for the ATM/VP network over the STM network increase when reserved sources are shared by more working connections, such as those in the self-healing mesh network.

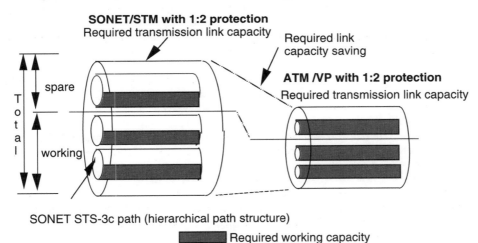

Figure 6-3
Figure 6-3: Spare capacity comparison between SONET and ATM Layer protection.

In terms of OAM capability, the transport overhead of the SONET transport network used to trigger SONET Layer protection switching is generated and transmitted every 125 ms. ATM transport allows for many OAM cells to be generated and transmitted within a 125 ms interval as long as the capacity is available. Unlike the STM transport's OAM fixed capacity, the ATM Layer OAM capacity can be assigned dynamically based on the particular maintenance activity or procedure invoked. Thus, ATM transport would be able to convey necessary network protection switching messages faster than its STM counterpart.

In addition to technology impact, a change in transport equipment cost model could change the criteria that are used to select appropriate survivable network architectures such as those described in Chapter 2. For example, in SONET transport networks, the cost of a SONET W-DCS is much higher than the cost of a SONET B-DCS, and the cost of a B-DCS is much higher than a SONET ADM. However, in ATM transport networks, the cost of a Virtual Channel Cross-Connect System (VCX) is higher than that of a VPX, but this cost difference is much less than for their SONET counterparts. A similar cost differential applies to the VPX and ATM/ADM. The implications of these cost model shifts on survivable network architecture selection remain to be studied.

6.2.3 Multilayer Survivable SONET-Based ATM Network Architectures

The ATM network protection layer depends on whether the ATM VC or VP transport network is considered, as depicted in Figure 6-4. The ATM VC transport network involves a large number of smaller ATM switches that terminate at VPs and VCs. The transmission path between two adjacent ATM switches is a SONET/STS path. The alternative transport system is the VP transport network that deploys a smaller number of larger ATM switches in strategic locations and uses less-expensive ATM VPX/ADMs to bring remote customer traffic to the strategically located ATM switch for switching and processing. The latter approach is sometimes referred to as the "Virtual CO" approach. It has been implemented in the present SONET network infrastructure by many LECs. For each type of ATM transport network, three survivable network architectures used in today's SONET networks (i.e., point-to-point systems, Self-Healing Rings, and mesh self-healing networks) may also be applicable to ATM transport networks. In this chapter, we will discuss the architecture and system designs for APS and SHRs in Sections 6.3 and 6.4, respectively. Self-healing schemes for ATM transport networks will be discussed in Chapter 7.

6.3 Key Design Issues in ATM Protection Switching

Protection switching at the ATM Layer is currently under study in a working group of ITU-T Study Group 13 and ANSI T1S1.5. Open technical issues associated with ATM protection switching include

- Protection switching communications mechanism (OAM cells, or ATM connections using AALs)
- Method of specifying a protected route (fragment or end-to-end)
- Unidirectional or bidirectional protection switching for point-to-point and point-to-multipoint systems (e.g., multicast video)
- 1:N, 1:1, and 1+1 protection switching
- Interlayer management for protection switching triggering (SONET to ATM Layer; including the protection control messages exchange protocol and its protection scheme)
- Protection switching system implementation and analysis

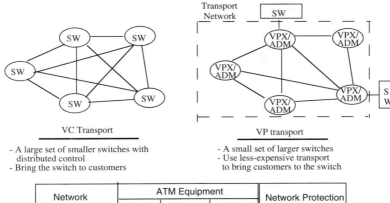

| Network | ATM Equipment | | | Network Protection |
Configuration	TM	ADM	VPX	Mechanism
Point-to-Point	✓	✓	✓	APS
Ring		✓	✓	SHR
Mesh			✓	SHN

VPX = Virtual Path Cross-Connect System
APS = Automatic Protection Switching System
SHR = Self-Healing Ring
SHN = Self-Healing Network (mesh)

Figure 6-4: ATM Layer protection options.

In addition to the foregoing system engineering issues, other application feasibility issues must be addressed before any practical ATM protection switching system can be realized. One of these issues involves how ATM protection switching can best interwork with existing SONET APS. The study of these application feasibility issues would help identify appropriate application areas for SONET/APS or ATM protection switching systems.

6.3.1 Protection Switching Communications Mechanism

There are two different design approaches for ATM protection switching systems: one is based on network management (i.e., fault management) capability; the other is based on signaling capability. The former design, currently under study in ITU-T, is independent of the routing design for working system, whereas the latter design uses the existing routing capability to implement the protection switching function. For the latter design, for example, B-ISUP signaling messages defined by ITU-T and PNNI signaling messages defined by the ATM Forum [5] can be used to trigger the ATM protection switching function [6,7]. Compared with the former design (i.e., using network management capability), the latter design may minimize development costs but may only be applicable to some particular networks using the same signaling messaging system. We discuss ATM/APS based on the network management capability in Section 6.4.2; the design based on the SVC capability will be discussed in Section 6.4.3.

The OAM cell has been suggested by some ITU-T contributors as a way to convey control messages for protection switching. However, it is also possible to use user cells with AAL Type 5 to transport control messages for protection switching, using a dedicated connection between switches. AAL Type 5 has been recommended by ITU-T as the format to carry B-ISDN signaling messages; it has also been considered for carrying restoration control messages for ATM self-healing networks [8]. Whether OAM cells or user cells with AAL Type 5 should be used for protection switching remains to be determined, and this choice depends on the definition of "fragment" for protection switching (which will be discussed in Section 6.3.2) and the delay requirement.

6.3.2 Protection Switching Fragmentation

ATM protection switching can be performed either at the VP or VC layer on a fragment or end-to-end connection basis, where the fragment is a portion of the end-to-end connection. Due to simplicity and protection switching time considerations, ATM protection switching systems at the VP layer on a fragment basis are likely to be implemented first. Thus, the following discussions are based on the VP protection switching on a fragment basis. However, these discussions can also apply to VC protection switching. Although it was suggested in the ITU-T SG13 working group that protection switching at the fragment level together with a fast protection switching process (<<1 second) may be a relevant measure for protection at the ATM Layer, these are issues needing further study.

As already mentioned, a fragment is a portion of the end-to-end connection. However, the definition of "fragment" for protection switching is not yet standardized and is under study in ITU-T SG13. There are three possible "fragment" definitions [9]. The first definition (which we will call the "OAM Definition") is consistent with the definition of ATM/OAM segments [10], as depicted in Figure 6-5(a). In this case, the endpoints of the segment generate and terminate OAM cells, and intermediate points of the segment pass through and copy OAM cells without modifying their contents. Using the OAM Definition for protection switching, the protection switching fragment (PS fragment, also called the protected fragment) is a concatenation of one or more VP links, where a VP link is a logical link between two adjacent VP Cross-Connect Systems (VPXs) or VP Add–Drop Multiplexers (ADMs). Segments may not overlap, as shown in the definition of OAM segment [see Figure 6-5(a)]. For this definition, the use of OAM cells for protection switching messaging may be a reasonable choice.

The second definition for the protection switching fragment (called the "Line Definition") is similar to the definition used in SONET line protection switching (see Chapter 2). In this definition, a protection switching fragment is defined as a single VP link as shown in Figure 6-5(b), when the VP layer is used for protection switching. In this case, the endpoints of the protection switching fragment will not, in general, correspond with the endpoints of the OAM segment. With this definition, the use of connections with AAL Type 5 for messaging may be a reasonable choice.

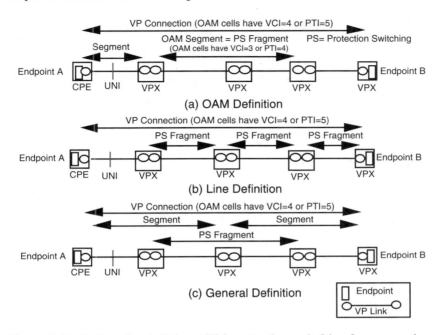

Figure 6-5: Options for defining ATM protection switching fragmentation.

In the third definition (called "General Definition"), shown in Figure 6-5(c), PS fragments can overlap OAM segments. We can consider the PS fragment in the OAM and Line definitions as special cases of the general PS fragment, which can be defined between any nodes and may allow network designers to choose any portion to be protected based on their network requirements. Note that the ATM protection switching procedure may enable services to recover from node failures when the OAM-like and general protected fragments are used.

When the OAM Definition is used, if the failure does not occur in the OAM segment endpoints, the detecting node should generate a protection switching message. It may use OAM cells, or a special connection using AAL Type 5, to notify the downstream OAM segment endpoint, which then generates protection switching messages toward the other end of the OAM segment to trigger protection switching, using either OAM cells or AAL-5 packets to disseminate the protection switching message. In this approach, VPs in an unaffected link would also be switched if that unaffected link is included in the OAM segment.

For the line approach, the node detecting the fault [e.g., Loss of Signal (LOS)] can generate the protection switching message/alarm using AAL-5 and send it to the other end (upstream) of the VP link to trigger protection switching. In this approach, using AAL-5 is required for protection switching messages, because these two VPX nodes do not necessarily terminate (i.e., extract) OAM cells (because they may not be at endpoints of OAM segments), which may trigger undesired actions downstream.

The AAL-5 format may also be used for the general approach, because the VPX node that is responsible for triggering protection switching may not be able to terminate OAM cells. Table 6-1 summarizes and compares approaches for defining ATM PS fragments [9].

Table 6-1: Comparison of approaches for defining ATM PS fragments.

System Factors	ATM Protected Fragment		
	OAM approach	Line approach	General approach
Protection switching unit	per VP	Line/VP	per VP
Spare capacity needed	Moderate	Most	Less
Engineering complexity	Moderate	Least	Complex
Node failure protection	Possible	No	Possible
Network design flexibility	Less	Less	Most
Protection switching cell	OAM/AAL 5	AAL 5	AAL 5

6.3.3 Multilayer Network Architecture Model and Interworking

Each survivable SONET-based ATM network architecture (see Figure 6-4) uses a multilayer model including SONET and ATM layers. In general, ATM Layer protection may require less spare capacity than SONET Layer protection at the expense of a slower restoration time and a more complex control system. How best to combine these two layer protection schemes in the same network is a challenging task for network planners and engineers who are seeking cost-effective solutions for SONET-based ATM network protection.

Figure 6-6 depicts a multilayer network survivability model for ATM transport networks [3]. This figure also shows an interworking relationship between survivable transport systems at each layer and the network management system across the layers. The network protection systems at each layer are managed and triggered by the layer management system at each layer; the process is then coordinated by the plane management system. The layer management system includes fault management for the "hard" network failures and performance management for "soft" failures (e.g., performance degradation). The fault and performance management systems at each layer are used to trigger either protection switching at the Optical Layer, the SONET Layer, or the ATM Layer or the rerouting scheme at the Network Layer that is service- and/or application-specific. The function of system (or plane) management is to collect the failure information, interpret the failure messages, identify and isolate the failure location(s), coordinate the timing of generating next higher layer AIS messages, and take necessary actions to recover from failures.

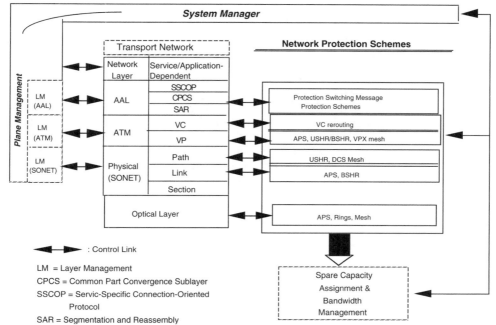

Figure 6-6: A multilayer network survivability model [3].

The following gives an example that shows the role of system management in network protection across layers. When a cable is cut, the SONET Line Terminating Equipment receives Loss of Signal (LOS). Here, there are two options: it starts to perform line protection switching (if any), or immediately generates the Path AIS and sends it to the downstream Path Terminating Equipment. When the PTE receives the Path AIS, it could initiate SONET path layer restoration (if any), or immediately generate the VP-AIS. Again, when the Virtual Path Terminating (VPT) equipment receives the VP-AIS, it could immediately trigger VP layer restoration and not generate the next higher layer AIS (i.e., VC-AIS) until it completely restores affected signals or a predetermined time-out period expires. Alternatively, it would immediately generate the VC-AIS and pass it to the equipment terminating the Network Layer function (in this case, STP and SCP if the application is for signaling). At the Network Layer, the system could choose to initiate network rerouting (if any) or simply do nothing. At this point, there is no standard to guide the timing for generating the next higher level AIS messages from the SONET Physical Layer to the Network Layer. However, Bellcore requirements [10] and T1S1.5 agreements in response to a Bellcore contribution have clarified that the VC-AIS is not sent until Physical Layer protection switching is given a chance to clear up the problem. Note that this scenario is based on an assumption that SONET/APS and SONET Self-Healing Rings would be universal protection systems for SONET/ATM networks. In reality, SONET/APS and SONET SHRs may not be applicable to DCS-based mesh networks. The DCS-based mesh network may use some self-configurable algorithms based on resource sharing to restore services [3,11]. Therefore, the waiting period of 50 ms before generating ATM layered alarms would become an unnecessary provision for

DCS-based mesh networks, which typically carry most demands in the core of LEC networks. In ITU-T, the working group responsible for ATM protection switching agreed at a meeting (March 1994) that the ATM Layer's AISs should be generated when needed without a waiting period of 50 ms. Thus, an alignment in this timing issue between ITU-T and T1S1 is needed.

If no interlayer coordination function exists (e.g., no control applied to timing for generating the next higher layer AIS messages), the AIS will be passed through each layer after it is generated, and that will trigger the protection switching or self-healing scheme at each layer (SONET, ATM, and the Network Layer) simultaneously. This may create a situation for network resource competition that causes network congestion or even network failures. Thus, some interlayer coordination function is needed to ensure that either the Network or ATM Layer performance remains at the acceptable level.

Thus, according to discussion of the preceding example, the main issues related to multilayer interworking (sometimes called "escalation" [12]) are defining in which layer the restoration process should start, when it should be escalated to another layer, and to which layer it should be escalated. Additional issues to be addressed include defining the relationship between the escalation strategy and the Network Management System, and the form of signaling to be used in the escalation mechanism.

The set of rules used to decide which mechanisms to activate, and when to halt mechanisms and to activate others, is called the escalation strategy. Two types of escalation strategies can be identified: activation of multiple restoration mechanisms in parallel, and sequential activation of restoration mechanisms.

In parallel strategies, different restoration mechanisms are activated at the same time to handle a single failure event. When one mechanism succeeds in restoring the failure, all activities are stopped. Although this achieves the fastest result, the individual mechanisms must be carefully coordinated so as not to obstruct each other or to compete for the same resources.

Sequential mechanisms lead to longer overall restoration times, but they are more easily managed. Individual mechanisms can then be optimized without risking contention problems. A sequential escalation strategy determines the order of activation of the mechanisms and coordinates between the mechanisms. There are two variables in sequential escalation. The first is the order in which the mechanisms are activated (i.e., bottom-up, top-down, or variable). The second variable involves those criteria used to decide when to escalate (i.e., time-out or diagnostics).

6.4 APS for SONET-Based ATM Networks

According to ITU Rec. I.311 [1], protection switching is defined as the establishment of a preassigned replacement connection using equipment but no network management control. For ATM services, protection can be performed either at the SONET Layer or at the ATM Layer. Examples of SONET/ATM protection switching systems have been proposed in [6,7,9,13–15]. In this section, we will discuss SONET Layer protection first and then discuss ATM Layer protection for point-to-point broadband network connections.

6.4.1 SONET Layer Protection for ATM Services

For the SONET-based ATM network, it is possible to use standardized SONET APS protocol to perform protection switching for ATM cells. However, the protection switching time of 50 ms may hit some ATM cells. The impact of this protection switching time on the service QoS depends on the service being supported. For example, real-time multimedia services may not tolerate ATM cell loss due to the 50 ms "hit" time. To avoid any ATM cell being lost due to protection switching, an ATM cell alignment is needed. In ATM networks, line protection switching is hitless if no ATM cells are dropped or duplicated upon switching. Figure 6-7 depicts an example of a hitless ATM protection switching system proposed by NTT [13]. This hitless protection switching system uses the SONET K1/K2 protocol to control the protection switching activity and uses an ATM cell alignment method to ensure that no cells are dropped when protection switching is performed. This specific ATM cell alignment method is also shown in the figure; the alignment is established and continuously maintained by a bit-by-bit comparison of user cells received from side 0 and side 1 (see Figure 6-7). This method does not use special OAM cells. Because alignment is continuously maintained, switching upon receiving Forced Switch (FS) or Reverse Request (RR) is always hitless.

6.4.2 ATM Layer Protection Switching Using OAM Cells or User AALs

This section discusses ATM Layer protection switching system design using OAM cells or user AALs (e.g., AAL-5). Such an ATM protection system can be performed at the VP or VC layer. For control system simplicity, VP protection has been proposed by major telecommunications equipment suppliers and service providers. For ATM Layer protection switching, ITU-T SG13 and T1S1 have initiated efforts to study its scheme and interworking with today's SONET networks.

6.4.2.1 ATM Protection Switching Mechanisms

This section describes briefly some general cases for ATM/VP protection switching that have been proposed in T1S1.5 94/94026 [14]. Possible system implementations will be discussed in Section 6.4.2.2. The proposed ATM/VP protection switching mechanism uses the following principles:

- Fault location needs to be specified in the protection switching cells sent by nodes in a protected route.
- A node that is a predefined protection endpoint must monitor every Protection Switching Cell it receives, and upon receiving a Protection Switching Cell with a fault location within the protected route, the node will immediately switch from the working VP (in the transmitting side) to the protection VP. It should be determined that the signal failure is within a protected route.
- A node that is a predefined protection endpoint will immediately switch the working VP to the protection VP upon detecting LOS, Loss of Frame (LOF), or Loss of Pointer (LOP).
- The predefined protection endpoint will send protection switching cells to the other predefined protection node whenever it receives LOS, LOF, LOP, or a Protection Switching Cell.

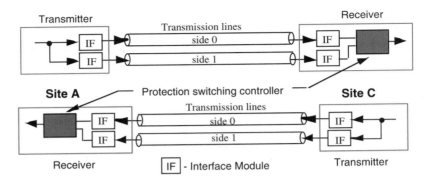

(a) 1+1 Line Protection Switching Configuration

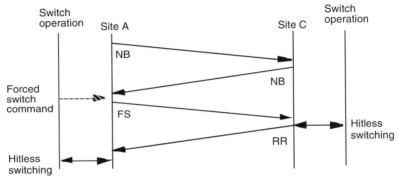

NB = No Bridge required, FS = Forced Switch, RR = Reverse Request

Note: Side 0 and side 1 are aligned by a bit by bit
comparison of received user cells.

(b) Protection Switching Control Sequence

Figure 6-7: Hitless protection switching for ATM services [13] (©IEEE 1993).

Figure 6-8 depicts an ATM 1:1 Protection Switching Cell transmission and restoration procedure [9,14]. The protection route is shown in the figure. When a link fails in one direction of the working link, Node D receives an LOS, LOF, or LOP signal in the case in which there is no intermediate node, or an OAM cell when there is an intermediate node X. If Node D determines that the fault is located within the protected route, it then switches the working VPs to the protection route and sends a Protection Switching Cell on the protection route to initiate protection switching at Node C. Upon receipt of the cell, Node C immediately switches the working VPs to the protection VPs. After the switchover at Node C, this node transmits data cells from Node A to Node D via the protection route. When Node D receives valid data cells, this node recognizes that the switch to the protection route has succeeded and stops sending Protection Switching Cells. User data cell flow in the B to A direction does not use the protection route in this case.

In the case of failures in both directions, for example in the side, Nodes C and D both receive a fault indication and send each other Protection Switching Cells along the protection route after each node determines that the failure is located within the protected route. Each node (C and D) immediately switches the working VPs to the protection VPs when it receives the Protection Switching Cell. After switching over at Nodes C and D, each node transmits data cells over the protection route. Then each node receives valid data cells, recognizes the switch to the protection route has succeeded, and stops sending Protection Switching Cells.

Figure 6-8(b) depicts a state diagram and a message flow model for the ATM protection switching procedure specified in Figure 6-8(a); this diagram describes an example of Protection Switching Cell transmission and restoration.

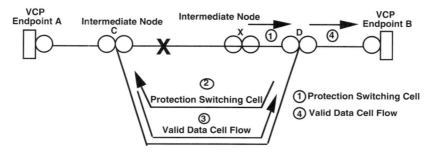

(a) 1:1 Protection Switching Procedure

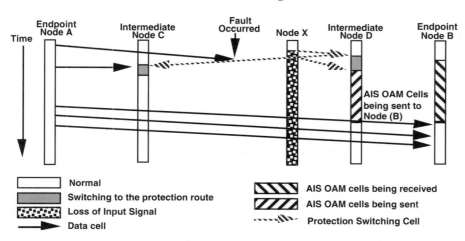

(b) State Diagram and Message Flow Model

Figure 6-8: 1:1 ATM protection switching procedure and state diagram [14].

For the 1:N protection switching scheme, switching from the working route to the protected route is done using the Protection Switching Cell with a failed link number.

However, further study is needed to specify the details of the protocol, especially for $1:N$ protected VPs.

Another ATM Layer protection switching algorithm for end-to-end restoration was reported in [16]. This algorithm preassigns the backup route and the backup VP for each target VP. The general procedures for this end-to-end restoration algorithm are summarized as follows and depicted in Figure 6-9. Figure 6-9(a) depicts a network configuration during normal conditions; Figures 6-9(b) and (c) depict restoration concepts during failure conditions.

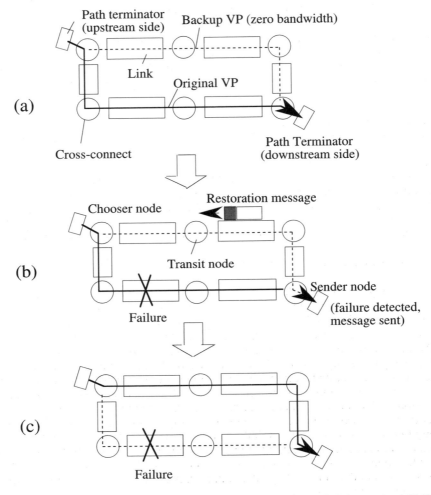

Figure 6-9: Algorithm for an end-to-end protection switching system [16]
(©1995 IEEE).

The restoration algorithm depicted in Figure 6-9 consists of three phases:

Phase 1 (Bandwidth capturing): When a VP failure occurs [Figure 6-9(b)], the Sender, the downstream node of the restoration pair, detects the failure and sends a restoration message along the backup VP. The Sender then switches from the failed VP to the backup VP. Each node along the backup route receives the restoration message, captures the appropriate bandwidth on the link and retransmits the message to the next node.

Phase 2 (Switching): When the Chooser, the upstream node of the restoration pair, receives the restoration message, it switches traffic from the failed VP to the backup VP, thereby completing failure restoration for this VP [Figure 6-9(c)].

Phase 3 (Dynamic route search): This phase is not normally needed, because backup VP routes and resources are designed for complete restoration. However, if a backup VP is not available due to multiple failures or for another reason, a new restoration route must be found dynamically using the control message transfer mechanism described in Section 7.3.

The achievement of restoration between path terminators gives rise to two benefits. One is the ability to recover from node failure. The line restoration scheme cannot generally perform node-failure restoration because node failure causes the loss of one of the restoration node pairs. Preplanned end-to-end restoration can recover from node failure, except for terminator node failure, by using the same process as that used for link failure. For terminator node failure, another restoration scheme, such as the dual-homing architecture, can also be used. The second benefit is a reduction in the required network resources, because the restoration route can be established more flexibly than with line restoration. This is an important factor in the design of spare capacity for alternate VPs.

6.4.2.2 Implementation Example of ATM Protection Switching Mechanisms

Figure 6-10 depicts an example block diagram of 1+1 ATM/VP APS that can switch every VP at the ATM Layer [9]. At the transmitter side, the functional blocks are the same as those used in SONET/APS. The duplicated SONET frames carrying ATM cells are transmitted. At the receiver side, there are two Optical-to-Electrical (O/E) converters and Overhead (OH) blocks that process the overhead in the SONET frame. The pure ATM cell signal is input to the Header Check block after the SONET overhead is processed. At the Header Check block, cell delineation is established and HEC bits in each cell are also checked so the error rate can be measured and any Out of Cell Delineation (OCD) anomaly can be detected. The Protection Switching Cell Detect block checks for the Protection Switching Cell that can trigger switching of any VP from the working line to the reserved line. At VPI Conversion 1 block, VPI values in the reserved line are converted so that the data cells after the ATM Multiplexer (MUX) block can be easily identified [i.e., which cells belong to which line (Working or Reserved)]. VPI values are converted according to predetermined rules, that is, converted VPI values may be equipment-proprietary values.

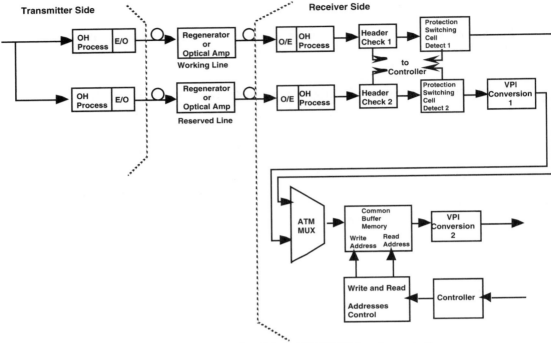

Figure 6-10: An example of 1+1 ATM/APS implementation.

At the ATM MUX block, the data cells in the working line and the reserved line are multiplexed by using ATM cell interleaving. In the case of multiplexing VBR cells, there is a possibility that some ATM cells could be discarded. The data cell exiting from the ATM MUX block is input to the Common Buffer Memory, where the order of the reading addresses is changed so that switching from the working to the reserved or from the reserved to the working line can be done without disruption of any cells. At VPI Conversion 2 block, converted VPIs are restored to the original values.

In the case of forced (i.e., manual) switching triggered by the Protection Switching Cell, the OAM cell or connection using AAL-5 may carry protection switching messages, depending on the definition of protected fragment being used. In this case, the reading cell speed may need to be controlled by leaving an idle cell (that is put in the Common Buffer Memory beforehand) as a space.

It is necessary to guard Protection Switching cells from error. The switching can certainly be done after receiving several consecutive Protection Switching Cells. One must consider the possibility that VPI values are not always converted because of value limits. However, in the present case we assume there are enough VPI values available, and we presume all VPI values can be converted regularly. The detailed timing diagram showing the process of hitless protection switching executed by the foregoing system implementation can be found in [9].

As depicted in Figure 6-10, when the cell just before a switching cell in the working line is read out from the Buffer, the Control block switches the cell stream that has the cell immediately after switching a cell in the reserved line. At this time, "hitless" switching is executed at every VP, and after this the cell information in the Buffer for the reserved line continues to be read out. In case the switching is to the reverse side, a similar procedure is followed. In unidirectional switching, it is necessary to give notice to the transmitter side that the VP switching has succeeded and which VP is on which line. The reverse switch to the working line after the network failure ends can also be done with the same procedure.

When a protection switch is initiated by a burst error in the working line, at Header Check blocks 1 and 2, HEC bits in both lines are always checked, and some data cells are output from this block when either the error rate exceeds the provisioned value or HEC anomalies are detected. Here we assume that seven consecutive HEC violations are detected and the state in the Cell Delineation is moved to the OCD Anomaly state. We can use this state as a switching trigger and switch the whole working line to the reserved line without any hit if we anticipate the phase difference between the working line and the reserved line and control the reading address from the Common Buffer Memory.

When Signal Degradation (SD) occurs, we can selectively switch VPs from the working line to the reserved line at the provisioned error rate. Suppose that the working line is carrying two VPs, one with voice traffic and the other with HDTV traffic. It is likely that HDTV applications cannot tolerate an error rate that may be acceptable for voice services. In this case we could only switch the VP carrying HDTV traffic. This is more useful in the protected network system of 1:N or M:N ($M < N$) where a more effective resource management scheme is needed. The nodal implementation described in Figure 6-10 for 1+1 protection switching can also be applied to 1:1 protection switching, except that the protection switching capability is needed at both the transmitting and receiving nodes. In addition, the proposed system implementation is applicable to the bidirectional scheme with the SONET K1, K2–like protocol.

The implementation of 1:1 ATM APS depicted in Figure 6-9 can be extended to a 1:N ATM APS that can switch every VP at the ATM Layer. In this design extension, at the transmitter side the functional blocks are the same as the usual blocks of the SONET Physical Layer, but the reserved line is made up from selectively multiplexed VPs by ATM interleaving in the working lines. The switching scheme is almost the same as a 1+1 scheme, but the priority control at the receiver side is also necessary (just as for the K1, K2 protocol used in SONET 1:N APS). In this case, there are several issues to be resolved including resource management of the reserved line, the abolition of cells in the ATM MUX block at the receiver side, and the optimal design of Common Buffer Memory [9].

6.4.3 ATM Layer Protection Switching Using SVC Signaling Capability

This section describes a restoration system for ATM networks that is built on top of a preplanned hop-by-hop routing system in the network and operates on a connection-by-connection basis, as described in [6]. A preplanned ATM protection switching method based on PNNI source routing can be found in Reference [7]. Given that ATM switches are being deployed in the network to support switched broadband services, the additional development cost and operations overhead associated with this restoration system are relatively insignificant.

The restoration system discussed in [6] consists of three key components: monitoring, detection, and recovery mechanisms. We will describe each of them briefly in the following.

6.4.3.1 The Monitoring and Detection Mechanisms

In the considered restoration system, the status of the signaling connection is used to derive the corresponding link status or neighboring ATM node status. At start-up, the assured mode data transfer on signaling connections is established between any two neighboring ATM switches across each interface connecting them. The Signaling ATM Adaptation Layer (SAAL) [17–20] polls the peer receiver at a specific time interval. The length of this time interval depends on the signaling connection's current activity level (actively sent, idle, etc.). A smaller polling time interval is used when the signaling connection is actively sending messages. If a response is not received within a predetermined time, the signaling connection is considered broken. The maximum time for an SAAL to detect a broken signaling connection depends on the values selected for the various protocol timers. After the SAAL detects a broken signaling connection (due to either link failures or node failures), it will notify the SAAL management, which triggers the recovery procedures via the local system management. If a link failure is detected locally, the recovery procedures will be triggered via local system management. Upon recovery from the failure, the SAAL management will trigger the start-up procedures to reestablish the assured mode data transfer on the signaling connections across the recovered interfaces.

6.4.3.2 The Recovery Mechanism

The recovery mechanism is built on top of the capabilities of a preplanned hop-by-hop routing scheme implemented in the switches. Each ATM switch selects, from a set of preplanned routes, the next ATM switch and the corresponding link to be used for the connection to reach the called party at the connection setup time. The same link is used to forward the setup message on the signaling connection to establish a user connection. The Connection Admission Control procedure examines the selected link to determine if it has sufficient bandwidth to accommodate the new call request. If there is not enough bandwidth available on the selected link, alternative routes will be examined. The call will be blocked if no preplanned alternative route can be found at any one of the traversed switches. The threshold used to determine whether there is sufficient bandwidth to accommodate a new connection should take into account the service quality to be delivered. At connection setup time, alternative routes will be used if congestion is detected on the selected route. However, after connections have been established, they will not be rerouted due to temporary traffic fluctuation.

In the remainder of this section, an ATM switch is called an upstream switch for a connection from a failure point if it is on the side toward the source switch (close to the calling party) of the connection. An ATM switch is called a downstream switch for a connection from a failure point if it is on the side toward the destination switch (close to the called party) of the connection. Figure 6-11 depicts examples of upstream and downstream switches.

From a given failure point, the same ATM switch can be an upstream switch for one connection and a downstream switch for another connection. Recovery procedures are activated at the two neighboring switches after a failure occurs.

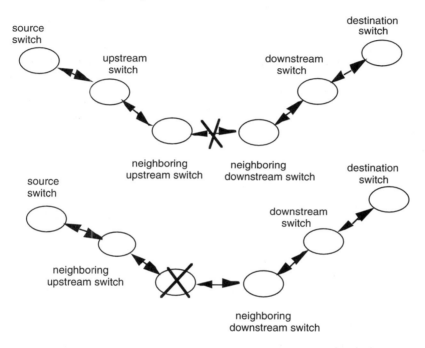

Figure 6-11: Upstream and downstream switches.

In the following, we describe a recovery procedure from the neighboring upstream switch that has been discussed in [6]. The recovery procedure from the downstream switch can be found in [6].

The neighboring upstream switches initiate the process of finding the alternative routes for the affected connections. When the recovery procedure is activated at the neighboring upstream switch, the affected connections are redirected to one of the preplanned alternative routes on a connection-by-connection basis. The Connection Admission Control (CAC) procedure implemented at each ATM switch is used to determine whether a reroute request for an affected user connection can be accepted on the selected alternative route. The CAC procedure may use a threshold value different from those values used for a new connection request to determine whether to accept a reroute request. If a reroute request cannot be accepted (due to the lack of either a route or the bandwidth), a reroute reject message is sent back to the corresponding source switch to release the connection and the associated network resources (e.g., the bandwidth, the VPI/VCI value, and the corresponding call record stored at the switch).

For each accepted reroute request, a reroute setup message is forwarded by the neighboring upstream switch toward the corresponding destination switch on the selected alternative route. At each switch, when a reroute setup message is received, the affected call identifier is extracted; the detailed actions that will take place can be found in [6].

6.4.3.3 Requirements for ATM Switches

The discussed restoration system uses the status of the signaling connection on each interface to detect a link failure or a neighboring node failure. There is no further requirement for an ATM switch with signaling to support the monitoring and detection mechanisms of the proposed restoration system.

The recovery mechanism is built on top of the capabilities of a preplanned hop-by-hop routing scheme implemented in the switches. Additional requirements for these switches to support the recovery mechanism of the proposed restoration system are summarized as follows:

- Local communications within a switch between the SAAL management, the system management, and the call control process are needed.
- To trigger the recovery procedures when a signaling connection transitions into unassured mode data transfer; to access the routing information and the resource allocation algorithm used by the call control process to reestablish affected connections.
- Additional communications between the neighboring switches are needed to release resources associated with affected existing connections along the no-longer-used segment of the original route due to rerouting.
- The recovery procedures described in Section 6.4.3.2 need to be implemented at each ATM switch.

6.4.3.4 The Performance Evaluation

The performance requirement of a restoration system in a network is closely related to the applications supported by the network services. For instance, VoD applications that utilize public carrier service to retrieve recorded video programs for entertainment, distance learning, information, and so forth, usually require long holding time and relatively high bandwidth [21]. From both the end-user and the network provider perspectives, restoring the existing connections when network failure occurs is more desirable than releasing the affected connections for this type of applications. The lost information could be retrieved again after the affected connection is restored. Furthermore, as long as the time to restore an affected connection is less than the connection time-out limit, to avoid unwanted connection release, a few seconds gap in the rare event of network failure in general is considered acceptable. In this subsection, the performance of the restoration system described in the previous section is evaluated from the VoD application perspective.

The performance studies are conducted via simulations at the call level. Signaling/control flows are modeled based on those defined in ITU-T draft Rec. I.375 [21] for VoD applications. Figure 6-12 depicts the functional components of the simulation model developed utilizing the OPNET network simulation tool.

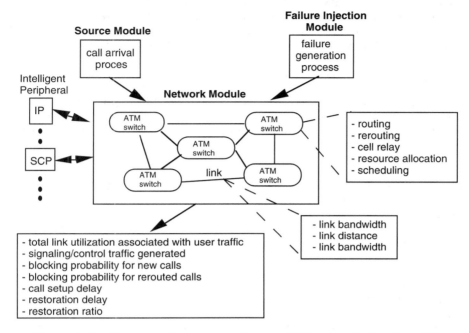

Figure 6-12: Functional component for an ATM network simulator [6].

The source module generates call arrivals and sends them to the corresponding source switches, which trigger the process of connection establishment at the switch. For VoD applications, communications between the source switch and the Service Control Point (SCP), and between the Set-Top Box (STB) and the Intelligent Peripheral (IP) (e.g., video gateway) during the call setup phase are simulated. The establishment of a connection between the STB and the video server is also simulated. The failure generation module generates link failure or node failure events, which trigger the restoration process at the switch.

The restoration blocking ratio, the restoration delay, and the average bandwidth utilization are measures used to evaluate the system's restoration capabilities. The new call blocking ratio and the setup delay are measures used to evaluate the impact of the restoration system on perceived network performance.

The network model used in the study has 23 ATM switches, 51 duplex links connecting these ATM switches, one SCP and one IP. The switches are numbered from 1 to 23. Both the SCP and the IP are attached to the same ATM switch, switch 10, in our study. All the links are assumed to have a link bandwidth of 150 Mbps (i.e., a STS-3c path) in each direction. The bandwidth required for each call in the direction to deliver a selected video program is 4 Mbps (i.e., assuming MPEG2 video transport). All signaling/control messages except one are assumed to be carried in one cell (53 bytes). The one exception is for the message sent from the IP to convey the program selection menu to the user. This message is assumed to be 870 cells, which is based on the regular color PC monitor resolution with a video compression ratio of 20:1. The node processing delay associated with each transmitted cell is assumed to be 2.83 ms (the time to transmit one cell at 150 Mbps).

The following inferences were derived based on observed simulation results reported in [6]:

- Noticeable video quality degradation is more likely due to failure detection latency than to restoration delay.
- Overhead generated by the protection switching procedure is insignificant compared with video connection bandwidth.
- The dominant component of the setup delay is the node transmission time of the menu sent from the IP to the source switch. Therefore, the number of hops of the route used between the source switch and the switch that the IP is attached to is a significant factor that affects the magnitude of the setup delay observed in the simulation experiments conducted.

6.4.4 Interworking between SONET and ATM Layer Protection Switching

Table 6-2 compares ATM protection switching systems and SONET APS systems in terms of design complexity, feature efficiency, and protection switching times.

SONET APS systems require duplicated electronics equipment [e.g., Optical Line Terminating Multiplexers (OLTMs)] to perform line layer protection switching capability. In contrast, the ATM protection switching system proposed here requires electronics duplication if either the OAM-like or general protected fragment approach is used, as the protected fragment is an ordered set of one or more VP links. If the node protection capability is needed, the ATM/VP protection switching system may provide a much more cost-effective solution than other alternatives (such as Self-Healing Rings or networks) in terms of equipment costs and control complexity.

The restoration time discussed in this section includes the failure/degradation detection time plus the protection switching time. Since a SONET APS system can perform line restoration functions under Signaling Failure (SF) and Signaling Degraded (SD) conditions, our comparative study here needs to consider both SF and SD scenarios. An SF is a "hard" failure caused by Loss of Signal (LOS), Loss of Frame (LOF), BER threshold exceeding 10^{-3}, line AIS, or some other protected hard failure. An SD is a "soft" failure caused by a BER exceeding a preselected threshold that usually ranges from 10^{-5} to 10^{-9}. SF is usually given priority over SD. Reference [22] reported detection times for a variety of BERs using ATM and STM technologies.

Table 6-2: Comparison between SONET/APS and ATM/APS.

System Characteristics	SONET/APS	ATM/APS
Protection switching unit	line	VP
Hardware failure ($>10^{-3}$) detection	fast	fast
Soft failure ($10^{-4} - 10^{-7}$) detection	slow (~2 seconds for 10^{-6} or worse)	fast (~1 ms)
Protection switching message exchange at a 125 ms interval	1	>>1
ATM Layer management capability needed	no	yes
Individual VP switching	not possible	possible
Unnecessary protection switching for VPs	possible	not possible
Protection switching control complexity	simpler	relatively complex
ATM node protection	not possible	possible

SONET/APS takes approximately 60 ms to restore services from hard failures where the hard failure is represented by LOS or LOF detection, rather than by poor BER. This restoration time includes 10 ms for LOS/LOF detection and 50 ms for protection switching. However, for the soft failure case, which is detected through poor BER performance (10^{-4} to 10^{-5}), it may take approximately 2 seconds to detect the system degradation with a BER of 10^{-4} [22]. Also, SONET/APS will switch all VPs in the degraded link to the protection link, even though this poor BER of 10^{-4}–10^{-7} may be caused by one or only a few VPs. This implies that there will be many unnecessary switches to and from the protection circuit if the SONET/APS is used to provide protection switching functions for ATM networks. False network management messages for those "good" VPs within the degraded link also would be sent to Operations Systems (OSs), and these would cause many unnecessary operations activities.

In contrast, ATM may detect the soft failure condition much more quickly than its SONET counterpart. For example, it may take approximately 1 ms to detect a BER of 10^{-4} or worse, compared with 2 seconds for its SONET counterpart [22]. This is due to a major difference between SONET transport overhead usage and VP/OAM/user cell usage. This difference is the frequency of signal generations and transmission during a time interval. In STM transport, transport overhead is generated and transmitted every 125 ms, and ATM transport allows many OAM/user cells to generated and transmitted within a 125 ms interval as long as the capacity is available. Unlike the STM transport's fixed OAM capacity, the ATM Layer OAM or user payload capacity can be assigned dynamically based on needs of the particular maintenance activity or procedure invoked.

Thus, ATM transport could convey necessary network management messages faster than its STM counterpart. This potentially allows for faster network response to "soft" failures for ATM transport. Thus, it has been suggested in [22] that performance defects due to line protection switching may be reduced by effecting line protection switches only for "hard" failures (i.e., BER > 10^{-3} or LOS, LOF), and using ATM VP OAM mechanisms for "soft" failure protection. Also, for the SD case, good VPs within the affected/degraded line will not be switched forward and backward as long as the VPXs have the capability to monitor the individual VP's QoS. Thus, many unnecessary switches to and from the protection circuit (along with unnecessary operations messages) can be avoided.

Considering the technology comparison shown in Table 6-2, the following hybrid protection switching system is proposed, as depicted in Figure 6-13 [9], for a network with existing APS systems available, as well as new ATM/VP protection switching systems. If LOS is detected (i.e., the entire line is dead), check if SONET/APS is available for protection switching. If it is, perform SONET protection switching; otherwise, perform ATM protection switching, if available. If a BER between 10^{-4} to 10^{-7} is detected, check the individual VP's performance measures. Perform ATM protection switching for those VPs with performance measures falling into the service-affected range.

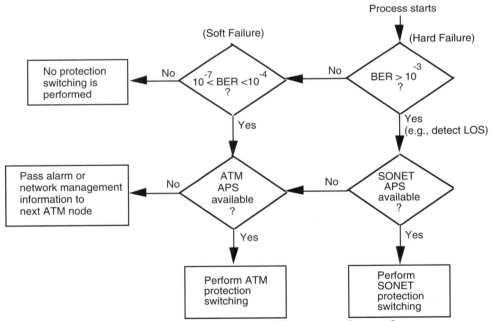

Figure 6-13: A proposed hybrid APS system and procedure.

6.5. ATM Self-Healing Rings
6.5.1 Potential Role of ATM Technology in Self-Healing Rings

To utilize the SONET ring bandwidth efficiently and cost-effectively, a bandwidth management system is needed because the major point-to-point demand in today's LEC networks is primarily at the DS1 (1.554 Mbps) level; present SONET rings transport signals at the STS-1 (51.84 Mbps) or STS-3c (155.52 Mbps) level. In the present SONET rings, grouping DS1 demands to STS-1 demands can be managed in a grooming or nongrooming manner. For the demand nongrooming system, the DS1 demand requirement for each node pair is directly converted to the STS-1 demand requirement by dividing the number of DS1s by 28. For the demand grooming system, grooming can be performed in a centralized or distributed manner, as depicted in Figure 6-14 [23,24]. Note that these SONET rings use only the centralized ring grooming or a nongrooming system, because the distributed ring grooming system at the DS1 level is too expensive to be implemented.

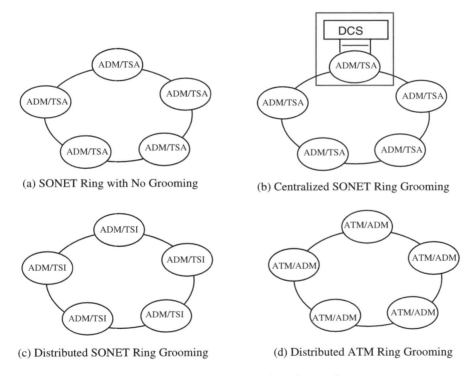

(a) SONET Ring with No Grooming

(b) Centralized SONET Ring Grooming

(c) Distributed SONET Ring Grooming

(d) Distributed ATM Ring Grooming

Figure 6-14: Ring grooming alternatives.

The centralized ring grooming system, as depicted in Figure 6-14(b), includes a Wideband DCS (W-DCS) (e.g., DCS 3/1 in this case) and a SONET ring with an ADM in each node. The DS1 demand grooming function is provided by the W-DCS (e.g., DCS 3/1), which is usually located at a major ring interconnection point. If the SONET ring evolves from today's fiber-hubbed networks, the W-DCS will probably be located at the hub. The function of this DCS grooming for SONET rings is similar to that for fiber-hubbed networks, in which DS1 demands between two offices will be routed via the W-DCS if these DS1 demands are not assigned to a direct (point-to-point) STS-1. The ADM used in this centralized ring grooming system can be a simple ADM because it does not need grooming capability. The signal add–drop in this case can be implemented by using a Time Slot Assignment (TSA) method that assigns dedicated time slots for each node, and those dedicated time slots can be dynamically assigned to DS1 ports. After the DS1 demand requirement is bundled into the STS-1 demand requirement, it can be transported by either an STS-1 ring or a Virtual Tributary (VT), where each STS-1 payload can accommodate up to 28 VT1.5 channels; use of the ring depends on relative economics. A SONET ring is called an STS (or VT) ring if it adds–drops signals at the STS (or VT) level.

On the other hand, the distributed ring grooming system, as depicted in Figure 6-14(c), distributes the demand grooming capability into each ring node by using a Time Slot Interchange (TSI) switching fabric within each ADM. The ADM with TSI (ADM/TSI) is functionally a mini-DCS that only serves the demand of one fiber system (i.e., a fiber pair). The TSI function in this distributed ring grooming architecture is performed at the VT (DS1) level, rather than at the STS-1 (DS3) level, as commonly used for self-healing control in the 2-fiber bidirectional Self-Healing Ring architecture (see Chapter 2). Compared with the centralized ring grooming system, the distributed ring grooming system generally requires less ring capacity for the same DS1 demand requirement, but at the expense of more complex and expensive ADMs. The relative economics of these two ring grooming systems depends on the relative costs of equipment (DCS, ADM/TSA, ADM/TSI) and the demand requirements.

To reduce SONET ring cost, an enhanced grooming system must combine the best features of centralized and distributed ring grooming systems. In other words, the new, more cost-effective SONET ring grooming system should have bandwidth allocation flexibility to reduce the ring capacity requirement, as does the distributed ring grooming system using ADM/TSIs, and should use simpler and less expensive ADMs, as does the ADM/TSA. This new system can be implemented through ATM technology as proposed in [23]. The conceptual diagram for this distributed ATM ring grooming architecture is depicted in Figure 6-14(d). This new ATM ring architecture is implemented at the VP layer; thus, it is also called an ATM/VP ring. We will discuss this ATM/VP ring architecture in the next two sections (Sections 6.5.2 and 6.5.3). An economic study comparing four ring grooming alternatives reported in [24] has suggested that the proposed ATM VP ring (along with ATM Layer protection) can achieve the same bandwidth usage efficiency as the SONET/TSI distributed grooming ring, but at a much lower cost. Please refer to [24] for a detailed economic study model and cost study results.

6.5.2 SONET-Based ATM Self-Healing Ring Architectures

The basic equipment supporting ATM rings is ATM Add–Drop Multiplexers (ADMs), the signal processing level of which can be at the VP or VC layer. Just like its SONET/ADM counterparts (i.e., ADM with TSA, and ADM with TSI), the ATM ADM may or may not have switching capability (i.e., switching fabric). For the former ring architecture, the self-healing function can be implemented at the SONET or ATM Layer;

for the latter ring architecture, only the ATM Layer is available for protection switching. Determining which ring architecture is appropriate depends on an application's QoS requirements and costs. We will discuss SONET-based ATM Self-Healing Rings in this section. The ATM Self-Healing Ring architecture (with switching capability) will be discussed in Section 6.5.3.

6.5.2.1 A SONET-Based ATM VP Ring Architecture

Figure 6-15 depicts an example of the SONET-based ATM Ring architecture using Point-to-point VPs (SARPVP), which was proposed in [23]. The VP used in the point-to-point VP add-drop multiplexing scheme carries VC connections between the same two ring nodes. In this SARPVP architecture, each ring node pair is preassigned a duplex VP. For example, in the figure, VP #2 and VP #2' (not shown in Figure 6-14) carry all VC connections from Node 1 to Node 3 and from Node 3 to Node 1, respectively. The physical route assignment for the VP depends on the type (unidirectional or bidirectional) of the considered SONET ring. If the considered ring is unidirectional, two diverse routes that form a circle are assigned to each VP. For example, in Figure 6-15, two physical routes, 1–2–3 and 3–4–1 are assigned to VP #2 and VP #2' (not shown in the figure) if the considered ring is a unidirectional ring. If the considered ring is bidirectional, only one route is assigned to each duplex VP (e.g., route 1–2–3 is assigned to both the VP #2 and VP #2'), and demands between Nodes 1 and 3 are routed through route 1–2–3 bidirectionally.

To avoid the VP translation at intermediate ring nodes of a VP connection, the VPI value is assigned on a global basis. The ATM cell add–drop or pass-through at each ring node is performed by checking the cell's VPI value. As the VPI value has global significance and only one route is available for all outgoing cells, it need not be translated at each intermediate ring node. Thus, no VP cross-connect capability is needed for the ATM/ADM of this ATM/VP ring architecture.

The STS-3c used in this ATM/VP ring always originates at one ring node and terminates at the next ring node for best network utilization. This type of STS-3c is sometimes referred to as "one-hop" STS-3c, as depicted in Figure 6-16. For ATM termination at the STS-3c level, the ATM/ADM can be upgraded from the present SONET/ADM by replacing the SONET STS-3 line cards with ATM line cards, thus minimizing the initial capital costs for evolving from the SONET/STM ring to the hybrid STM/ATM ring, and eventually to the ATM/VP ring. Of course, operations for processing ATM cells will be different from those for processing SONET channels. The requirements and design for SONET rings require no change. Thus, the ATM/ADM with STS-3c terminations could be the first candidate for transition from the SONET ring to the ATM ring for early deployment of ATM technology on the SONET infrastructure.

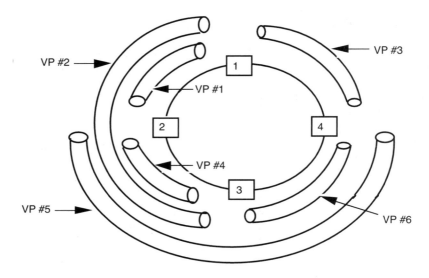

VP #2 → → VP #3

VP #1

VP #4

VP #5 → → VP #6

Working - ATM/VP for bandwidth management
Protection - SONET path or ATM/VP for cost-effective protection

Figure 6-15: An ATM/VP ring concept [23].

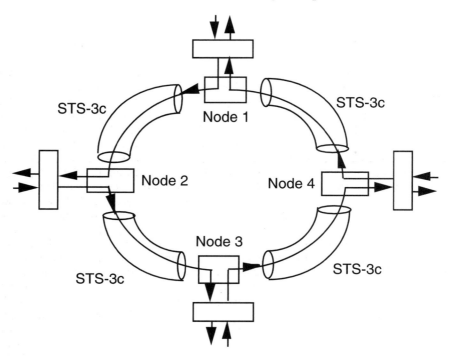

STS-3c

Node 1

STS-3c

Node 2

Node 4

STS-3c

Node 3

STS-3c

Figure 6-16: One-hop STS-3c placement for ATM/VP ring.

6.5.2.2 ATM/VP ADM Functional Diagram

Figure 6-17 depicts a possible ADM configuration with STS-3c terminations for an ATM/VP ring implementation that supports DS1 services via circuit emulation. In this architecture, the hardware design for signal terminations above STS-3c is the same as that for SONET/ADMs for ring applications. This ATM/VP ADM is designed to be upgradable from existing SONET/ADMs.

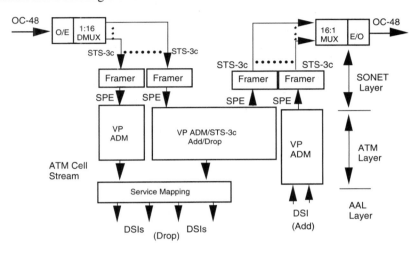

Figure 6-17: An example of an ATM/VP ADM functional diagram.

Figure 6-18 depicts a possible STS-3c chip configuration used in the ATM/VP ADM described in the previous figure. The ATM/VP add–drop function, which is performed at the STS-3c level, requires three major modules. The first module is the ATM/SONET interface, which converts the STS-3c payload to an ATM cell stream and vice versa. The functions performed in this module include cell delineation, self-synchronization, and scrambling. The scrambling process here increases the security and robustness of the cell delineation process against malicious users or unintended simulations of a cell header followed by a correct Header Error Control (HEC) in the information field. The second module performs header processing, which includes cell addressing (VPI in this case) and HEC. To perform cell add–drop/pass-through, this module checks the VPI value of each cell to determine if it should be dropped or passed through, and it identifies an idle cell that can be used to insert cells from the considered office (i.e., signal adding) via a simple sequential access protocol. This access protocol can be implemented by the third module that passes through each nonidle cell and inserts the added cells into outgoing idle cells sequentially. The third functional module also includes a service mapping module that maps ATM cells to their corresponding DS1 cards based on VPI/VCI values of ATM cells. This service mapping module first multiplexes all ATM cells from different STS-3c payloads into a single ATM cell stream, and then it distributes ATM cells to corresponding DS1 groups according to their VPI values. For each DS1 group, the ATM cells are further divided and distributed to the corresponding DS1 cards by checking their VCI values. This service mapping module essentially just performs a simple VPI/VCI comparison function.

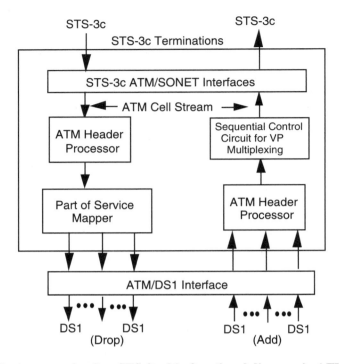

Figure 6-18: An example of an STS-3c chip functional diagram in ATM VP/ADM.

For ATM termination at the STS-3c level, the ATM/ADM can be upgraded from the present SONET/ADM by replacing the SONET STS-3 line cards with ATM line cards, thus minimizing the initial capital costs for evolving from the SONET/STM ring to the hybrid STM/ATM ring, and eventually to the ATM/VP ring. Of course, operations for processing ATM cells will be different from those for processing SONET channels. The requirements and design for SONET rings require no change. Thus, the ATM/ADM with STS-3c terminations could be the first candidate for transition from the SONET ring to the ATM ring for early deployment of ATM technology on the SONET infrastructure.

6.5.2.3 Self-Healing Control Schemes

The SONET-based ATM/VP ring architecture can be implemented as either a bidirectional or unidirectional ring architecture. For the bidirectional line-switched Self-Healing Ring, the self-healing function can be implemented at the SONET Layer in the same way as already defined by the T1X1.5 standards group. For the unidirectional ATM/VP ring, the self-healing function can be implemented at either the SONET Layer (STS-3c, STS-12c) or the VP layer. If the SONET Layer is used for protection, the path-switched self-healing system implemented in present SONET SHRs remains unchanged. In this case, the working ring uses the VP layer for bandwidth management and the protection ring uses the SONET Layer for protection.

If the VP layer is used for protection, the chip for adding–dropping VPs within the STS-3c is used for both the working and protection STS-3c's. The VP ring with the VP layer protection may use the same self-healing protocol as defined in the SONET path-switched SHR [18], except that the path selection may be triggered by the VP-AIS (Alarm Indication Signal) rather than by the STS (or VT) AIS. Compared with SONET layer protection, the ATM/VP ring with VP layer protection may be more flexible and simpler, but the VP self-healing protocol needs to be standardized. It has been recommended in Bellcore GR-2837-CORE [25] that bidirectional rings use SONET Layer protection and unidirectional rings use ATM Layer protection due to the spare capacity assignment characteristics [26].

6.5.2.4 Spare Ring Capacity Engineering

The protection layer could significantly affect the working and spare capacity requirements for unidirectional and bidirectional Self-Healing Rings. In the example shown in Figure 6-19, the SONET ring having the centralized demand pattern (i.e., all demands concentrate to one node) usually favors the unidirectional SHR (see Chapter 2). However, due to one-hop SONET structure used in ATM/VP rings (see Figure 6-16), the unidirectional ring will need an OC-48 ring when the SONET Layer protection is used but can still use an OC-12 ring if the ATM Layer protection is implemented. On the other hand, the impact of the one-hop SONET structure for ATM/VP rings on bidirectional ring capacity requirements may not be as significant as on its unidirectional counterpart. This is partially because, even in SONET case, the one-hop case is similar to point-to-point demand, which naturally favors the bidirectional SHR (see Chapter 2). In this example, the SONET/ATM bidirectional SHR needs an OC-12 ring for both the SONET Layer protection and the ATM Layer protection, although the spare capacity requirement for the ATM Layer protection is much less than for its SONET Layer protection counterpart.

6.5.3 ATM Self-Healing Rings with Switching Capability

ATM Self-Healing Rings having ATM switching capability typically use the ATM Layer for ring protection switching. One such ring was proposed by NTT and reported in [27]. Figure 6-20 shows an ATM self-healing control scheme used in an ATM/VP SHR architecture described in [27]. The proposed ATM/ADM has an ATM switching module and additional preplanned VPI table (i.e., VPI-F table) for failure recovery. The self-healing procedure from failure detection to recovery of in-service VPs is as follows. In the case of a unidirectional transmission line failure, the first downstream ADM detects

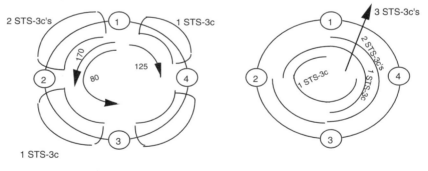

Centralized Demand Pattern:

Demand (1,2) --> 170 Mbps
Demand (1,3) --> 80

(a) Demand Example

3 STS-3c's

80

125   170

2 STS-3c's

1 STS-3c

1

2   4

3

Required Working Capacity: 3STS-3c's
Required Protection Capacity using
ATM Layer Protection: 3 STS-3c's

6 STS-3c's

3 STS-3c's
2 STS-3c's

1 STS-3c

1

2   4

3

Required Protection Capacity
using SONET Layer Protection:
3 + 2 + 1 = 6 STS-3c's

===> Need an OC-48 Unidirectional Self-Healing Ring
   (Protection at SONET Layer)
===> Need an OC-12 Unidirectional Self-Healing Ring
   (Protection at ATM Layer)

(b) Working and Spare Capacity Requirements for Unidirectional Self-Healing Ring

2 STS-3c's   1 STS-3c

170
125
80

1 STS-3c

1

2   4

3

Required Working Capacity: 2 STS-3c's
Required Protection Capacity using
ATM Layer Protection: 2 STS-3c's

3 STS-3c's

2 STS-3c's
1 STS-3c

1 STS-3c

1

2   4

3

Required Protection Capacity
using SONET Layer Protection:
2 + 1 = 3 STS-3c's

===> Need an OC-12 Bidirectional Self-Healing Ring
   (Protection at SONET Layer)
===> Need an OC-12 Bidirectional Self-Healing Ring
   (Protection at ATM Layer)

(c) Working and Spare Capacity Requirements for Bidirectional Self-Healing Ring

Figure 6-19: An example of spare capacity engineering for ATM/VP rings.

the Line/Path alarm. If the alarm satisfies the condition of Far End Receiving Failure (FERF) generation, the downstream ADM sends FERF to the ADM upstream of the failure. If the upstream ADM receives FERF in excess of the guard time, the ADM confirms that the downstream transmission line has failed. It then switches all working VPs destined for the downstream ADM to their respective protection VPs. Path restoration uses the preplanned VPI-F table, which indicates the paths to be activated after transmission line failure. These actions recover all affected VPs on the failed line. The downstream ADM waits for the guard time and then connects all affected working VPs to their respective protection VPs. In this proposal, all working VPs between two adjacent ADMs will be rerouted, even if only one transmission line fails. For bidirectional line failure, the procedure is similar to the one just described, except that the adjacent ADMs now cannot receive FERF. However, rerouting is still performed, because both ADMs detecting the Line/Path alarm connect working VPs to protection VPs after the guard time expires.

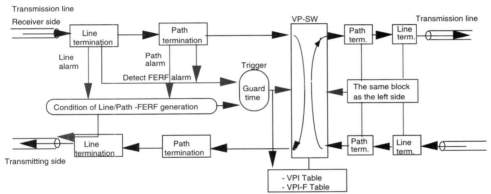

Figure 6-20: Self-healing control at the ATM Layer [27] (©1992 IEEE).

6.5.4. Broadband Self-Healing Ring Evolution

Since SONET rings have been widely deployed in the U.S. telecommunications industry, evolution to SONET/ATM rings has become a crucial issue for network providers desirous of remaining competitive. The evolution strategy may depend on business plans, network infrastructure, economics, and timing of equipment availability. A cost-effective three-phase network evolution path that uses the proposed SONET/ATM VP ring architecture has been proposed in [28]. This three-phase evolution path was designed primarily for network providers that have already deployed SONET rings and are looking for business opportunities in the emerging broadband data service areas. This evolution path, as depicted in Figure 6-21, evolves from the present SONET/STM ring transport through a hybrid SONET STM/ATM ring transport, and eventually to a SONET/ATM VP-based ring transport for supporting both switched and nonswitched voice and data services. This three-phase network evolution path may yield a potential opportunity for early deployment of SONET/ATM networks, thus facilitating the introduction of future B-ISDN services.

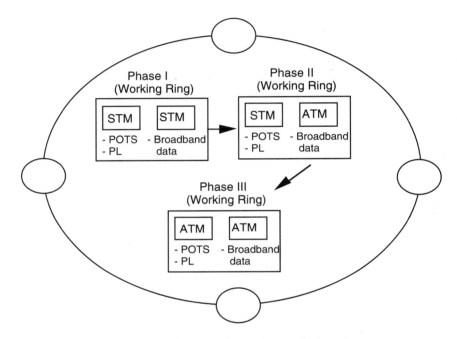

Figure 6-21: A three-phase ring evolution plan.

In Phase I, emerging broadband data service is added to the existing SONET/STM ring that heretofore supported only nonswitched DS1 and DS3 services. The DS1 and DS3 carrying broadband data services are mapped to SONET VT1.5 and STS-1 paths, respectively, and are then transported to the ADM on the ring.

Phase II is to upgrade the ring transport format to a hybrid STM/ATM format by replacing the STS-3 line cards carrying high-speed data services with STS-3c line cards implementing the ATM VP add–drop scheme, as described in Section 6.5.2 or [23,24]. In this phase, the STS-3 cards carrying existing nonswitched DS1 and DS3 services remain unchanged (i.e., still using the STM format), and new STS-3c cards carry high-speed data services. The broadband data service's DS1/DS3s access the network via the cell format of the ATM Adaptation Layer (AAL) to support the format conversion between the service and the signal transport. Thus, the ring architecture forms a hybrid STM/ATM transfer mode architecture. The Self-Healing Ring system remains unchanged, as in

Phase I. It is expected that the Phase II ring transport will provide a much more efficient and cost-effective transport for high-speed data services than the SONET/STM ring in Phase I, due to inherent ATM technology characteristics that include a shared bandwidth and a simple ADM design for the SONET/ATM VP ring architecture. In this phase, Operations Systems may need to upgrade to ATM/VP-based OSs. The savings on transport for supporting high-speed data services in this phase may justify the OS upgrade investment. This phase also offers a good opportunity for network operations personnel to gain experience on ATM/VP-based network operations that will make the last transition to the fully ATM/VP ring easier.

When the operations experience gained from the high-speed data services that use the ATM/VP technology in Phase II is adequate, Phase III can then be introduced to replace the remaining STM portion (that supports nonswitched DS1 and DS3 services) by the ATM/VP add–drop scheme. In this phase, nonswitched DS1s and DS3s are converted to ATM cells via circuit emulation (i.e., AAL Class A).

A case study based on a five-node LEC ring model network indicated that the SONET-based ATM/VP ring had either significant cost savings or more spare capacity [24], as compared with an STM ring. An integrated operations system for the SONET-based ATM/VP ring for both switched and nonswitched broadband services in this phase should further reduce the OAM costs. Thus, network evolution from Phase II to III may be justified by its potential economic and operations benefits. However, these potential benefits still require more detailed studies. Again, as with the evolution from Phase I to II, the self-healing system remains unchanged when the ring evolves from Phase II to III.

In Phase III, when the ATM technology is mature, broadband data service access on customer premises can be converted to the ATM cell format, which then can be directly placed into the STS-Nc payload in the ADM of the ring. This can further reduce network cost, because the DS1/DS3-to-ATM (or VT/STS-to-ATM) conversion is no longer needed. After completing this three-phase network evolution, the underlying ring transport system is a true SONET/ATM B-ISDN transport system.

6.5.5 Applications to Mobility Management

6.5.5.1 PCS Hand-off Feature

The ATM Self-Healing Ring architecture may provide reliable and very fast hand-off capability to support existing cellular services or emerging Personal Communications Service (PCS). Hand-off is the most time-critical feature needed to support PCS mobility management, because the process needs to ensure call continuation service. Three types of hand-off procedures are available: network controlled (e.g., current analog cellular system); portable-assisted [e.g., GSM, Interim Standards (IS)-54, IS-95]; and portable controlled [e.g., Digital European Cordless Telephone (DECT), Personal Access Communications System (PACS)]. For both network-controlled and portable-assisted hand-offs, the network [e.g., the serving Wireless Service Center (WSC)] tracks the actual transmission quality of the call in progress. For the network-controlled hand-off, when the transmission quality falls below a certain level, the serving WSC requests neighboring WSCs to measure the signal strength of the call in progress and report the measurement results back to the serving WSC. For the case of portable-assisted hand-off, the serving WSC requests the portable to perform measurements on the other channels and reports results to the serving Radio Control Unit (RCU) via the deteriorating radio link. The serving RCU then decides on a target WSC based on the results reported from the WSCs or a portable. For the portable-controlled hand-off, the portable continuously monitors the signal strength and quality from the access base station and several hand-off candidate base stations. When the quality starts falling below a certain level, the portable accesses the "best" candidate base stations for an available traffic channel and launches a hand-off request. The network (target WSC) then initiates the hand-off process to the serving WSC if the best candidate base station is outside the serving WSC area.

For all three types of hand-offs, after some signaling message exchanges between the serving WSC and the target WSC, the network then routes the call to the target WSC. It is desirable that direct communication channels for signaling messages between the serving and the target WSCs be used to minimize the hand-off delay [29]. The associated signaling mode serves this purpose well, and the use of SONET-based ATM/VP Self-Healing Rings [23–25] connecting those WSCs would provide not only fast transport but also fast protection in case of failure of signaling links and/or the WSC.

In the ATM transport platform, there are three approaches to route the call and signaling messages for hand-off completion. They are anchor routing, dynamic point-to-point connection rerouting, and the preestablished point-to-multipoint routing. Figure 6-22 shows two classes of rerouting methods that may be implemented using ATM technology. In the following discussions, it is assumed that the transmission of ATM cells is terminated at the base stations.

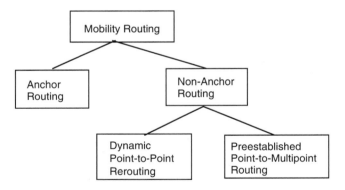

Figure 6-22: Classes of rerouting methods for hand-off.

6.5.5.2 Anchor Routing

An anchor routing is used to designate the initial serving switch of a PCS call as a fixed connection point between the wireless network and the wireline network for the entire duration of the call. This switch is called the "anchor switch." During interswitch hand-offs, all bridging is performed in this anchor switch by establishing connections between the target switch and the anchor switch. No matter how many hand-offs occur during the entire call duration, the anchor switch always retains the connection to the wireline network and bridges the connection to the new switch that is currently providing service to the portable. This approach has been adopted in the GSM and IS-41. Figure 6-23 illustrates an example of an anchor routing during hand-off, where it is assumed that a call is established between wireline and wireless users. As shown in this figure, the portable moves from base station coverage area A served by the WSC-A to base station coverage area B served by the WSC-B while the call is in progress. After hand-off, the WSC-A bridges the call connection to WSC-B.

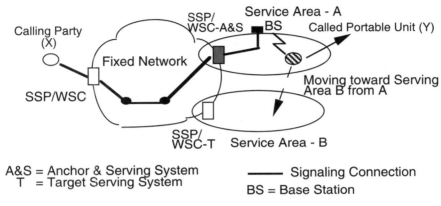

(a) Initial Connections

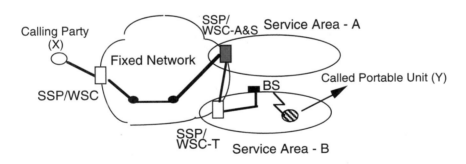

(b) Resulting Connections

Figure 6-23: An example of anchor routing.

Figure 6-24 shows an example of the call routing path before and after hand-off. From the wireline network to the WSC-A and to the base station BS1, the connection (VC) is via VCI1, VCI3 and the reverse path is via VCI4, VCI2. Normally, the same VCI can be used in both directions. The use of different VCIs is a generalized case for services that may require distinct QoSs in each direction. After hand-off, the route of the call between the wireline network and the WSC-A remains the same. The connection between the

WSC-A and WSC-B is established, and the corresponding virtual paths are VCI9 and VCI6. The WSC-B routes the call to the base station BS2 via VCI7 and via VCI8 for the reverse direction. In this example, it is assumed that the WSCs are Virtual Path ATM switches. Notice that the anchor routing is suitable for all three types of hand-off procedures, namely, network-controlled, portable-assisted, and portable-controlled hand-offs.

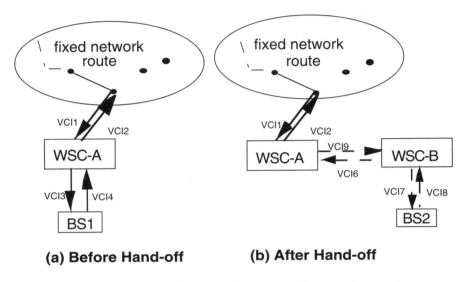

(a) Before Hand-off   **(b) After Hand-off**

Figure 6-24:   Routing path before and after hand-off for anchor routing scheme.

6.5.5.3 Ring Architecture for Reliable and Fast Anchor Rerouting

The most significant advantage of the anchor rerouting method is that the signaling connection rerouting remains at the local level among WSCs. Two primary concerns for this approach are switch processing capacity and survivability of the anchor WSC. However, these two concerns may be alleviated by using a SONET ring or a SONET-based ATM/VP Self-Healing Ring [23–25] to interconnect local WSCs, as shown in Figure 6-25. As specified in [25], each SONET or SONET-based ATM/VP SHR is required to have dual-access capability in order to protect the network from the single-node failure. Thus, in the proposed configuration shown in Figure 6-25, the WSC ring will have two gateway nodes interworking with the wireline signaling and service backbone network, which could be a mesh network, an interconnection network of rings, or a network using other architectures. The typical ring reconfiguration time due to a node failure is approximately 50 ms [23], probably not enough to disconnect a PCS call.

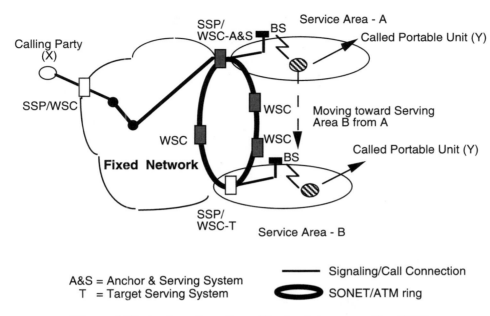

Figure 6-25: Anchored routing with ring interconnecting WSCs.

6.6 Summary and Remarks

In this chapter, we have discussed technical issues and several proposals and architectures for Automatic Protection Switching systems and Self-Healing Ring architectures for SONET-based ATM networks. Readers who are interested in more detailed technology and architecture designs may refer to [3,11,30,31,32]. Some protection switching architectures have become commercially available, and some are still in the standardization process. Evolution is crucial if network providers are to remain competitive in terms of service deployment and capital investment. In addition, multiple-layer networking and coordination for SONET-based ATM networks become challenging problems not only for network transport systems, but also for the Network Management System. These are challenges to be resolved before any practical, large-scale survivable SONET-based ATM networks can be realized.

References

[1] ITU-T Recommendation I.311, "B-ISDN General Network Aspects," January 1993.

[2] K. Sato, H. Ueda, and N. Yoshikai, "The Role of Virtual Path Crossconnection," *IEEE Magazine of Lightwave Telecommunications Systems,* Vol. 2, No. 3, pp. 44-54, August 1991.

[3] T.-H. Wu, "Emerging Technologies for Fiber Network Survivability," *IEEE Commun. Mag.*, pp. 58-74, February 1995.

[4]   R. Rawamura, K.-I. Sato, and I. Tokizawa, "Self-Healing ATM Networks Based on Virtual Path Concept," *IEEE J. Selected Areas in Commun.*, pp. 120–127, January 1994.

[5]   The ATM Forum, *Private Network-Network Interface Specification Version 1.0 (PNNI 1.0)*, March 1996.

[6]   D. Hsing, L. Kant, and B. Cheng, "A Restoration System for ATM Networks," *Proc. IEEE MILCOM*, pp. 483–489, November 1996.

[7]   D. Hsing, B-C. Cheng, G. Concu, and L. Kant, "A Restoration methodology based on Pre-Planned Source Routing in ATM Networks," *Proc. ICC'97*, Montreal, Canada, June 1997.

[8]   N. Yoshikai, and T-H. Wu, "Control Protocol and Its Performance Analysis for Distributed ATM Virtual Path Self-Healing Networks," *IEEE J. Selected Areas in Commun.*, pp. 1020-1030, August 1994.

[9]   Y. Fujita, T-H. Wu, and H. Fowler, "ATM Protection Switching Systems and Applications," *Conference Records of International Switching Symposium (ISS'95)*, pp. 234–238, April 1995.

[10] Bellcore GR-1248-CORE, "Generic Requirement for Operations of ATM Network Elements," Issue 3, August 1996.

[11] T.-H. Wu, *Fiber Network Service Survivability* Artech House, May 1992.

[12] L. Nedenlof, K. Struyve, C. O'Shea, H. Misser., Y. Du, and B. Tamayo, "End-to-End Survivable Broadband Netyworks," *IEEE Commun. Mag.,* Vol. 33 No. 9, pp. 63-70, September 1995.

[13] H. Ohta, and H. Ueda, "Hitless Line Protection Switching Method for ATM Networks," *Proc. ICC,* pp. 272-276, May 1993.

[14] H. J. Fowler, and Y. Fujita, "VP Protection Switching at the ATM Layer," Contribution T1S1.5/94-026, 1994.

[15] R. Kawamura, K. Sato, and I. Tokizawa, "Self-Healing Techniques Utilizing Virtual Paths," *Proc. the 5th International Network Planning Symposium*, pp. 129–134, Kobe, Japan, May 1992.

[16] R. Kawamura, and I. Tokizawa, "Self-healing Virtual Path Architetcure in ATM Networks," *IEEE Commun. Mag.,* Vol. 33, No. 9, pp. 72- 79, September 1995.

[17] ITU-T Draft Recommendation Q.2100, "B-ISDN ATM Signaling Adaptation Layer Overview Description," 1994.

[18] ITU-T Draft Recommendation Q.2010, "Broadband Integrated Service Digital Network Overview - Signaling Capability Set 1, Release 1," 1995.

[19] ITU-T Draft Recommendation Q.2110, "B-ISDN ATM Adaptation Layer Service Specific Connection Oriented Protocol (SSCOP)," 1994.

[20] ITU-T Draft Recommendation Q.2140, "B-ISDN ATM Adaptation Layer Service Specific Coordination Function for Signaling at the Network Node Interface (SSCF at NNI)," 1995.

[21] TU-T Draft Recommendation I.375, "Network Capabilities to Support Multimedia Services," November 1994.

[22] J. Anderson, B. T. Doshi, S. Dravida, and P. Harshavardhana, "Fast Restoration of ATM Networks," *IEEE J. Selected Areas in Commun.*, pp. 128-138, January 1994.

[23] T.-H. Wu, D. Kong, and R. C. Lau, "A Broadband Virtual Path SONET/ATM Self-Healing Ring Architecture and its Economic Feasibility Study," *Proc. IEEE GLOBECOM*, pp. 834–840, December 1992.

[24] T.-H. Wu, D. Kong, and R. C. Lau, "An Economic Feasibility Study for A Broadband Virtual Path SONET/ATM Self-Healing Ring Architecture," *IEEE J. Selected Areas in Commun.,* Vol. 10, No. 9, pp. 1459-1473, December 1992.

[25] Bellcore GR-2837-CORE, "ATM Virtual Path Functionality in SONET Rings - Generic Criteria," Issue 3, October 1996.

[26] J. Sosnosky, "Potential Network Applications for Virtual Path SONET ATM Rings," *Conference Records of National Fiber-Optic Engineers Conference (NFOEC),* Vol. 3, pp. 375-387, June 1994.

[27] Y. Kajiyama, N. Tokura, and K. Kikuchi, "ATM Self-Healing Ring," *Proc. IEEE GLOBECOM,* pp. 639–643, December 1992.

[28] T.-H. Wu, "Cost-Effective Network Evolution," *IEEE Commun. Mag.,* pp. 64-73, September 1993.

[29] T.-H. Wu, and L-F. Chang, "Architectures for PCS Mobility Management on ATM Transport Networks," *Proc. International Conference on Universal Personal Communications,* Tokyo, Japan, November 1995.

[30] T.-H. Wu, J. C. McDonald, K. Sato and T. P. Flanagan (eds.), "Integrity of Public Telecommunications Networks," *IEEE J. Selected Areas in Commun.,* January 1994.

[31] T. V. Landegem, (Editor), "Self-Healing Networks for SDH and ATM," *IEEE Commun. Mag.,* Vol. 33, No. 9, September 1995.

[32] C. A. Siller, Jr. and M. Shafi, *SONET/SDH: A Sourcebook of Synchronous Networking,* IEEE Press, New York, 1996.

Chapter 7

ATM Self-Healing Mesh Networks

7.1 Introduction

The point-to-point systems using Automatic Protection Switching and Self-Healing Rings discussed in Chapter 6 are cost-effective survivable network architectures for small-scale ATM networks. As with the SONET/SDH networks, these two architectures may not be cost-effective when the network becomes large, due to the expensive network equipment and operations costs associated with interworking of these architectures (see Chapter 2). The self-healing mesh network architecture is an alternative that provides a cost-effective solution for the required level of network survivability in large broadband networks. This is accomplished by sharing spare capacity for restoration. According to ITU-T Rec. I.311 [1], self-healing is the establishment of a replacement connection by the network without a network management control function, and when a connection failure occurs, the replacement connection is found by the network elements and rerouted depending on the network resources available at that time. Therefore, a self-healing network sometimes is referred to as a dynamically restorable network. Equipment typically used for mesh networks is Digital Cross-connect Systems (DCSs) for SONET/SDH networks or Virtual Path Cross-Connect Systems (VPXs), or Virtual Channel (VC) switches for ATM networks. The use of ATM VPXs or VC switches depends on the transport network architectures being considered (see Chapters 2 and 3).

As discussed in Chapter 2, SONET/SDH self-healing systems may not be able to restore services completely within 2 second's in large metropolitan Local Exchange Carrier (LEC) networks [2]. This is due primarily to serial processing and slow, serial cross-connect characteristics of the DCS system. ATM/VP may be one of the potential technologies that would meet such a 2-second service restoration objective, due to its fast cell routing scheme on the self-routing switching fabric and its inherent parallel processing and switching (cross-connection) capability.

To implement a cost-effective ATM self-healing mesh network, the designers need to understand alternative self-healing mesh network architectures and associated trade-offs in costs and survivability. Furthermore, if the underlying transport network uses a multilayer structure (e.g., ATM on SONET/SDH, or IP on ATM), the potential of each of the layer technologies and how they are differentiated need to be clearly understood before a cost-effective solution can be realized. This chapter provides background information that can help readers to investigate these questions.

In this chapter, we discuss some potential ATM technology roles for self-healing mesh networks in Section 7.2. Sections 7.3 and 7.4 review several proposed ATM self-healing mesh network architectures and associated restoration control message transfer mechanisms, respectively. ATM switching systems that potentially could be used to implement self-healing mesh networks are discussed in Section 7.5. Spare capacity designs and some case studies for ATM self-healing networks are discussed in Section

7.6. Section 7.7 discusses VP tracing methods needed for monitoring and triggering ATM self-healing VP networks. Section 7.8 discusses escalation issues and some proposals that would address the multilayer self-healing mesh network designs. Summary and remarks are given in Section 7.9.

7.2 Potential Role of ATM in Self-Healing Networks

7.2.1 Network Restoration Using Virtual Paths

A layered transport network architecture, constructed of circuit, path, and transmission-media layers, is an effective concept for paving the path toward Broadband ISDN, because it simplifies the network design and its OAM functions. In particular, the path layer provides flexibility and controllability. Several network restoration schemes based on the path layer have been studied [3].

The ATM cells in an ATM network are transported and cross-connected either on the Virtual Channel (VC) layer or on the Virtual Path (VP) layer, depending on application, traffic patterns, and network size. For survivability, VP restoration is simpler and faster than VC restoration and has thus been the focus of the currently proposed ATM restoration systems. Thus, we will primarily focus in this chapter on VP restoration. The VC self-healing scheme is reported to improve the multigrade reliability of multilayer self-healing B-ISDN networks; it will be briefly discussed in Section 7.8.

The VP concept has already been standardized on ITU-T Rec. I.311 [1]. A VP is a labeled path established by defining a VPI for the path segment and updating the VPI tables in the cross-connect nodes. The VPI is the number in the cell header identifying the assigned path of the cell. A VP is different in several ways from the digital path in STM networks. Its single-path layer structure is a simpler path layer than existing hierarchical structures. Moreover, path bandwidth flexibility and independence in route and bandwidth establishment of VPs are attractive. In STM networks, a digital path is established by assigning a time slot of the TDM frame at each cross-connect of the path. Path route establishment and bandwidth assignment are, therefore, not independent: only a digital path with a fixed bandwidth can be established. In contrast, the route and bandwidth of a VP are defined independently because the route is defined in the VPI tables in the cross-connect nodes and the bandwidth is logically defined and managed in a database. Consequently; a VP route can be established without assigning its bandwidth along the path connection; in other words, assigning zero bandwidth is possible.

7.2.2 Digital Path Restoration vs. VP Restoration

Currently proposed ATM self-healing networks [4–7] use both a centralized protection control architecture with a preplanned method and a distributed protection control architecture with a divisive reconfiguration method. Although these architectures are similar in design to SONET/SDH DCS self-healing networks [8–11], ATM technology provides several potential benefits.

Table 7-1 compares the principle characteristics of VP restoration in ATM networks with those of digital path restoration in STM networks. One of the most important advantages is the ability to preestablish backup paths using zero-bandwidth VPs, as mentioned in Section 7.2.1. Independence between alternate route selection and capacity assignment for ATM VP networks helps simplify the design of the self-healing protocol.

ATM technology significantly reduces the cross-connect time for network reconfiguration in distributed control mesh-type restoration systems, due to the use of a labeled routing scheme on a self-healing switching fabric and to ATM's inherent parallel processing and switching (cross-connect) capability [12]. In addition, due to separation of physical route establishment and capacity assignment, the restoration protocol used in the ATM self-healing network may be more efficient than its SONET/SDH counterpart (which requires a three-phase protocol to complete the restoration process; see Chapter 2).

Table 7-1: Comparison of digital path and VP restoration.

System Attribute	Digital Path (STM)	Virtual Path (ATM)
Restoration unit	digital path (STS, STM, VT-1.5, VC-3, VC-4)	Virtual Path
Number of restoration layers	multiple path layer	single path layer
Preestablished backup path	not cost-effective	flexible (zero-bandwidth protection VP)
Built-in path OAM mechanism	path overhead (POH)	OAM cells
Restoration protocol	less efficient (three phases)	more efficient (fewer than three phases)
Number of restoration paths	few (less than 50 paths/link)	can be many (up to 4096 paths/link)
Traffic type	one (CBR)	several (CBR, VBR, UBR, ABR)
Managed QoS parameters	transmission delay, bit error	bit error, Cell Transfer Delay, CDV, cell loss/misinsertion, throughput (ABR), etc.
Bandwidth dimensioning scheme	not required	required (to handle multiple QoS parameters and traffic types)
Managed resources	bandwidth (time slot)	bandwidth, number of VPIs

The signaling channels, which use either user information cells or OAM cells for the ATM/VP networks, have a much higher bandwidth for restoration message exchange than that for SONET/SDH networks. This results in a shorter control-message transfer time.

A powerful VP layer failure management scheme using OAM cells is also a significant advantage [13]. VP Alarm Indication Signal (AIS) cells are generated immediately after receiving a defect indication and are transmitted every second, as long as the defect indicating an interruption of cell transfer capability at the VP level exists. A VP Remote Detect Indication (RDI) cell is sent to the far end from a VP termination point when a VP-AIS state has been declared. This enables cross-connect nodes along the failed VP to detect the failure very rapidly, which allows the self-healing scheme to quickly trigger restoration.

7.3 ATM Self-Healing Network Architectures

7.3.1 Class of ATM Self-Healing Network Protocols

As with SONET/SDH self-healing networks, ATM VP self-healing networks may use a common generic restoration algorithm but the self-healing protocols provoking this restoration algorithm may differ, depending on timing, control location, restoration point, and direction (unidirectional or bidirectional) of generating restoration paths when the network is in the restoration process. Figure 7-1 depicts such as a class of ATM self-healing network protocols:

1. Self-healing control scheme: centralized restoration versus distributed restoration
2. Timing for restoration route generations: guided restoration (prior failure) versus unguided restoration (at the time of failure)
3. Signal restoration level: local (line) restoration versus end-to-end (path) restoration
4. Bidirectional restoration methods: coordinated bidirectional restoration versus uncoordinated bidirectional restoration

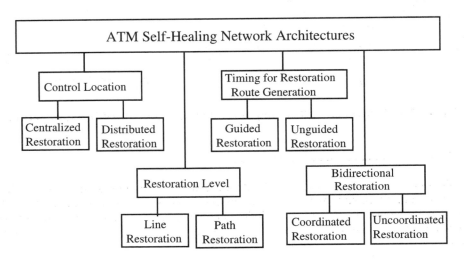

Figure 7-1: A class of ATM self-healing networks.

7.3.1.1 Centralized Restoration vs. Distributed Restoration

Figure 7-2 depicts examples of centralized and distributed self-healing control systems. In a centralized ATM self-healing network, an end node of a failed facility detects the failure and informs the central controller via a separate data communications network. The central controller then computes the rerouting paths based on the topology and network resource data it has at that time. Once the rerouting path(s) is obtained, the centralized controller instructs the corresponding ATM equipment (e.g., VPX or VC switches) along the new path to perform resource allocation for traffic rerouting. In a centralized ATM self-healing network, path rerouting and acknowledgment messages are sent using OAM cells. An ATM VP bandwidth management system proposed in [14] may be enhanced to serve as a centralized control self-healing system.

In contrast, in the distributed control environment, the route computation and resource allocation functions are performed at the local node, which is responsible for triggering the restoration protocols. These functions are performed through exchanges of restoration messages in a distributed manner. We will discuss a distributed self-healing system in Section 7.3.2.

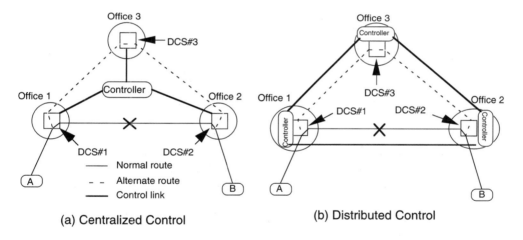

(a) Centralized Control   (b) Distributed Control

Figure 7-2: Centralized self-healing control vs. distributed self-healing control.

7.3.1.2 Line Restoration vs. Path Restoration

Services carried by an ATM transport network can be restored on a physical line (line restoration) or on an end-to-end VP (path restoration) basis, as shown in Figure 7-3. In the example shown, the link between Nodes Q and R fails. If the line restoration method is used, all VPs in link Q–R are replaced by VPs in link Q–T–R. If the path restoration method is used, each VP affected by the link failure selects a new route for restoration. In the example shown in Figure 7-3(b), all VPs in route P–Q–R–U are replaced by VPs in route P–S–T–U.

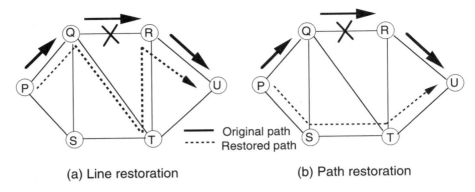

(a) Line restoration (b) Path restoration

Figure 7-3: Types of route restoration.

Line restoration reestablishes VPs for all affected VPs carried over the failed link. Standby VPs can be established on a preplanned or dynamic basis. When the preplanned method is used with the ATM distributed self-healing control architecture, all preassigned VPs for each link can either be stored in each node connected by that link or in the central controller and then downloaded to the affected node's ATM VPX when needed. For a VP that requires protection routing, protection routes can be preestablished without reserving capacity. When a network failure is detected, necessary path capacities are assigned along the preestablished protection path routes. This process can be performed using distributed or centralized processing. When the dynamic line restoration method is used with the ATM distributed self-healing control architecture, standby VPs for all affected VPs carried over the failed link are dynamically computed when needed [15]. In this case, the restoration path-finding process is similar to the one for SONET/DCS self-healing networks.

Path restoration establishes new VPs for each end-to-end working VP. These standby VPs can be preplanned or computed dynamically whenever needed. If the alternate routes are found using the dynamic rerouting method (i.e., the flooding-based rerouting method), restoration may be slower than that using the preplanned method, because the area searched for the alternate routes may be large.

7.3.1.3 Guided Restoration vs. Unguided Restoration

The difference between guided restoration and unguided restoration is the timing involved in providing rerouting information to the restoration process. For guided restoration, rerouting information is computed periodically. Thus, rerouting information (including rerouting paths) may be ready when the network component has failed and the restoration process is triggered. In contrast, in the unguided restoration method, rerouting paths are searched when the restoration process is provoked via the exchanges of route searching messages and acknowledgment messages. Compared with unguided restoration, guided restoration may restore service faster but may have a relatively high probability of incomplete restoration (due to less adaptation to the network status at that time) and require a local topology database for route computation to minimize the message exchange overhead. One example of guided restoration can be found in [16].

7.3.1.4 Coordinated Bidirectional Restoration vs. Uncoordinated Bidirectional Restoration

As the ATM connection is unidirectional in nature, the ATM connection restoration is typically performed unidirectionally. Bidirectional restoration may be executed in a coordinated or uncoordinated manner. For uncoordinated bidirectional restoration, the bidirectional ATM connection can be restored from both the downstream and upstream directions simultaneously and independently. However, because the paths for the downstream and upstream restoration for bidirectional restoration may be different, that would result in more extensive switchback operations. Such inefficient operations may be avoided by coordinating the restoration process from the downstream and upstream directions. We will discuss a coordinated bidirectional restoration process (also called double-search process) in Section 7.3.3.

7.3.2 Distributed Self-Healing Protocol

Figure 7-4 depicts a distributed ATM self-healing network protocol proposed in [15]. The example in Figure 7-4 uses a unidirectional, line, unguided restoration method. In this example protocol, the downstream node of the failed link is designated as the Sender node, which will broadcast restoration messages to the other end node of the failed link (called the Chooser node). The Chooser node decides the alternate routing paths for restoration based on information included in the received restoration messages. In the node, the restoration messages are logically distinguished from user information messages.

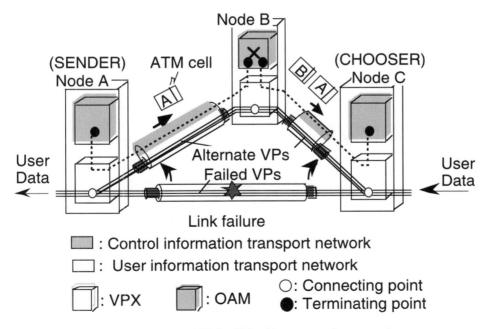

Figure 7-4: An ATM self-healing network protocol.

The self-healing algorithms discussed in [15] use the same basic concept as the SONET distributed DCS restoration algorithms (see Chapter 2), except that restoration time and restoration ratio in the Chooser are recorded immediately after the search for alternate routes is completed. Figure 7-5 presents a diagram of the time flow of this algorithm. When the Sender detects the failures, it generates and broadcasts route-search messages to its adjacent nodes. When a route-search message arrives at an intermediate node, it is checked to determine if the received restoration message has errors or missequencing. If no error is detected, the intermediate node adds the restoration information (e.g., available VP capacity, VP identifier, and node ID) and rebroadcasts the message to adjacent nodes. If an error is detected, the message is discarded and the intermediate node sends a retransmission request to the node that sent the message. This process continues until all messages are transmitted correctly between the sending and receiving nodes.

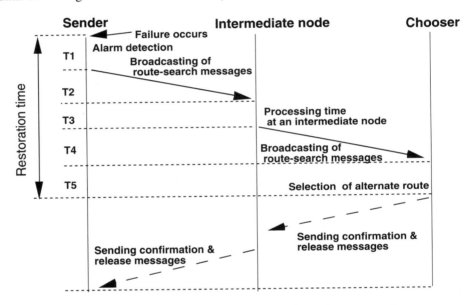

Figure 7-5: Time flow model of a self-healing protocol.

When the Chooser receives the first route-search message, it compares the available spare capacity of the alternate route against that of the failed VPs. The Chooser has enough information about the alternate route and its spare capacity to decide which failed VPs are to be replaced by the alternate VPs. How the Chooser decides to restore which failed VPs depends on the restoration route selection algorithm. The objective of the restoration route selection algorithm in the Chooser node may be to minimize delay, maximize restoration ratio, or the like.

The Chooser selects the failed VPs that can be restored using the alternate route, and it sends the user information in the failed VPs through the newly selected VPs. If all failed VPs cannot be completely restored in the first run, the second route-search message received by the Chooser is examined in the same way. This process continues until all failed VPs have been reassigned or no available spare capacity or route can be found. After all alternate VPs are determined, the Chooser records the affected capacity that has been restored; it then sends a confirmation message along each selected alternate VP to request that the nodes along the alternate route update the VPI routing table so that traffic going through the failed VP can be rerouted to the alternate route. The Chooser also broadcasts a message to the Sender to release the unused protected VPs and their spare capacity.

If the reliability of the restoration messages is a concern, the link-by-link error control mechanism for confirmation messages and release messages along with an end-to-end error control mechanism for both messages may be considered [17].

If the failed VP is bidirectional, the restoration algorithm can perform the VP restoration for each direction simultaneously. Since the upstream nodes (i.e., Chooser node) for both directions select the alternate route independently, the alternate route selected for one VP direction may be different from that for the other VP direction. Having the bidirectional VPs follow different alternate routes is operationally inefficient and so the restored VPs should be switched back as soon as the failed transmission link is restored.

7.3.3. Coordinated Bidirectional (Double-Search) Restoration

The self-healing algorithm described in Section 7.3.2 restores the VPs unidirectionally. For a bidirectional VP, the route-search process for each VP direction needs to start from the downstream node (i.e., Sender node) to the upstream node (i.e., Chooser node) for both directions. The overhead generated by searching alternate routes to restore a bidirectional VP is almost twice the overhead needed to restore a unidirectional VP. The overhead can be reduced significantly if the bidirectional VP restoration works in a coordinated manner, rather than independently. In this subsection, we discuss a coordinated bidirectional VP restoration method, called the double-search algorithm, as proposed in [18,19]. The concept of this double-search algorithm may be used not only for ATM networks, but also for the SONET/SDH networks.

7.3.3.1 Double-Search Restoration Algorithm

Figure 7-6 depicts the procedures of a double-search self-healing algorithm proposed in [18]. This algorithm consists of five phases using three self-healing control messages for route search, response, and acknowledgment. This five-phase self-healing algorithm is summarized as follows.

Phase 1 (Detection Phase)   When a transmission link fails [Figure 7-6 (a)], each node (X and Y) at the end of the failed link receives an alarm (e.g., Loss of Signal, AIS) and detects the link failure. Nodes X and Y then become Senders and broadcast route-search messages to the neighboring nodes.

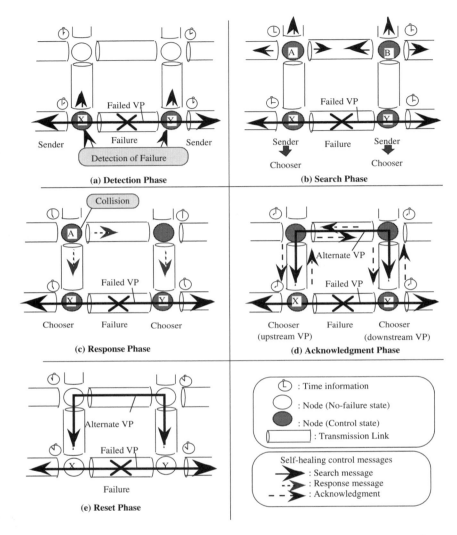

Figure 7-6: Procedures of a double-search self-healing algorithm [18].

Phase 2 (Search Phase) After broadcasting the search messages, the two Senders become Choosers [Figure 7-6(b)]. In this search phase, every node that receives a search message begins to search for alternate VP routes. At this time, the state in each node changes from a no-failure state to a control state.

When a node (A and B in our example) receives a search message, it reserves capacity for use as alternate VPs and broadcasts new search messages to its neighboring nodes. As each node rebroadcasts this search message, it increments the hop number in the message by one. If the hop number has already reached the hop count limit, the node does not rebroadcast the message.

Phase 3 (Response Phase)   If during the search phase, one node (e.g., A) [Figure 7-6(c)] receives search messages from both Senders, this is called a collision. If a collision occurs, the affected node enters a response phase. In this phase, the affected node combines the contents of the two search messages. This combined message, called a response message, contains all the information needed to identify the alternate VPs, including their routes and available VP capacity. The node sends this response message to the two Choosers according to the route information in the message. When the Choosers receive this response message, they can identify the alternate VP candidates. If collisions occur at several intermediate nodes, the Choosers receive several response messages.

Phase 4 (Acknowledgment Phase)   After receiving one or more response messages, both Choosers enter an acknowledgment phase in which they select alternate VPs [Figure 7-6(d)]. When a bidirectional VP fails, the Choosers select the alternate upstream and the downstream VPs independently. In our example, node X selects the alternate upstream VP and node Y selects the alternate downstream VP.

The Choosers select the alternate VPs as follows. Each Chooser knows the capacity of each failed VP beforehand. Using this data, they select alternate VPs and send the acknowledgment messages back along the route of the alternate VPs, restoring the failed VPs. If some affected VPs remain unrestored, the responsible Chooser selects another alternate VP. The Choosers repeat this process until all failed VPs have been restored or until there are no more alternate VP candidates.

Phase 5 (Reset Phase)   After the restoration is completed, the node state needs to be returned to the no-failure state. Each node performs this process by using the time of failure information [20] in the search message [Figure 7-6(e)]. Each node calculates the time elapsed since the failure occurred. After a predetermined fixed time, the node returns to the no-failure state.

Many restoration messages are broadcast in a self-healing network. When nodes receive these messages, they have to reserve available VP capacity for use as alternate VPs. If the reserved VPs are not used for restoration, an invalid process will occur in the nodes. For example, when a multiple transmission link failure occurs, this invalid process may cause an invalid allocation of available VP capacity. The area affected by restoration messages should thus be reduced to achieve efficient operation and management.

In self-healing algorithms, the hop count limit (the maximum number of nodes that the restoration messages can pass through) generally restricts the search area. Conventional self-healing algorithms have to send the restoration information from a Sender to a Chooser to find alternate VPs. The double-search self-healing algorithm, however, gets this information when the restoration messages from two Senders meet at an intermediate node. The hop count limit can thus be set to almost half that of conventional algorithms. For example, the proposed algorithm with a hop count limit of four can obtain the same information about alternate VPs as a conventional algorithm with a hop count limit of eight. This double-search algorithm certainly reduces the restoration time by reducing the search area.

In the same way as its unidirectional counterpart, because the two Choosers in the double-search algorithm select the alternate routes independently, the route of the selected alternate upstream VP route may be different from that of the selected alternate downstream VP route. Thus, for operational efficiency, the restored VPs should be switched back as soon as the failed transmission link is stored.

7.3.3.2 Case Study

Figures 7-7 and 7-8 depict some performance results of the proposed double-search restoration algorithm that can be compared with results obtained using a unidirectional restoration algorithm [8]. This performance analysis was reported in [18] and was based on a lattice network model. As depicted in Figure 7-7, the number of invalid nodes with the proposed double-search algorithm may be reduced to one-fifth of that with a conventional algorithm [8]. Here, valid nodes are those belonging to the alternate VP candidates; invalid nodes are those receiving a search message but not belonging to the alternate VP candidates. This result suggests that the coordinated bidirectional searching method would be more efficient than uncoordinated bidirectional searching methods.

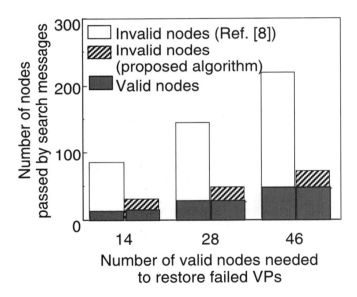

Figure 7-7: Reduction in the number of invalid nodes.

Figure 7-8 depicts the restoration ratio (the capacity of restored VPs to the capacity of failed VPs) for failed bidirectional VPs. The average restoration ratio shown here was the result of all possible simulation cases of a transmission link failure, which may statistically characterize the self-healing algorithm. In general, this result shows that the restoration ratio and the restoration time for a coordinated bidirectional restoration method are at least comparable to those for an uncoordinated bidirectional restoration method. Please refer to [18] for a detailed simulation model and assumptions.

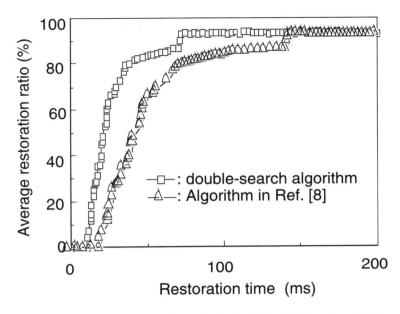

Figure 7-8: Restoration ratio analysis for failed bidirectional VPs.

7.3.4 Triggering Methods for Self-Healing Networks

In the ATM network, the control information for restoration has generally been considered to be transferred by an OAM cell. The information for dynamic restoration may need a longer bit length than an OAM cell, as will be discussed in the next section. This occurs primarily because, in the self-healing protocol, each intermediate node needs to add a new node and capacity information into the routing list carried within the restoration message. Assuming an average of seven hops for an alternate rerouting path for a large network, it is impossible for the restoration message to be carried on a single OAM cell. To transfer all the restoration information at the same time, a mechanism for allocating multiple OAM cells for the information needs to be specified. We will discuss one example of this multicell mechanism in Section 7.4.

Alternatively, the signaling protocol used for PNNI [21] potentially may be used to trigger the dynamic restoration process in self-healing networks. PNNI routing is based on source routing that determines desired paths using hierarchical topology state information. When the source node detects a connection failure, it computes a new path based on the current topology, taking into account any useful information that may become available at the time of reroute. Although PNNI Phase 1 Routing [21] does not support an automatic rerouting of an exiting connection after the detection of a certain type of network failure, the crankback mechanism that is specified in PNNI automatic recovery from failure during a connection setup may be used to trigger connection rerouting. When the call cannot be processed according to the designated path during the call setup process, it is cranked back to the creator of that designated path (i.e., source node), with an indication of the problem. An example of the crankback process is depicted in Figure 7-9.

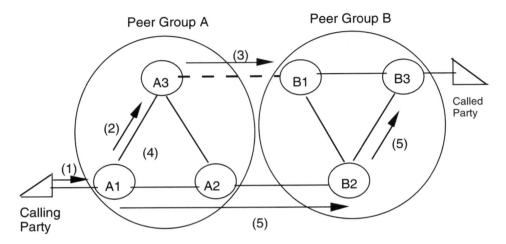

(1) Setup of UNI signaling
(2) Node A1 decides the route, and orders Node A3 to connect to
    Peer Group B
(3) Breaking occurs between Node A3 and Node B1
(4) Node A3 sends a release message to Node A1
(5) Node A1 decides the route again, and orders Node A2 to connect
    to Peer Group B.

Figure 7-9: An example of the PNNI crankback procedure.

To carry out crankback, all nodes in the network must exchange the information on network topology (such as port IDs in the node, address information) by using the signaling protocol. Thus, this mechanism potentially may be used as a new triggering method for self-healing networks. However, the current version of PNNI needs to be enhanced to provide such a network restoration capability. For example, we consider two disjoint paths from the source to the destination, which can support the QoS required by the call. To avoid setting up two erroneously overlapping paths across the Peer Group, each route has to be aware of the presence of the other one. An additional information exchange needed to support this mutual awareness of paths is not supported by the current version of the PNNI Phase I specification [22]. This additional information exchange feature for alternate rerouting following path failure needs to be added to the PNNI Phase I specification.

7.4 Reliable Control Message Transfer for Restoration

The message transport control protocol in the self-healing algorithms discussed in the previous section potentially may cause a miscount of the network restoration ratio, because an alternative route and its available capacity may not be available due to line errors or ATM cell loss in the route-search messages. Thus, restoration message protection for the self-healing protocol needs to be considered.

This section first reviews the restoration message structure and its transfer mechanism through OAM cells or user/signaling AAL. Then the reliable restoration message transfer protocols, including link-by-link retransmission protocols, will be discussed.

7.4.1 Restoration Message Structure

There are, in general, three restoration messages in a self-healing protocol: route-search, confirmation, and release. To analyze the performance of the message transfer protocol, the information contained in these messages must be identified. Table 7-2 shows examples of the information that may be contained in the restoration message. The message and the usage types indicate the management and application types, respectively. Note that if we focus only on the self-healing function, this information may not be needed. This protocol, however, can be adopted not only for fault management like self-healing control, but also for other management functions, such as configuration management and performance management. The flow type is used to distinguish the message type. The message index is used to identify each message. In a self-healing protocol, many route-search messages are sent for the same failure. In some cases, a message may return to the originating node. These messages should be deleted as soon as possible, because they are not needed to restore the network. To do this, a message index may need to be assigned to the control message bytes.

In a route-search message, the Sender and Chooser node IDs indicate where the Sender and Chooser are physically located. The failed link ID identifies the failed link between the Sender and the Chooser. The capacity of failed link is used when the nodes reserve VP capacity. The hop count limit indicates the maximum number of links on an alternate route. The intermediate-node ID indicates the nodes passed through by a route-search message. The available VP capacity and the available number of VP connections are used for selecting alternate VPs. These three parameters related to alternate VPs are added each time a message passes through an intermediate node.

The confirmation message carries the selected node ID and the selected link IDs and their capacities; the release message carries the release command. We optionally add a time stamp and other useful information for multiple-failure restoration. Details of the way this information is used can be found in [15,20].

Table 7-2: Example contents in self-healing control messages for line restoration in ATM networks.

Item	Purpose	bytes/bits	
Message type	Indication of management type: configuration, performance, fault, etc.	4 bits	
Usage type	Indication of function type: self-healing, signaling, etc.	1 byte	
Flow type	Indication of self-healing flow: broadcast, acknowledge (confirmation), cancel	4 bits	
Message index	Self-Healing Control (SHC) message no. (indication of message trace)	1 byte	
Node ID (Sender ID)	Indication of node that recognized failure	1 byte	
Failed link ID	Indication of failed link no. (VPI)	1 byte	(): Option
Capacity of failed link	Indication of a failed link capacity (peak bit rate)	2 bytes	N: Number of intermediate nodes
Hop count limit	Limit of broadcasting area	1 byte	
(Time stamp)	(Measurement of delay time)	(6 bytes x N)	
(Priority value) (Judgment message)	(Priority value assigned by sender to an SHC message) (Control for ordering of restoration)	(1 byte) (1 byte)	Multiple failure restoration
(Chooser (candidate) ID) (Required capacity of each sender–chooser pair)	(Multiple destinations)	(1 byte) (2 bytes)	
Node ID (intermediate-node ID) Available alternate VP ID Available VP capacity	Basic data for selecting optimum alternate route	2 bytes x N 2 bytes x N 2 bytes x N	N: Number of intermediate nodes
Selected node ID Selected link ID (and capacity)	Acknowledgment of alternate route(s)	1 byte 1 byte	
Release message	Release of unused VP capacity in each node	1 byte	

7.4.2 Transfer Mechanism for Control Messages

Generally, the restoration message for the ATM VP self-healing mechanism has been considered to be transferred by an OAM cell [13]. Because a large amount of information is required in a restoration message, as shown in Section 7.4.1, it may be impossible to transfer all of it in a single OAM cell. For example, by the time a route-search message passes through seven nodes, the information in the message takes up to 50 bytes. This means the message may require more than one OAM cell for message exchange in the ATM Layer [18,23].

Moreover, the definition of OAM cells may need to be revisited for restoration. A physical-layer OAM cell is used in and terminated at the Physical Layer [13]. A performance-monitor cell for traffic measurement is one such cell. An ATM-Layer OAM cell, for example, a VP-AIS cell, is generated in and terminated at the ATM Layer. For the self-healing control mechanism, another OAM cell type must be assigned to the upper layer for control information transfer. This cell is called a Self-Healing Control (SHC) cell to distinguish it from the other OAM cells.

The restoration message can be divided into SHC cells by using the ATM Adaptation Layer (AAL), as depicted in Figure 7-10.

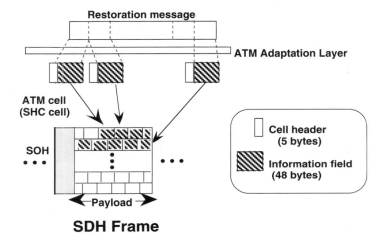

Figure 7-10: Transfer mechanism using ATM cells.

7.4.3 Reliable Restoration Message Transfer Protocol

One approach to provide reliability for restoration message transfer uses link-by-link retransmission for protecting the route-search messages. Figure 7-11 depicts the time flow for the protocol that protects the message signals. The basic protocol is the same as that described in [15], except that retransmission using an error recovery mechanism is applied to the Phase 1 operation (Option 1) or both the Phase 1 and Phase 2 operations (Option 2). Since the reliability of messages can be guaranteed by this link-by-link error control mechanism, the selection of a restoration route and its establishment are done using messages passed in only one direction.

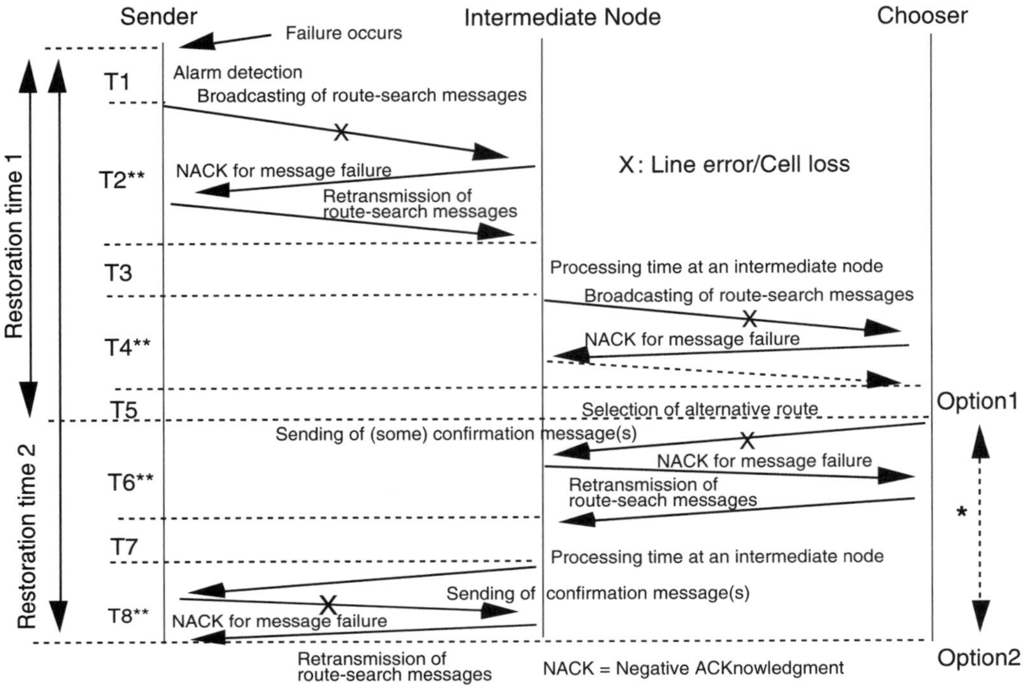

*: Also consider confirmation process between Sender and Chooser.

**: These values are statistical, and they depend on error conditions in the message link.

Figure 7-11: Time flow for ATM VP self-healing network with error recovery technology.

7.4.4 Error-Recovery Mechanism in ATM Adaptation Layer

There are two possible recovery mechanisms in the AAL: Go-Back-N and Selective Retransmission (SR). This section briefly summarizes these two error recovery methods [23].

The Go-Back-N retransmission method is commonly used in most existing protocols for error recovery in terrestrial networks. The basic operation of Go-Back-N is shown in Figure 7-12. A receiver buffers only those packets that are correctly received and in sequence. If one packet in a sequence of transmitted packets is lost, then this packet, plus subsequent packets that have already been sent, must be retransmitted. This sometimes results in a large number of packet retransmissions for only a small number of packets lost.

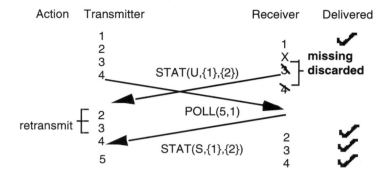

Figure 7-12: Error recovery in Go-Back-*N* retransmission mode.

Unnecessary retransmission of packets is avoided by using the Selective Retransmission method, in which only the corrupted or lost packets are retransmitted. The basic operation of Selective Retransmission is shown in Figure 7-13. The transmitter resends only the "known" lost packets while continuing to send new data within the constraint of its end-to-end window. Achieving this objective requires the receiver to inform the transmitter of the reception status of each packet (received, delivered to higher layer, missing, received in error, etc.). The control messages from the receiver therefore need to carry more status information than is required for the Go-Back-*N* method.

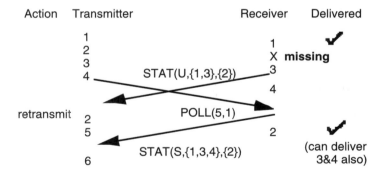

Figure 7-13: Error recovery in selective retransmission mode.

In general, Selective Retransmission has better performance than Go-Back-N in terms of overhead. Error recovery in the AAL in general may be considered to be based on Selective Retransmission; however, few precise analyses based on Selective Retransmission have been reported. For those reported, the parameters are limited and the models are greatly simplified. For example, [24] reported a throughput characteristics analysis of Selective Retransmission based on a very simplified model. Therefore, to evaluate the worst-case scenario in terms of delay, the Go-Back-N method may be used for error recovery at the data link layer. The case study results reported in [17] suggested that the delay caused by using link-by-link error control may be insignificant, compared with the 2-second objective for complete restoration [25]. In [17], the average delay penalty due to the added link-by-link error recovery was estimated to be 7.5 ms for a 100 percent network restoration ratio. The queuing delay in each node was found to have more effect on network restoration time than the link delay introduced by the error-recovery mechanism. The window size in the data link level should thus be optimized to obtain faster restoration.

7.5 Self-Healing Switching Systems

As already discussed, the VPs play an important role in ATM self-healing mesh networks. A VP is a logical, point-to-point connection between two exchanges or users; it has a fixed route and a fixed capacity [26,27]. To provide a more efficient and less expensive ATM self-healing mesh network, the ATM VP Cross-Connect System (VPX) must be designed to be cost-effective. The ATM VPX must have the following characteristics and capabilities [12]:

1. Nonblocking characteristics
2. No restrictions on using the intermediate link in a multistage switching network
3. Cell-sequence integrity
4. Large system throughput (e.g., greater than tens or hundreds of gigabits per second)
5. Minimum cell-loss probability
6. Cross-connect capability for various VP speeds
7. VP management capabilities

The second requirement means that the switching fabric should be determined only by the input and output port connection relationships and should not be restricted by intermediate link connections. Even when VP capacity is dynamically altered in the network, blocking stemming from the utilization of the inner links should be avoided. For example, blocking may occur in a non-self-routing three-stage switching network if the VPs are set up through intermediate links and the VP bandwidths are dynamically changed. Examples of switching fabrics that meet requirements 1–3 are one-stage switching networks and self-routing multistage networks.

To operate an ATM network effectively, the cross-connect system should support such VP management mechanisms as VP maintenance signal transfer, VP transmission quality monitoring, VP rerouting against line failure, and transmission line protection switching.

Various switching fabric structures have been studied [27–29], each with different features and application areas, as shown in Table 7-3. The shared buffer type may be suitable for smaller switches and is robust against bursty traffic. The input buffer type with Random-In, Random-Out (RIRO) queuing and contention control may be more suitable for large switching networks. The output and cross-point buffer types have intermediate characteristics. Additional functions, such as broadcast connection and priority control, can be implemented easily with the output buffer and cross-point buffer types.

Table 7-3: ATM Switching Architectures.

Switch types	Configuration	Feature
Output buffer		FIFO control Bus bottleneck
Shared buffer		RIRO control Smaller buffer size
Input buffer		RIRO control Contention control to increase throughput
Cross-point buffer		FIFO control Larger buffer size

B: Buffer,   MUX: Multiplexer,   DMUX: Demultiplexer,

FIFO: First-In, First-Out Logic,   RIRO : Random-In, Random-Out Logic

The functional architecture of the ATM VPX proposed in [29] is depicted in Figure 7-14. It consists of I/O ports, a self-routing or shared buffer switch, a contention-control module, and a VP management module. Each input port has its own buffer, VPI tables, and input controller. The VPI table provides the designated output port number and the new VPI for the incoming cells. The input controller and contention control module use a parallel-processing architecture that dynamically assigns the send time of input cells to prevent cell collision in the routing switch. The self-routing or shared buffer switch receives the cells, which are tagged with their designations, from the input ports and transfers them to the designated output ports using self-routing control.

Note that most of the ATM VPX systems or prototypes reported today that directly terminate 2.4 Gbps ATM cell streams (i.e., STS-48c or STM-16 terminations) use shared memory design to support a Cell Loss Rate of 10^{-9} [30–32] (also see Figure 3-18 for an example of an ATM VPX system configuration).

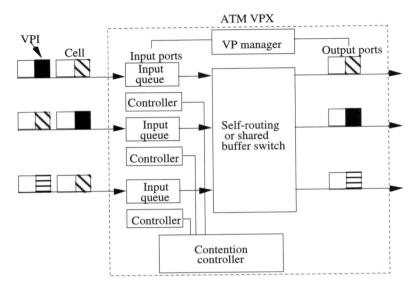

Figure 7-14: An ATM VPX functional architecture [29].

Recently developed network elements such as switches and cross-connects offer strong computational power, reflecting the rapid advances in CPU power and cost reduction. Self-healing implies failed-path restoration using distributed network element control mechanisms. Network failure management schemes for a distributed network element control environment have been studied [33–36] from the viewpoints of parallel processing, fast intranetwork element message transmission, and robustness compared with centralized Operations System control. Each network element in a self-healing network includes a Network Element Function (NEF) and a Synchronous Equipment Management Function (SEMF) [37]. The distributed control process and local databases for self-healing are implemented in the SEMF, which is separated from NEF physically as depicted in Figure 7-15. This scheme provides independence between the SEMF and the NEF, and simplicity in modifying or upgrading the control process or databases.

Although distributed network element control is an effective scheme for network control, it is not suitable for all network control applications. For example, the operator control functions for a Wide-Area Network are best provided by conventional centralized control. Therefore, distributed network element control schemes and centralized control schemes must coexist, and they require interaction.

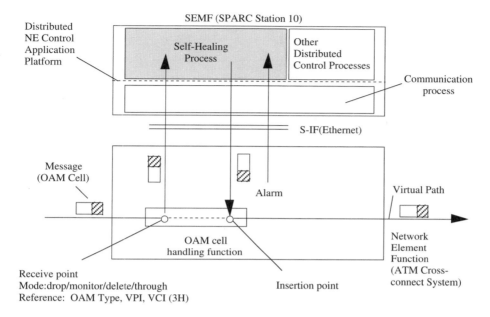

Figure 7-15: A platform for distributed NE control application.

7.6 Backup VP Assignment and Spare Capacity Design and Analysis

7.6.1 A Spare Capacity Design Algorithm for ATM Self-Healing Networks

To guarantee adequate restoration probability with minimum spare resources, a spare resource design algorithm is needed. This algorithm for the line-restoration scheme is almost the same as that for the path-restoration one. It is assumed that the network topology and the failed VP's route and bandwidth are already known. The route of a backup VP must be determined so that the failure of any one node does not lead to the failure of both the failed VP and its backup VP. This means that the routing of the backup VP must be completely independent from that of the failed VP. The amounts of spare resources distributed on each link are designed by computing the maximum resources needed in each link to handle all single-node failures in the network.

Figure 7-16 depicts the flow of the spare resources distribution algorithm [7]. It is very difficult to find the absolute minimum solution by searching all combinations of all possible routes for all backup VPs, because the number of combinations is huge. As an initial condition, a shortest route among all possible routes that satisfy the routing requirement is chosen, and then the spare resource amount is minimized by subsequently assigning the backup route. This permits determination of a cost-effective backup VP route and its spare capacity. In addition to [7], readers who are interested in virtual path routing on survivable ATM networks may refer to [38] for details.

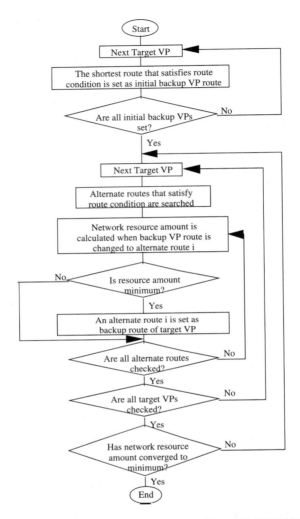

Figure 7-16: A spare resources distribution algorithm [7] (©1994 IEEE).

7.6.2 Spare Capacity Analysis for Path Restoration vs. Line Restoration

The spare capacity needed for using the spare capacity allocation algorithm described in Section 7.6.1 for both line restoration and path restoration has also been reported in [7], and these results are summarized in Figure 7-17. The results shown in Figure 7-17 were based on a Japanese long-haul network of 56 nodes and 200 links. The number of working VPs in the network is 2350, and the spare capacity ratio is assumed to be 0.35. In addition, it is assumed that the spare capacity design is based on the requirement of 100 percent restoration for any single-link failure. As expected, path restoration requires 37.5 percent less spare capacity than line restoration (35 percent for path restoration vs. 56

percent for line restoration). Note that the proposed spare capacity allocation algorithm may be applied to both dynamic restoration (i.e., self-healing) and preplanned restoration (i.e., protection switching).

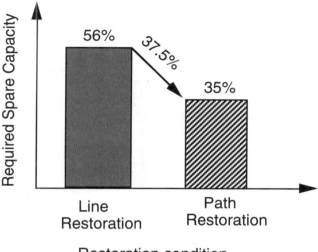

Figure 7-17: Required spare resources for line restoration and path restoration [7] (©1994 IEEE).

The number of VPI values available for protection is crucial when designing a backup VP connection. Although zero-bandwidth backup VPs do not consume spare capacity resources, VPI numbers must be reserved for all of the VPs that may fail. If a large number of backup VPs are defined for the same link, link capacity may become unusable due to a shortage of VPI numbers. The magnitude of this problem can be calculated by evaluating the ratio between the total backup VP capacity in a link and the number of backup VPIs in the same link. Reference [7] calculates the VPI Capacity Ratio (VCR) as follows [see Figure 7-18(a)]:

$$VCR = \frac{C1 / (C1 + C2)}{N1 / (N1 + N2)}$$

where C1 is total occupied VP capacity in a link; C2 is the total spare VP capacity in that link; N1 is the number of working VPs; and N2 is the number of backup VPs.

This ratio indicates the likelihood of insufficient VPI numbers for a link. This parameter should be used in addition to the spare resource algorithm when designing backup VPs, because the algorithm does not consider the availability of VPI numbers.

Figure 7-18(b) shows the VCR values in the proposed self-healing scheme under the conditions that were used for simulating the backup capacity [7]. The values were calculated as the average over all links. When designing for path restoration, because the VCR value (1.25) is near 1, the target VPs are not seriously affected by a lack of VPI numbers. On the other hand, when designing for line restoration, the lack of VPI numbers may be experienced in links accommodating narrow-bandwidth VPs, because the VCR value (2.2) is comparatively large. Therefore, the available VPI numbers may be of more concern in line restoration than in path restoration.

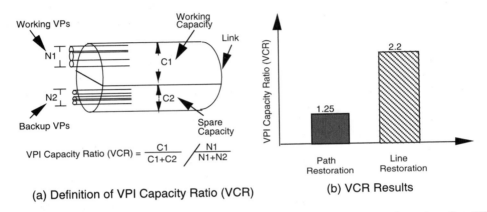

(a) Definition of VPI Capacity Ratio (VCR)   (b) VCR Results

Figure 7-18: VPI Capacity Ratio (VCR) for line restoration and path restoration [7] (©1994 IEEE).

7.6.3 Impact of VP Assignment on Network Restoration Ratio

One broadband network evolution issue involves evaluating how the ATM VP technology works under the network capacity (both the working and spare capacities) designed for SONET/SDH networks. This section summarizes some results reported in [39] that may provide insight for this network evolution question.

Given the end-to-end STS-1 demand requirement, the physical routing path for each node pair for working traffic, and working and spare capacities on each link, the Network Restoration Ratio (NRR) for the ATM VP distributed self-healing network depends on how the end-to-end demand is assigned to VPs due to the single-VP-restoration constraint. The single-VP-restoration constraint, which is the fundamental principle of VP restoration, means that only one affected VP can be restored via one protection VP on one alternate route. This VP assignment problem for each node pair includes finding the number of VPs and their corresponding sizing.

The VP considered here is the point-to-point VP that carries Virtual Channels (VCs) (e.g., STS-1s in our case here) with the same source and destination pair. There are two options for point-to-point VP assignment for each node pair: multiple VP assignment (denoted M-VP assignment) and single VP assignment (denoted S-VP assignment). The M-VP assignment provisions multiple point-to-point VPs for the same node pair to carry the end-to-end demands; in contrast, the S-VP assignment allocates only one VP carrying all end-to-end demands (without service separation) for that node pair. The M-VP assignment is usually used when the end-to-end traffic includes different types of services with different classes of service requirements (e.g., voice, video, and SMDS). In other words, different types of services are assigned to different VPs in the M-VP assignment. In addition to service separation, an appropriate bandwidth sizing for each VP may help improve the restoration ratio. The VP assignment for the end-to-end demand needs to be determined during the service provisioning process. Compared with the M-VP assignment, the S-VP assignment may make VP management simpler (due to fewer VPs needed), but it may have a lower restoration ratio and a more stringent QoS requirement for all services carried on that VP.

Table 7-4 depicts the impact of VP assignment on the NRR of the network for which working and spare capacities are optimized with respect to SONET/SDH DCS distributed restoration [39]. The model network used in this study is a metropolitan LEC network (see Figure 7-19), which includes 15 nodes and 28 links. The working and spare capacities for each link, as shown in Figure 7-19, are based on a spare capacity assignment algorithm [40], which was designed primarily for their SONET DCS distributed restoration system. The SONET/DCS and ATM/VP distributed restoration algorithms used here are based on the algorithms proposed in [40] and [15], respectively. The demand requirement in the model network consists of special service circuits for private line voice/data services and message circuits for public switched voice services. Thus, the M-VP assignment requires two point-to-point VPs for each node pair, in which each VP carries one type of circuit. The S-VP assignment requires only one point-to-point VP for each node pair in which this VP carries two types of circuits on it. In Table 7-4, for the ATM/VP network, the hop count limit is nine; the bandwidth used for restoration messages is 10 percent of the OAM bandwidth for each link, where the OAM bandwidth is 1 percent of the link capacity.

Table 7-4: Network restoration ratios for SONET and ATM/VP distribution restoration.

Link	Affected STS-1s	Network Restoration Ratio (NRR)		
		SONET	Multi-VP Assignment	Single-VP Assignment
(8–9)	108	100	100	63.0
(8–11)	146	100	99.7	65.7
(10–11)	134	100	99.9	99.0
(10–15)	88	100	100	63.6
(11–15)	69	100	100	21.7
(12–13)	71	100	100	99.8
(14–15)	55	100	100	94.8
Other links	see Fig. 7-19	100	100	100.0
NOTE: The spare capacity here is optimized for SONET STS-1 networks.				

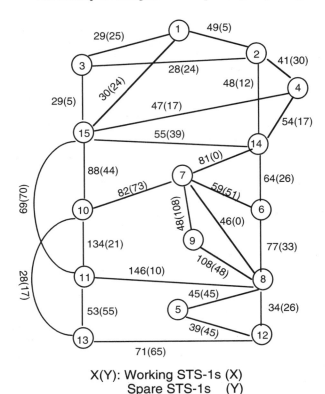

X(Y): Working STS-1s (X)
      Spare STS-1s   (Y)

Figure 7-19: A model network and capacity assignment for VP assignment analysis.

As shown in Table 7-4, the SONET/DCS distributed restoration system may restore 100 percent of the affected services for all link failures; the spare capacity shown here is designed for such a 100 percent STS-1 restoration. The ATM/VPX distributed restoration system with the M-VP assignment may completely restore services for all link failure cases except for two links, (8–11) and (10–11), for which restoration ratios are 99.7 percent and 99.9 percent, respectively. These non–100 percent restoration ratios occur because the spare capacity in the model network is designed for the STS-1 restoration, rather than VP (a group of STS-1s) restoration.

As also shown in Table 7-4, the ATM/VPX distributed restoration system using S-VP assignment cannot completely restore services for some single-link failure scenarios. In particular, for the case of link (11-15) failure, the system can only restore 21.7 percent. This low restoration ratio occurs because link (11–15) carries 12 VPs with 1 VP having 54 STS-1s for node pair (11,15). According to the single-VP-restoration constraint and the spare capacity design primarily for STS-1 restoration, the protocol simply cannot find an alternate route with at least 54 available spare STS-1s to restore this particular VP. (Note that this example shows a significant, inherent difference between STS-1 restoration and VP restoration. For example, 54 STS-1s belonging to node pair (11,15) are affected by the failure of link (11–15); these affected 54 STS-1s can be rerouted via several alternate routes for STS-1 restoration, but they can only be rerouted via a single alternate route for VP restoration.) This example implies that the higher the working VP capacity, the less likely the VP will be restored, if the spare capacity is not designed for VP restoration (e.g., the spare capacity shown in Figure 7-18 is designed for SONET/STS-1 restoration). This also implies the M-VP assignment is more efficient and cost-effective than the S-VP assignment for ATM/VPX distributed restoration system in terms of the NRR and the spare capacity engineering. In fact, when the M-VP assignment is used for this example: 5 VPs with sizes of 5, 6, 10, 16 and 17 STS-1s for node pair (11,15), the restoration ratio increases from 21.7 to 100 percent when link (11–15) fails. The foregoing results imply that the optimal spare capacity assignment designed for SONET/STS-1 restoration may not be an optimal spare capacity assignment for the ATM/VPX distributed restoration system due to the difference between STS-1 restoration and VP restoration. The spare capacity assignment for the S-VP assignment scheme may be more expensive than that for the M-VP assignment, because the latter scheme has a more balanced spare capacity assignment over the network.

In summary, this case study suggests that the VP assignment for end-to-end demands may significantly affect the Network Restoration Ratio, assuming the network spare capacity is not planned for the ATM/VP distributed self-healing network. The result also indicates that the smaller the VP capacity, the more likely a VP can be restored. This also implies that assigning different types of services to different point-to-point VPs (for each node pair) and an appropriate bandwidth sizing for each VP would help improve the NRR as well as engineer more cost-effective spare capacity planning. Furthermore, the spare capacity assignment optimal for SONET/DCS distributed networks may not be optimal for ATM/VP distributed restoration networks, due to the inherent difference between SONET/STS-1 restoration and ATM/VP restoration.

7.6.4 Spare Capacity Analysis for VP vs. SONET/STS Restorations

To quantify the advantages of using ATM technology rather than the existing SONET/SDH technology for network restoration, [5] has reported a case study comparing total capacity needed for both SONET/SDH networks and ATM/VP networks. This case study is based on a 40-node lattice network model with assumptions of 64 percent utilization for SDH networks (due to the multistage multiplexing effect) and 90 percent utilization for ATM/VP networks. Table 7-5 summarizes the results of total capacity needed, in terms of OC-48 units, for an SDH network and an ATM//VP network that was reported in [5]. The total capacity needed in Table 7-5 consists of working capacity and spare capacity. In addition, the path restoration method and the basic restoration unit of DS3 (equivalent to STS-1) are used to obtain results in the table.

As shown in Table 7-5, ATM saves approximately 28 percent in total capacity uniformly across all the scenarios. Note that the basic restoration unit considered here is DS3 (equivalent to STS-1). Thus, if the SDH network uses a larger unit for restoration such as OC-12, as suggested in [5], the spare capacity needed for the SDH network will be higher; hence, the total capacity required will be larger than the figures shown in Table 7-5 for the SDH case. Therefore, the difference in spare capacity requirement between the ATM/VP network and the SDH network will be even larger than reported here.

Table 7-5: Comparison of total capacity required for ATM/VP networks and SDH networks.

Traffic/Failure Scenario	Total capacity for SDH networks (in OC-48 units)	Total capacity for ATM/VP networks (in OC-48 units)
Symmetric traffic under single failure	1402	1009
Symmetric traffic under multiple failures	1498	1079
Asymmetric traffic under single failure	600	438
Asymmetric traffic under multiple failures	648	461

7.7 ATM VP Trace Methods

To construct an ATM self-healing network, especially one with the centralized control scheme, the actual network configuration must always be known by the Operation System (OS) in real time. For VP route reconfiguration to be made correctly and quickly, the VP configuration stored in the OS database should always match the current network configuration. When a VP is reconfigured or some inconsistency is detected, the database must be updated. This section describes the VP trace function, which is indispensable for precise administration of the ATM network configuration. With this function the OS can automatically collect the VP configuration information, including the system number, interface number, and the VPI number by using OAM cells. This function can also be useful for SDH path trace and VC tracing [41].

7.7.1 VP Trace Function

The VP trace function required for the ATM self-healing networks must meet the following OAM requirements:

1. The performance of all in-service VPs must be constantly monitored to ensure that VP performance degradation is detected immediately.

2. Any failure in the ATM network must be detected. When a failure is detected, all related VP handling systems and the OS must be informed immediately. Moreover, line blocking or protection switching must be performed so as to minimize the effect on service.

3. When a VP is setup or recovered, its connectivity and performance must be tested. When a failure is detected, the location of the failure point must be specified.

4. It must be possible to confirm the current network configuration quickly and accurately. The network configuration database must be updated quickly to ensure efficient and precise network administration.

The VP OAM functions that satisfy these requirements are discussed in [13]. Reference [42] explains how the VP maintenance signal transfer function can be implemented in an ATM network. The VP trace function that meets these four requirements is discussed in the following.

In the ATM network, the VPs can be reconfigured dynamically and flexibly. The VP reconfiguration may be caused by VP setup, hitless protection switching, and/or self-healing following a failure. Furthermore, the adaptive VP route and bandwidth re-allocations imposed by network customer demands may also change the VP configuration. To administer the network efficiently, the following requirements must be considered by the OS:

1. The VP configuration data should be stored in databases managed by the OS, which remotely manages the VP configuration.

2. Real-time operation:

   • When the network is to be reconfigured, the configuration data in the database should be confirmed to correspond to the actual VP configuration.

   • When the network has been reconfigured, the final VP configuration should be confirmed to be correct. The configuration data in the database corresponding to the new VP configuration should be updated accurately.

The VP trace function is one of the schemes that may satisfy the foregoing requirements. The VP configuration information, called the SYS-IDs, is collected remotely and automatically by the OS to confirm the current VP configuration in real time. The collected SYS-IDs are used to update the database. The SYS-IDs are uniquely defined by each VP handling system in advance to identify all VPs. The SYS-IDs represent the components of the VP configuration, office number, system number, interface number, and VPI number. An OAM cell (trace cell) is defined to collect the SYS-IDs, where this OAM cell may be arbitrarily dropped or inserted from or into in-service VPs.

7.7.2 Alternative VP Trace Methods and Evaluation

Three methods for implementing the VP trace function are possible [42]. Figure 7-20 depicts the first method (Method A), which enables unidirectional VPs to be traced. When a VP is to be or has already been reconfigured, the OS designates both the start and terminating equipment. The start equipment generates a trace cell containing its own SYS-ID, and inserts it into the corresponding VP. The intermediate equipment detects and drops the trace cell. They append their own SYS-ID to the end of the SYS-IDs already present in the trace cell and reinsert this new trace cell. The terminating equipment drops and collects the trace cells, which now contain the accumulated SYS-IDs. These cells are passed to the OS as the trace result. The OS compares the trace results with the corresponding data and collates them as needed. The current VP configuration can thus be easily determined by the OS. The OS can arbitrarily designate the start and terminating equipment. If VP terminators are designated, an end-to-end VP trace is performed.

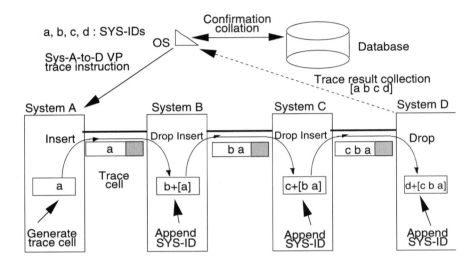

Method A: Each system appends its SYS-ID to trace cell in order (unidirectional VP)

Figure 7-20: VP tracing, Method A [42] (©IEEE 1992).

In Method B (see Figure 7-21), the SYS-IDs are appended to the trace cell in order, in the same way as for Method A, except that the OS also sends a loop-back instruction to the terminating equipment (Equipment D) to initiate VP trace.

In Method C (see Figure 7-22), the designated start equipment generates a SYS-ID request cell; the equipment that detects it then generates and sends back a trace cell that includes its own SYS-ID. The SYS-IDs are accumulated by the starting equipment, which then transmits them to the OS.

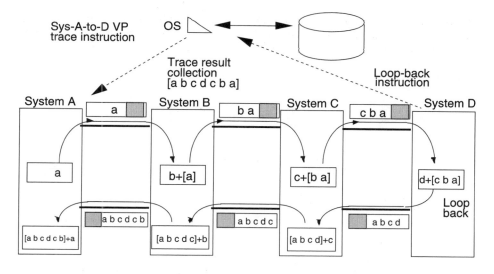

Method B: Each system appends its SYS-ID
to trace cell in order (bidirectional VP)
Figure 7-21: VP tracing, method B [42] (©IEEE 1992).

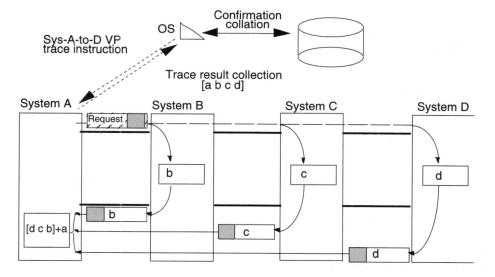

Method C: Each system generates a trace
cell with its own SYS-ID

Figure 7-22: VP tracing, Method C [42] (©IEEE 1992).

Table 7-6 depicts the evaluation results among three VP tracing methods [42]. With Method C, the number of trace cells increases with the quantity of intermediate equipment, and thus this method negatively affects user traffic. The loop-back function is required in Methods B and C, and so the processing time of Method A is the shortest. Therefore, from the viewpoints of traffic sensitivity, functional simplicity, and time taken, Method A is the best.

Table 7-6: Comparison of VP trace method alternatives [42].

	Method A	Method B	Method C
Number of required trace cells	$(N-1)/n$	$(2N-1)/n$	N
Loop-back function	not required	required	required
Required trace Time	T	$2T$	$2T$

N: number of devices to be traced
n: maximum number of SYS-IDs allowed in one cell
T: total processing time for trace by Method A

The VP trace function takes full advantage of the basic ATM features, such as nonhierarchical path structure and the ability to directly multiplex/demultiplex cells. Thus, the VP trace function can be accomplished much more simply than can the methods used to gather path-configuration data in SONET/SDH networks.

Details of possible system implementation of the VP trace function and a proposed VP trace cell structure can be found in Reference [42].

7.8 Multilayer B-ISDN Self-Healing Networks

The ATM-based transport functions are composed of several layers (SONET-section/line/path, ATM-VP/VC, AAL, and Network Layer); these layers may have similar capabilities for managing and controlling the network, as shown in Figure 4-14 (see Section 4.4.5). For example, rerouting function under a network failure can be performed on the SONET Layer (e.g., SONET rings), or the ATM/VP Layer (e.g., ATM/VP self-healing network), or on the Network Layer (e.g., MTP-3 in the case of signaling) [43]. It is thus necessary to minimize the functional redundancies across the layers to reduce network and operation costs.

Because current telecommunication networks usually support one level of availability, different services can accept different levels of network availability. Offering different levels of availability not only saves the network resources, it also allows the network operator to provide its customers with services at the most appropriate cost. When the VP is used as the only network restoration unit, the network cost might become high because high availability is provided to all services even if most traffic does not explicitly require it. Reference [44] proposes a multiple-availability-level ATM network architecture based on VC-route self-healing schemes. This architecture enables recovery from transit-switch failures as well as cross-connect or transmission-link failures, and it can also be used in conjunction with physical-level and VP-level restoration. With this architecture, QoS requirements, including network availability and cell loss rate and delay, can be achieved by VP selection or by VC routing; multiple levels of availability are supported by using Connection Admission Control (CAC).

When a failure occurs in the network, for each logical configuration (VC network) and pair of origin–destination switches (VP network access and VP termination points) affected by the failure, the self-healing scheme finds alternate routes (VC routes and VP connections) to replace the failed ones. When a failure occurs, the destination switch becomes the Sender and the origin switch becomes the Chooser. To find an alternate VC route for the failed VC route, the Sender sends recovery message cells using a flooding mechanism or selects predetermined backup routes. In the case of flooding, the recovery message cells are broadcast throughout the network; with predetermined backup routes, the cells are sent to the Chooser through those routes. In either case, when the Chooser receives the recovery message cells, it identifies the backup candidate routes and decides which one should be used as the new VC route. The unoccupied bandwidth of an outgoing VP is calculated by a switch located along each backup candidate route. If the outgoing VP's unoccupied bandwidth is lower than the minimum unoccupied bandwidth of a recovery message cell, the values in the message are updated at the switch. Upon reception of the recovery message cells, the Chooser determines the VP with the largest unoccupied bandwidth.

Each QoS level (network availability) is associated with a different logical configuration of the VC network. For a connection that belongs to a given service class, the CAC procedure can use only the VC route and the origin and destination switches predetermined for that service class. Reference [44] has suggested that the optimum configuration knowing the user's requested QoS after the failure can be provided to a VC self-healing network in which the CAC functions take statistical gain, required VP bandwidth, and network efficiency into account.

7.9 Summary and Remarks

We have reviewed key technologies for ATM self-healing mesh networks and their network architecture. Although these technologies have not yet been deployed commercially, there may be a demand for such ATM self-healing networks if the network provider is seeking a cost-effective survivable scalable ATM network with flexible and global restoration capabilities. Before such robust networks can be achieved, however, several issues remain to be resolved. The optimum interworking scheme among multiple layers, such as between the Network Layer and the ATM Layer, is an open issue for network planners and engineers. The hardware and operations needed to provide both ATM self-healing and ATM congestion control need to be investigated. The synchronization between centralized and distributed OS control remains an open task for designing a practical self-healing network. It is hoped that this chapter provides necessary background information for those who are interested in investigating these open questions.

References

[1] ITU-T Recommendation I.311, "B-ISDN General Network Aspects," Temporary Document 5G (XVIII), January 1993.

[2] T.-H. Wu, H. Kobrinski, D. Ghosal and T. V. Lakshman, "The Impact of SONET DCS System Architecture on Distributed Restoration," *IEEE J. Selected Areas in Commun.*, pp. 79-87, January 1994.

[3] T.-H. Wu, *Fiber Network Service Survivability*, Artech House, 1992.

[4] K. Sato, T. Hadam, and I. Tokizawa, "Network Reliability Enhancement with Virtual Path Strategy," *Proc. IEEE GLOBECOM'90*, pp. 403.5.1- 403.5.6, San Diego, CA, December 1990.

[5] J. Anderson, B. T. Doshi, S. Dravida, and P. Harshavardhana, "Fast Restoration of ATM Networks," *IEEE J. Selected Areas in Commun.*, Vol. 12, No. 1, pp. 128-138, 1994.

[6] E. Ayanoglu and M. Veeraraghavan, "Advanced Topics in Broadband ATM Networks," *Tutorial – IEEE INFOCOM '94*, Toronto, Canada, June 1994.

[7] R. Kawamura, K. Sato, and I. Tokizawa, "Self-Healing ATM Networks Based on Virtual Path Concept," *IEEE J. Selected Areas in Commun.*, Vol. 12, No. 1, pp. 120-127, January 1994.

[8] W. D. Grover, "The Self-Healing Network," *Proc. IEEE GLOBECOM'87*, pp. 1090-1095, November 1987.

[9] H. R. Amirazizi, "Controlling Synchronous Networks with Digital Cross-Connect Systems," *Proc. IEEE GLOBECOM'88,* pp. 1560-1563, 1988.

[10] H. Komine, T. Chujo, T. Ogura, K. Miyazaki, and T. Soejima, "A Disributed Restoration Algorithm for Multiple-Link and Node Failures of Transport Networks," *Proc. IEEE GLOBECOM'90*, pp. 0459-0463, 1990.

[11] S. Hasegawa, Y. Okanoue, T. Egawa, and H. Sakauchi, "Control Algorithm of SONET Integrated Self-Healing Networks," *IEEE J. Selected Areas in Commun.*, Vol. 12, No. 1, pp. 110-119, 1994.

[12] K. Sato, H. Ueda, and N. Yoshikai, "The Role of Virtual Path Cross Connection," *IEEE Mag. Lightwave Telecommunications Systems,* Vol. 2, No.3, pp.44-54, August 1991.

[13] ITU-T Recommendation I.610, "B-ISDN Operation and Maintenance Principles and Fuctions," 1993.

[14] R. Kawamura, H. Hadama, K. Sato, and I. Tokizawa, "Fast VP-Bandwidth Management with Distributed Control in ATM Networks," *IEICE Trans. on Commun.*, Vol. E77-B, No. 1, pp. 5-14, January 1994.

[15] R. Kawamura, K. Sato, and I. Tokizawa, "Self-Healing Techniques utilizing Virtual Paths," *Proc. 5th International Network Planning Symposium,* Kobe, Japan, pp. 129–134, May 1992.

[16] M. M. Slominski and H. Okazaki, "Guided Restoration of ATM Cross-Connect Networks," *Proc. IEEE ICC'94*, pp. 466-470, May 1994.